Biosensors - Micro and Nanoscale Applications

Biosensors - Micro and Nanoscale Applications

Editor

Rodolph Donovan

scitus academics

Biosensors - Micro and Nanoscale Applications
Edited by **Rodolph Donovan**

ISBN: 978-1-68117-314-6
Library of Congress Control Number: 2016937624

www.scitusacademics.com
Box No. 4766, 616 Corporate Way,
Suite 2, Valley Cottage,
NY 10989

Printed in United States of America on Acid Free Paper

Preface

Today, the implementation of new technological platforms in biosensor-based development focuses primarily on the miniaturization of analytical systems and the reduction of detection limits. Rapid scientific and technological advances allow the use of biosensors online to detect minute concentrations of various chemical compounds in a wide range of matrixes and to monitor extremely low levels of biomarkers even in living organisms and individual cells. A biosensor is an analytical device used with a physicochemical detector to detect a biological component. The bioelement interacts with the test analysis and the biological response is converted into an electric signal by the transducer. Biosensors, depending on their specific application, are also known as immunosensors, optrodes, resonant mirrors, chemical canaries, biochips, glucometers and biocomputers. Biosensor samples are increasingly sophisticated, mainly because of the combination of advances in two technological areas: microelectronics and biotechnology. Biosensors are very useful tools for measuring a wide variety of analytes, including organic compounds, gasses, ions and bacteria. Many optical biosensors are based on the phenomenon of surface resonance (SPR). The book embodies the current state of the art and potential micro and nanoscale performance options in biosensor construction and application.

Table of Contents

Chapter 1

Nano-Sized Elements in Electrochemical Biosensors

Joanna Cabaj, Jadwiga Sołoducho*

Faculty of Chemistry, Wroclaw University of Technology, Wrocław, Poland

ABSTRACT

The emerging nanotechnology has opened novel opportunities to explore analytical applications of the fabricated nano-sized materials. Recent advances in nano-biotechnology have made it possible to realize a variety of enzyme electrodes suitable for sensing application. In coating miniaturized electrodes with biocatalysts, undoubtedly the most of the potential deposition processes suffer from the difficulty in depositing process and reproducible coatings of the active enzyme on the miniature transducer element. The promising prospects can concern to the obtaining of thin protein layers by using, i.e. electrochemical deposition, electrophoretic deposition as well as monolayer methods (Langmuir-Blodgett procedure, Layer-by-Layer—LbL). Many aspects dealing with deposition of enzyme by techniques employing electric field are considered, including surface charge of enzyme, and its migration under applied electric filed. The using of nanoscale materials (i.e. nanoparticles, nanowires, nanorods) for electrochemical biosensing has seen also explosive increase in recent years following the discovery of nanotubes. These structures offer a promise in the development of biosensing, facilitating the great improvement of the selectivity and sensitivity of the current methods. Finally, the perspectives in the further exploration of nanoscaled sensors are discussed.

Keywords: Enzymes, Immobilization, Nanobiosensors, Nanoparticles

1. INTRODUCTION

To date, there is an increasing necessity for mighty analytical tools with high sensitivity, fast response, selectivity, accuracy and low cost of production/operation. Notably, biosensors have found comprehensive adoption in the area of environmental control as well as pharmaceutics and medical diagnostics. Consequently, the main objective in biosensors design is the sufficient development of a biosurface, firmly sensitive and selective for a respective analyte, which may be able to generate measurable signals coupled to an adequate transducer.

Thus, many groups of researchers tend also to combine nanoparticles into the materials used for biosensors in order to improve the sensitivity of the system in potential sensing applications. Most recent studies show that biosensors composed with nanoparticles do take on rapid, simple, and accurate measurements, which offers exciting new opportunities for the development of biosensor capabilities. Owing to the emerging roles that nanoparticles are playing in the improvement of biosensors in recent years, it is necessary and meaningful for us to investigate the researches of nanoparticle-based biosensors from the point of view of management of technology. As a vital part of management of technology, grasping the latest development of technology and identifying the emerging characters can help us get competitive advantages in the future.

Several kinds of nanoparticles, including metal nanoparticles, oxide nanoparticles, semiconductor nanoparticles, and even nanodimensional conducting polymers have been used in biosensing systems. Owing to these unique properties, different kinds of nanoparticles always play different roles in different sensing systems. Generally, metal nanoparticles are always used as elements of “electronic wires”. Oxide nanoparticles are often applied to immobilize biomolecules, while semiconductor nanoparticles are often used as labels or tracers [1].

The development of high performance and reliable miniaturized enzyme electrodes is a crucial objective worth pursuing in the nanosized biotechnology area. The miniaturization bids numerous advantages [2]. Miniaturized enzyme electrodes are useful in analysis of small sample volumes, hence practical if only small amounts of biological fluids are provided or, waste saving if larger quantities are available. These small systems can be also integrated into new technologies like microarray sensor or microfluidic systems. When used in vivo, miniaturized systems create less damage of the tissues and hence quick healing.

Certainly, there are several methods (i.e. Langmuir-Blodgett techniques, LbL) with which enzymes can be deposited including but not limited to entrapment and crosslinking [3]. But an important problem in the application of enzymatic proteins for the development of miniaturized electrodes is the difficulty in depositing uniform and reproducible layer coatings of enzymes on the transducer [4].

Electrochemical and electrophoretic deposition are offered as techniques, which can employ electric field to produce apparently uniform and reproducible layer coatings of biocatalysts over very small areas. Electrochemical deposition is known for several decades, but by the contrast electrophoretic deposition is rather recent technique [5]. However, further development work needs to be done according to optimize the parameters for a broader use of especially in the fabrication of miniaturized enzyme systems [4].

2. IMMOBILIZATION OF ENZYMES IN MINIATURIZED SYSTEMS

2.1. Electrochemical Deposition

A study of literature indicates that electrochemical deposition is one of the most techniques employed for enzyme immobilization, because of simple, cost effective apparatus. Electrochemical deposition can be employed to drive enzymes alone to deposit on the support, as well as with other components including i.e. collagen [6] , noble metal salts (Pt, Pd) [7] , monomers such as pyrrole, 1, 3-diaminobenzene [8] , some redox mediators (i.e. Prussian blue) or redox centers [9] , nanomaterials like carbon nanotubes and metal nanoparticles [10] . The final goal of all these efforts is to fabricate enzyme electrodes with appropriate characteristics in terms of preserved activity, enhanced kinetics and stability to fit with the specific application.

Generally, the activity of enzyme electrode prepared by electrochemical deposition depends primarily on thickness of the enzyme layers [11].

The latter yields deposition of thin enzyme coatings because only enzymes present nearby vicinity of the electrode surface precipitate. Matsumoto et al. [12] observed that only few tens of nanometer are produced. The thickness of the enzyme coating can be increased to reach much thicker layers if i.e. a surfactant is added to enzyme solution prior deposition.

According to kinetics, it was observed, that enzymes deposited under electrochemical deposition may or may not keep similar kinetics as nativeproteins [13] [14]. This fact depends on the environment of the deposited enzyme coating and presence of other components. In case of stability, enzyme electrodes produced by electrochemical technique have usually moderate stabilities [4].

2.2. Electrophoretic Deposition

Electrophoretic deposition is carried on from low conductivity aqueous solutions/suspensions (Figure 1). The technique requires high strength electric fields, which can reach several hundreds of volts to move the charged biocatalysts from bulk of the solution to the electrode. Both parameters of elevated zeta potential and high strength electric field yield significant migration of enzymatic proteins under electrophoretic deposition [4]. As a result, more enzymes reach the surface of the electrode and precipitate to form thick-deposited coatings [15] . It is weighty, that when high strength electric fields are employed in electrophoretic deposition, continuous direct current can no longer be utilized because it can induce heat and extensive water electrolysis [4]. This situation causes changes in temperature and pH in the environment of electrodes which maylowered the activity of the protein [16]. To avoid the problem it is possible to apply pulsed direct current and alternating current [4].

Alternating current is no fixed anode or cathode but the polarization of each electrode changes continuously between the positive and negative signs. Due to the fact, that a negatively charged enzyme is subjected to asymmetrical alternating current signal, it should only oscillate in one location because the migration achieved during the first half cycle when one of electrodes is positively charged should be neutralized during the second half process when the other electrode becomes positively charged [4]. According to that, the migration of charged enzyme under symmetrical signals is zero. In results, only thin enzyme coatings can be deposited [5] [17]. Whereas, under unsymmetrical alternating current field is applied to negatively charged protein, large amount of enzymes accumulate nearby the surface of electrode and the enzymes precipitate to yield thick deposited layers. Furthermore, because alternating current fields generate the minimum of water electrolysis as well as heat, the maximum enzymes activity could be preserved after deposition [5] [17]. There is found a several enzymes (i.e. glucose oxidase, peroxidase), which have been successfully immobilized by this method [4] [5] [17].

2.3. Thin Layer Methods

Adsorption of organic molecules on solid conducting supports to produce thin nanostructured films are one of the most employed architectures and represents an important approach in the field of nano-manipulation. Langmuir-Blodgett (LB) technique promotes a high control of the physical and chemical properties of nanostructured organic films and plays an important role in the production of miniaturized devices applicable as platforms for enzyme immobilization [18].

Other pathways to prepare platforms based on nanomaterials, aiming the fabrication of electrochemical biosensors are dispersion in solvents, adsorption (e.g. LbL), formation of covalent bonding.

As well, the utilization of hybrid organic/inorganic thin films can, in a simple manner, be employed in solid conductor electrodes. The possibility to incorporate hybrids containing nanostructured materials for enhance electrochemical properties makes these techniques much attractive in the field of bionanoelectrochemistry [19] [20] .

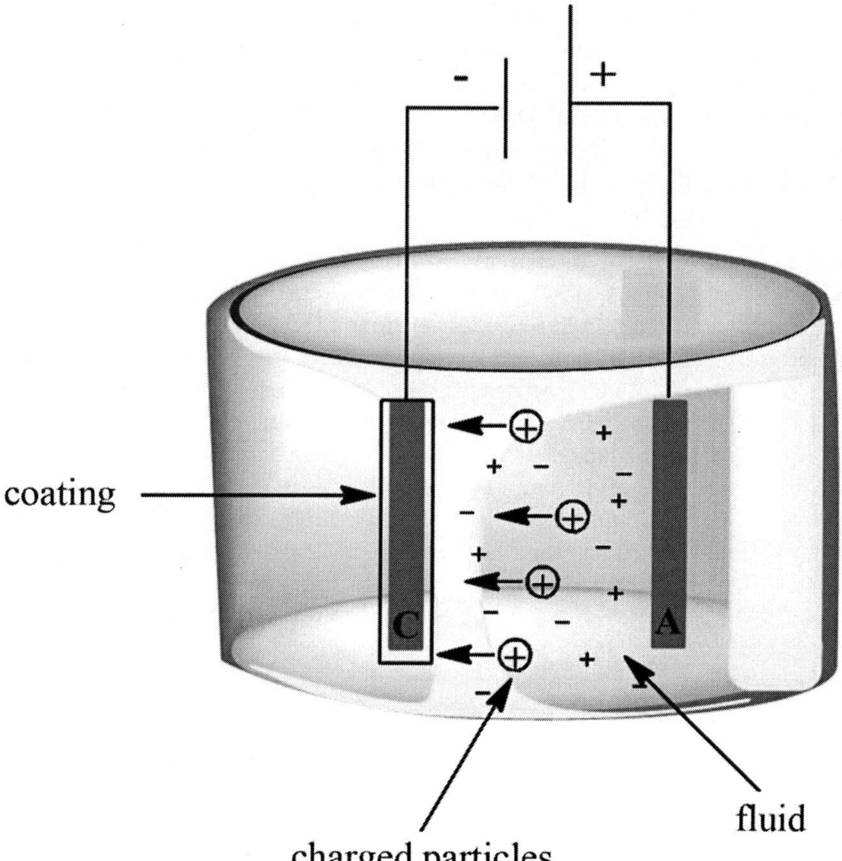

Figure 1. Electrophoretic coating.

The aim is to preserve the native enzyme molecular conformation and to arrange it in a suitable position for the molecular recognizing of an external molecule of a solution put on contact with the Langmuir-Blodgett device. As defended in a review by Girard-Egrot [21], the successful incorporation of enzymes on a preformed Langmuir monolayer depends strongly on the methodology employed. The most one commonly used is the adsorption of the enzyme from the subphase, avoiding direct adsorption of the macromolecule present at the water surface. This strategy was used to produce electrochemical sensors containing i.e. phytic acid [22], horseradish peroxidase [23], hemoblogin [24], and urease [25], tyrosinase [26], to detect a diversity of substances such as including phytic acid, hydrogen peroxide, glucose, choline, urea, and phenols.

In these types of sensors the sensor sensitization can be achieved by i.e. an amphiphilic heterocyclic semiconducting structures admixed into the film [27] or other, more sophisticated architectures have been developed in order to enhance the performance of LB-based enzyme electrochemical sensors. For instance, Sun et al. [28] used pyriduylthio-modified carbon nanotubes as Langmuir-Blodgett films to support hydrogenase added in a subsequent adsorption from solution.

Alternatively, a modern concept of self-assembly was introduced by Decher and co-workers [29][30] at 90 decade as a low-cost and simple method to obtain nanostructured thin films under controlled conditions (pH, temperature, polyelectrolyte concentration, ionic strength, etc.). For this purpose, a large variety of materials for electrochemical sensing and biosensing can be obtained [31] -[33] . Basically, the processes of film fabrication by LbL technique (Figure 2) is governed by the adsorption of organic polyelectrolytes with opposite charges present on their molecular structure, in such a way that film roughness, thickness, porosity and morphology can be controlled at molecular level [34]. Important advantage in the use of LbL technique to construct biosensors is the possibility to incorporate organic/inorganic composite materials that contributes for the maximization of the biodevices electrochemical signal [35]. Also, it is important to emphasize that most hybrids based on nanomaterials has been utilized to detect electrochemical signal from biochemical reactions.

Proteins have also been used in LbL method to construct alternate multilayer of ceramic nanotubes (halloysite), spherical particles leading to an array of new ordered nanoparticles-tubules, which were applied to load co-enzymes (NAD) for the development of enzymatic nanoreactors. Decher and co-workers [36] also reported the use of protein/polyelectrolyte hybrid films via specific recognition. One of the main challenges is to maintain the integrity of the native protein structure to promote their utilization for technological applications.

Ram and co-workers [37] reported the utilization of LbL technique to produce nanostructured films of poly (ethylene imine) and poly(sodium polystyrene sulfonate), cholesterol oxidase and cholesterol esterase. The strong stability of multilayer films was also evaluated and contributes directly for the electrochemical properties of the film warranting the glucose oxidase immobilized on solid conductor supports remained active on the electrode surface.

3. SUMMARY AND FUTURE TRENDS

Electrochemical nanobiosensors offer without doubts an important step toward development of selective, down to few target molecules sensitive biorecognition device for medical and security applications. In their case, very high amplification of signal could be reached, i.e. using high diameter carbon nanotubes filled with nanoparticles and their following electrochemical stripping.

The utilization of nano-manipulation techniques has also become an interesting approach to fabricate electrochemical devices with high specificity and molecular order. Moreover, the sensitivity and overall behavior of biosensors has grown rapidly as an outcome of incorporating different nanomaterials in their construction.

Electrochemical nanobiosensors consisting from single carbon nanotube are clear examples of future path of biosensor development. These strategies waits for exploration. There is high expectation that such devices will develop toward reliable point-of-care diagnostics of cancer and other diseases, and as tools for intra-operation pathological testing, proteomics and systems biology.

ACKNOWLEDGEMENTS

Authors are gratefully acknowledged for financial support of NCN-Grant no. 2012/05/B/ST5/00749 and Wrocław University of Technology.

REFERENCES

1. Luo, X., Morrin, A., Killard, A.J. and Smyth, M.R. (2006) Application of Nanoparticles in Electrochemical Sensors and Biosensors. Electroanalysis, 18, 319-326.
2. Dukhin, A.S. and Dukhin, S.S. (2005) Aperiodic Capillary Electrophoresis Method Using an Alternating Current Electric Field for Separation of Macromolecules. Electrophoresis, 26, 2149-2153.
3. Sheldon, R.A. and van Pelt, S. (2013) Enzyme Immobilisation in Biocatalysis: Why, What and How. Chemical Society Reviews, 42, 6223-6235.

4. Amman, M. (2014) Electrochemical and Electrophoretic Deposition of Enzymes: Principles, Differences and Application in Miniaturized Biosensor and Biofuel Cell Electrodes. Biosensors and Bioelectronics, 58, 121-131.

5. Ammam, M. and Fransaer, J. (2009) AC-Electrophoretic Deposition of Glucose Oxidase. Biosensors and Bioelectronics, 25, 191-197.

6. Wang, S.S. and Vieth, W.R. (1973) Immobilization of Whole Cells in a Membraneous Form. Biotechnology and Bioengineering, 15, 93-115.

7. Johnston, D.A., Cardosi, M.F. and Vaughan, D.H. (1995) The Electrochemistry of Hydrogen Peroxide on Evaporated Gold/Palladium Composite Electrodes. Manufacture and Electrochemical Characterization. Electroanalysis, 7, 520-526.

8. Gao, Z.Q., Binyamin, G., Kim, H.H., Barton, S.C., Zhang, Y.C. and Heller, A. (2002) Electrodeposition of Redox Polymers and Co-Electrodeposition of Enzymes by Coordinative Crosslinking. Angewandte Chemie International Edition, 41, 810-813.

9. Ackermann, Y., Guschin, D.A., Eckhard, K., Shleev, S. and Schuhmann, W. (2010) Design of a Bioelectrocatalytic Electrode Interface for Oxygen Reduction in Biofuel Cells Based on a Specifically Adapted Os-Complex Containing Redox Polymer with Entrapped Trametes Hirsuta Laccase. Electrochemistry Communications, 12, 640-643.

10. Chiu, J.Y., Yu, C.M., Yen, M.J. and Chen, L.C. (2009) Glucose Sensing Electrodes Based on a Poly(3,4-Ethylenedioxythiophene)/Prussian Blue Bilayer and Multi-Walled Carbon Nanotubes. Biosensors and Bioelectronics, 24, 2015-2020.

11. Almeida, N.F., Beckman, E.J. and Ataai, M.M. (1993) Immobilization of Glucose Oxidase in Thin Polypyrrole Films: Influence of Polymerization Conditions and Film Thickness on the Activity and Stability of the Immobilized Enzyme. Biotechnology and Bioengineering, 42, 1037-1045.

12. Matsumoto, N., Chen, X.H. and Wilson, G.S. (2002) Fundamental Studies of Glucose Oxidase Deposition on a Pt Electrode. Analytical Chemistry, 74, 362-367.

13. Iwuoha, E.I., Saenzde Villaverde, D., Garcia, N.P., Smyth, M.R. and Pingarron, J.M. (1997) Reactivities of Organic Phase Biosensors. 2. The Amperometric Behaviour of Horseradish Peroxidase Immobilised on a Platinum Electrode Modified with an Electrosynthetic Polyaniline Film. Biosensors and Bioelectronics, 12, 749-761.

14. Vidal, J.C., Garcia, E. and Castillo, J.R. (1999) In Situ Preparation of Overoxidized PPy/oPPD Bilayer Biosensors for the Determination of Glucose and Cholesterol in Serum. Sensors and Actuators B, 57, 219-226.

15. Ammam, M. and Fransaer, J. (2010) Alternating Current Electrophoretic Deposition of Saccharomyces cerevisiae Cells and the Viability of the Deposited Biofilm in Ethanol Production. Electrochimica Acta, 55, 9125-9131.

16. Ammam, M. and Fransaer, J. (2011) Effects of AC-Electrolysis on the Enzymatic Activity of Glucose Oxidase. Electroanalysis, 23, 755-763.

17. Ammam, M. and Fransaer, J. (2010) Micro-Biofuel Cell Powered by Glucose/O2 Based on Electro-Deposition of Enzyme, Conducting Polymer and Redox Mediators: Preparation, Characterization and Performance in Human Serum. Biosensors and Bioelectronics, 25, 1474-1480.

18. Schnorr, J.M. and Swager, T.M. (2011) Emerging Applications of Carbon Nanotubes. Chemistry of Materials, 23, 646-657.

19. Baughman, R.H., Zakhidov, A.A. and de Heer, W.A. (2002) Carbon Nanotubes—The Route toward Applications. Science, 297, 787-792.

20. Hrapovic, S., Liu, Y., Male, K.B. and Luong, J.H.T. (2004) Electrochemical Biosensing Platform Using Platinum Nanopartyicles and Carbon Nanotubes. Analytical Chemistry, 76, 1083-1088.

21. Girard-Egrot, A.P., Godoy, S. and Blum, L.J. (2005) Enzyme Association with Lipidic Langmuir-Blodgett Films: Interests and Applications in Nanobioscience. Advances in Colloid and Interface Science, 116, 205-225.

22. Caseli, L., Moraes, M.L., Zucolotto, V., Ferreira, M., Nobret, T.M., Zaniquelli, M.E.D., Pereira, U. and Oliveira Jr., O.N. (2006) Fabrication of Phytic Acid Sensor Based on Mixed Phytase-Lipid Langmuir-Blodgett Films. Langmuir, 22, 8501-8508.

23. Schmidt, T.F., Caseli, L., Viitala, T. and Oliveira Jr., O.N. (2008) Enhanced Activity of Horseradish Peroxidase in Langmuir-Blodgett Films of Phospholipids. Biochimica et Biophysica Acta, 1778, 2291-2297.

24. Yin, F., Kafi, A.K.M., Shin, H.K. and Kwon, Y.S. (2005) Human Serum Albumin-Octadecylamine Langmuir-Blodgett Film Formed by Spreading Human Serum Albumin Solution Directly on Subphase's Interface Covered with a Layer of Octadecylamine. Thin Solid Films, 488, 223-229.

25. Caseli, L., Nobre, T.M., Zaniquelli, M.E.D., Zucolotto, V. and Oliveira Jr., O.N. (2008) Using Phospholipid Langmuir and Langmuir-Blodgett Films as Matrix for Urease Immobilization. Journal of Colloid and Interface Science, 319, 100-108.

26. Apetrei, C., Alessio, P., Constantino, C.J.L., de Saja, J.A., Rodriguez-Mendez, M.L., Pavinatto, F.J., Fernandes, E.G.R., Zucolotto, V. and Oliveira Jr., O.N. (2011) Biomimetic Biosensor Based on Lipidic Layers Containing Tyrosinase and Lutetium Bisphthalocyanine for the Detection of Antioxidants. Biosensors and Bioelectronics, 26, 2513-2519.

27. Cabaj, J., Chyla, A., Jedrychowska, A., Olech, K. and Soloducho, J. (2012) Detecting Platform for Phenolic Compounds-Characteristic of Enzymatic Electrode. Optical Materials, 34, 1677-1681.

28. Sun, Q., Zorin, N.A., Chen, D., Chen, M., Lieu, T.X., Myake, J. and Qian, D.J. (2010) Langmuir-Blodgett Films of Pyridyldithio-Modified Multiwalled Carbon Nanotubes as a Support to Immobilize Hydrogenase. Langmuir, 26, 10259-10265.

29. Decher, G. and Hong, J.D. (1991) Buildup of Ultrathin Multilayer Films by a Self-Assembly Process: II. Consecutive Adsorption of Anionic and Cationic Bipolar Amphiphiles and Polyelectrolytes on Charged Surfaces. Berichte der Bunsengesellschaft für physikalische Chemie, 95, 1430-1434.

30. Decher, G. (1997) Fuzzy Nanoassemblies: Toward Layered Polymeric Multicomposites. Science, 277, 1232-1237.

31. Siqueira Jr., J.R., Gasparoto, L.H.S., Crespilho, F.N., Carvalho, A.J.F., Zucolotto, V. and Oliveira Jr., O.N. (2006) Physicochemical Properties and Sensing Ability of Metallophthalocyanines/Chitosan Nanocomposites. Journal of Physical Chemistry B, 110, 22690-22694.

32. Jeong, R.A., Hwang, J.Y., Joo, S., Chung, T.D., Vark, S.P., Kang, S.K., Lee, W.Y. and Kim, H.C. (2003) In Vivo Calibration of the Subcutaneous Amperometric Glucose Sensors Using a Non-Enzyme Electrode. Biosensors and Bioelectronics, 19, 313-319.

33. Jie, G., Liu, B., Pan, H., Zhu, J.J. and Chen, H.Y. (2007) CdS Nanocrystal-Based Electrochemiluminescence Biosensor for the Detection of Low-Density Lipoprotein by Increasing Sensitivity with Gold Nanoparticle Amplification. Analytical Chemistry, 79, 5574-5581.

34. Shi, G.Y., Qu, Y.H., Zhai, Y.Y., Liu, Y., Sun, Z.Y., Yang, J.G. and Jin, L.T. (2007) {MSU/PDDA}n LBL Assembled Modified Sensor for Electrochemical Detection of Ultratrace Explosive Nitroaromatic Compounds. Electrochemistry Communications, 9, 1719-1724.

35. Hou, Y., Cheng, Y., Hobson, T. and Liu, J. (2010) Design and Synthesis of Hierarchical MnO2 Nanospheres/Carbon Nanotubes/Conducting Polymer Ternary Composite for High Performance Electrochemical Electrodes. Nano Letters, 10, 2727-2733.

36. Cassier, T., Lowack, K. and Decher, G. (1998) Layer-by-Layer Assembled Protein/Polymer Hybrid Films: Nasnoconstruction via Specific Recognition. Supramolecular Science, 5, 309-315.

37. Ram, K., Bertoncello, M.P., Ding, H., Paddeu, S. and Nicolini, C. (2001) Cholesterol Biosensors Prepared by Layerby-Layer Technique. Biosensors and Bioelectronics, 16, 849-856.

38. Musameh, M., Wang, J., Merkoci, A. and Lin, Y.H. (2002) Low-Potential Stable NADH Detection at Carbon-NanotubeModified Glassy Carbon Electrodes. Electrochemistry Communications, 4, 743-746.

39. Dequaire, M., Degrand, C. and Limoges, B. (2000) An Electrochemical Metalloimmunoassay Based on a Colloidal Gold Label. Analytical Chemistry, 72, 5521-5528.

40. Sadik, O.A., Aluoch, A.O. and Zhou, A.L. (2009) Status of Biomolecular Recognition Using Electrochemical Techniques. Biosensors and Bioelectronics, 24, 2749-2765.

41. Zhang, S., Wang, N., Yu, H., Niu, Y. and Sun, C. (2005) Tailoring the Surface Potential of Gold Nanoparticles with Self-Assembled Monolayers with Mixed Functional Groups. Biochemistry, 67, 15-22.

42. Yogeswaran, U. and Chen, S.M. (2008) A Review on the Electrochemical Sensors and Biosensors Composed of Nanowires as Sensing Material. Sensors, 8, 290-313.

43. Ajayan, P.M. (1999) Nanotubes from Carbon. Chemical Reviews, 99, 1787-1800.

44. Lynam, C., Gilmartin, N., Minett, A.I., O'Kennedy, R. and Wallace, G. (2009) Carbon Nanotube-Based Transducers for Immunoassays. Carbon, 47, 2337-2343.

45. Zhao, Y.D., Zhang, W.D., Chen, H., Luo, Q.F. and Li, S.F.Y. (2002) Direct Electrochemistry of Horseradish Peroxidase at Carbon Nanotube Powder Microelectrode. Sensors and Actuators B, 87, 168-172.

46. Yamamoto, K., Shi, G., Zhou, T.S., Xu, F., Xu, J.M., Kato, T., Jind, J.Y. and Jin, L.T. (2003) Study of Carbon Nanotubes-HRP Modified Electrode and Its Application for Novel On-Line Biosensors. Analyst, 128, 249-254.

47. Wang, J.X., Li, M.X., Shi, Z.J., Li, N.Q. and Gu, Z.N. (2002) Direct Electrochemistry of Cytochrome c at a Glassy Carbon Electrode Modified with Single-Wall Carbon Nanotubes. Analytical Chemistry, 74, 1993-1997.

48. Davis, J.J., Coles, R.J., Allen, H. and Hill, O. (1997) Protein Electrochemistry at Carbon Nanotube Electrodes. Journal of Electroanalytical Chemistry, 440, 279-282.

49. Guiseppi-Elie, A., Lei, C.H. and Baughman, R.H. (2002) Direct Electron Transfer of Glucose Oxidase on Carbon Nanotubes. Nanotechnology, 13, 559-564.

50. Dechakiatkrai, C., Chen, J., Lynam, C., Min, S.K., Kim, S.J., Phanichphant, S. and Wallace, G.G. (2008) Direct Ascorbic Acid Detection with Ferritin Immobilized on Single-Walled Carbon Nanotubes. Electrochemical and SolidState Letters, 11, K4-K6.

51. Chen, R.J., Zhang, Y.G., Wang, D.W. and Dai, H.J. (2001) Noncovalent Sidewall Functionalization of Single-Walled Carbon Nanotubes for Protein Immobilization. Journal of the American Chemical Society, 123, 3838-3839.

52. Rubianes, M.D. and Rivas, G.A. (2003) Carbon Nanotubes Paste Electrode. Electrochemistry Communications, 5, 689-694.

53. Perez, B., Pumera, M., del Valle, M., Merkoci, A. and Alegret, S. (2005) Glucose Biosensor Based on Carbon Nanotube Epoxy Composites. Journal of Nanoscience and Nanotechnology, 5, 1694-1698.

54. Pumera, M., Merkoci, A. and Alegret, S. (2006) Carbon Nanotube-Epoxy Composites for Electrochemical Sensing. Sensors and Actuators B, 113, 617-622.

55. Wang, J. and Musameh, M. (2004) Carbon Nanotube Screen-Printed Electrochemical Sensors. Analyst, 129, 1-2.

56. Sanchez, S., Pumera, M., Cabruja, E. and Fabregas, E. (2007) Carbon Nanotube/Polysulfone Composite Screen-Printed Electrochemical Enzyme Biosensors. Analyst, 132, 142-147.

57. Armada, M.P.G., Losada, J., Cuadrado, I., Alonso, B., Gonzalez, B., Casado, C.M. and Zhang, J.B. (2004) Preparation of Biosensors Based in a Siloxane Homopolymer with Interacting Ferrocenes for the Amperometric Detection of Peroxides. Sensors and Actuators B, 101, 143-149.

58. Liu, G.D. and Lin, Y.H. (2006) Amperometric Glucose Biosensor Based on Self-Assembling Glucose Oxidase on Carbon Nanotubes. Electrochemistry Communications, 8, 251-256.

59. Liu, G.D. and Lin, Y.H. (2006) Carbon Nanotube-Templated Assembly of Protein. Journal of Nanoscience and Nanotechnology, 6, 948-953.

60. Qu, F.L., Yang, M.H., Jiang, J.H., Shen, G.L. and Yu, R.Q. (2005) Amperometric Biosensor for Choline Based on Layer-by-Layer Assembled Functionalized Carbon Nanotube and Polyaniline Multilayer Film. Analytical Biochemistry, 344, 108-114.

61. Zhao, H.T. and Ju, H.X. (2006) Multilayer Membranes for Glucose Biosensing via Layer-by-Layer Assembly of Multiwall Carbon Nanotubes and Glucose Oxidase. Analytical Biochemistry, 350, 138-144.

62. Yu, X., Chattopadhyay, D., Galeska, I., Papadimitrakopoulos, F. and Rusling, J.F. (2003) Peroxidase Activity of Enzymes Bound to the Ends of Single-Wall Carbon Nanotube Forest Electrodes. Electrochemistry Communications, 5, 408-411.

63. Patolsky, F., Weizmann, Y. and Willner, I. (2004) Long-Range Electrical Contacting of Redox Enzymes by SWCNT Connectors. Angewandte Chemie International Edition, 43, 2113-2117.

64. Pumera, M., Sanchez, S., Ichinose, I. and Tang, J. (2007) Electrochemical Nanobiosensors. Sensors and Actuators B, 123, 1195-1205.

65. Cui, Y., Wei, Q.Q., Park, H.K. and Lieber, C.M. (2001) Functional Nanoscale Electronic Devices Assembled Using Silicon Nanowire Building Blocks. Science, 293, 1289-1292.

66. Zhou, X.T., Hu, J.Q., Li, C.P., Ma, D.D.D., Lee, C.S. and Lee, S.T. (2003) Silicon Nanowires as Chemical Sensors. Chemical Physics Letters, 369, 220-224.

67. Shao, M.W., Yao, H., Zhang, M.L., Wong, N.B., Shan, Y.Y. and Lee, S.T. (2005) Fabrication and Application of Long Strands of Silicon Nanowires as Sensors for Bovine Serum Albumin Detection. Applied Physics Letters, 87, Article ID: 183106.

68. Shao, M.W., Shan, Y.Y., Wong, N.B. and Lee, S.T. (2005) Silicon Nanowire Sensors for Bioanalytical Application: Glucose and Hydrogen Peroxide Detectionadv. Advanced Functional Materials, 15, 1478-1482.

69. Janata, J. and Blackburn, G.F. (1984) Immunochemical Potentiometric Sensors. Annals of the New York Academy of Sciences, 428, 286-292.

70. Janata, J. (1989) Principles of Chemical Sensors. Plenum Press, New York.

71. Klabunde, K.J. (2001) Introduction to Nanotechnology. In: Klabunde, K.J., Ed., Nanoscale Materials in Chemistry, John Wiley and Sons, New York.

72. Du, J., Yu, X.P. and Di, J.W. (2012) Comparison of the Direct Electrochemistry of Glucose Oxidase Immobilized on the Surface of Au, CdS and ZnS Nanostructures. Biosensors and Bioelectronics, 37, 88-93.

73. Zheng, B.Z., Xie, S.P., Qian, L., Yuan, H.Y., Xiao, D. and Choi, M.M.F. (2011) Gold Nanoparticles-Coated Eggshell Membrane with Immobilized Glucose Oxidase for Fabrication of Glucose Biosensor. Sensors and Actuators B, 152, 49-55.

74. Hanefeld, U., Gardossi, L. and Magner, E. (2009) Understanding Enzyme Immobilization. Chemical Society Reviews, 38, 453-468.

75. Brogan, K.L., Wolfe, K.N., Jones, P.A. and Schoenfisch, M.H. (2003) Direct Oriented Immobilization of F(ab') Antibody Fragments on Gold. Analytica Chimica Acta, 496, 73-80.

76. Rao, S.V., Anderson, K.W. and Bachas, L.G. (1998) Fundamental Review, Oriented Immobilization of Proteins. Microchimica Acta, 128, 127-143.

77. Liu, Z.M., Liu, J., Shen, G.L. and Yu, R.Q. (2006) A Reagentless Tyrosinase Biosensor Based on 1,6-Hexanedithiol and Nano-Au Self-Assembled Monolayers. Electroanalysis, 18, 1572-1577.

78. Nakanishi, K., Sakiyama, T., Kumada, Y., Immamura, K. and Imanaka, H. (2008) Recent Advances in Controlled Immobilization of Proteins onto the Surface of the Solid Substrate and Its Possible Application to Proteomics. Current Proteomics, 5, 161-175.

79. Snabe, T., Rode, G.A., Neves-Petersen, M.T., Buus, S. and Petersen, S.B. (2006) Oriented Coupling of Major Histocompatibility Complex (MHC) to Sensor Surfaces Using Light Assisted Immobilization Technology. Biosensors and Bioelectronics, 21, 1553-1559.

80. Madoz-Gúrpide, J., Abad, J.M., Fernández-Recio, J., Vélez, M., Vázquez, L., Gómez-Moreno, C. and Fernández, V.M. (2000) Modulation of Electroenzymatic NADPH Oxidation through Oriented Immobilization of Ferredoxin: NADP+ Reductase onto Modified Gold Electrodes. Journal of the American Chemical Society, 122, 9808-9817.

81. Ha, T.H., Jeong, J.Y. and Chung, B.H. (2005) Immobilization of Hexa-Arginine Tagged Esterase onto Carboxylated Gold Nanoparticles. Chemical Communications, 48, 3959-3961.

82. Brown, K.R., Fox, A.P. and Natan, M.J. (1996) Morphology-Dependent Electrochemistry of Cytochrome c at Au Colloid-Modified SnO2 Electrodes. Journal of the American Chemical Society, 118, 1154-1157.

83. Lin, J., Qu, W. and Zhang, S. (2007) Disposable Biosensor Based on Enzyme Immobilized on Au-Chitosan-Modified Indium Tin Oxide Electrode with Flow Injection Amperometric Analysis. Analytical Biochemistry, 360, 288-293.

84. Tangkuaram, T., Ponchio, C., Kangkasomboon, T., Katikawong, P. and Veerasai, W. (2007) Design and Development of a Highly Stable Hydrogen Peroxide Biosensor on Screen Printed Carbon Electrode Based on Horseradish Peroxidase Bound with Gold Nanoparticles in the Matrix of Chitosan. Biosensors and Bioelectronics, 22, 2071-2078.

85. Shumyantseva, V.V., Carrara, S., Bavastrello, V., Jason Riley, D., Bulko, T.V., Skryabin, K.G., Archakov, A.I. and Nicolini, C. (2005) Direct Electron Transfer between Cytochrome P450scc and Gold Nanoparticles on Screen-Printed Rhodium-Graphite Electrodes. Biosensors and Bioelectronics, 21, 217-222.

86. Abad, J.M., Velez, M., Santamaria, C., Guisan, J.M., Matheus, P.R., Vazquez, L., Gazaryan, I., Gorton, L., Gibson, T. and Fernandez, V.M. (2002) Immobilization of Peroxidase Glycoprotein on Gold Electrodes Modified with Mixed Epoxy-Boronic Acid Monolayers. Journal of the American Chemical Society, 124, 12845-12853.

87. Challa, S.S.R.K. (2010) Nanocomposites: Nanomaterials for the Life Sciences. Wiley-VCH, Weinheim.

88. Yang, W.W., Wang, J.X., Zhao, S., Sun, Y.Y. and Sun, C.Q. (2006) Multilayered Construction of Glucose Oxidase and Gold Nanoparticles on Au Electrodes Based on Layer-by-Layer Covalent Attachment. Electrochemistry Communications, 8, 665-672.

89. Xiao, Y., Patolsky, F., Katz, E., Hainfeld, J.F. and Willner, I. (2003) "Plugging into Enzymes": Nanowiring of Redox Enzymes by a Gold Nanoparticle. Science, 299, 1877-1881.

90. Gao, F.X., Yuan, R., Chai, Y.Q., Chen, S.H., Cao, S.R. and Tang, M.Y. (2007) Amperometric Hydrogen Peroxide Biosensor Based on the Immobilization of HRP on Nano-Au/Thi/Poly (p-Aminobenzene Sulfonic Acid)-Modified Glassy Carbon Electrode. Journal of Biochemical and Biophysical Methods, 70, 407-413.

91. Yang, G., Yuan, R. and Chai, Y.Q. (2008) A High-Sensitive Amperometric Hydrogen Peroxide Biosensor Based on the Immobilization of Hemoglobin on Gold Colloid/L-Cysteine/Gold Colloid/Nanoparticles Pt-Chitosan Composite Film-Modified Platinum Disk Electrode. Colloids and Surfaces B, 61, 93-100.

92. Tang, L., Zeng, G.M., Shen, G.L., Li, Y.P., Zhang, Y. and Huang, D.L. (2008) Rapid Detection of Picloram in Agricultural Field Samples Using a Disposable Immunomembrane-Based Electrochemical Sensor. Environmental Science & Technology, 42, 1207-1212.

93. Turner, M., Golovko, V.B., Vaughan, O.P., Abdulkin, P., Berenguer-Murcia, A., Tikhov, M.S., Johnson, B.F. and Lambert, R.M. (2008) Selective Oxidation with Dioxygen by Gold Nanoparticle Catalysts Derived from 55-Atom Clusters. Nature, 454, 981-983.

Chapter 2

New Materials for the Construction of Electrochemical Biosensors

Robert Săndulescu[1], Mihaela Tertiş[1], Cecilia Cristea[1] and Ede Bodoki[1]

[1] Analytical Chemistry Department, "Iuliu Haţieganu" University of Medicine and Pharmacy Cluj-Napoca, Romania

ABSTRACT

The development of electrochemical sensors has attracted great interest due to these sensors' high sensitivity and selectivity. Here, we present the general concept and the classification of biosensors, their advantages and drawbacks, the main strategies in electrochemical biosensor technology and the materials used in electrochemical sensors, such as electrodes and supporting substrates, materials for improved sensitivity and selectivity, materials for bioreceptor immobilization, and biological recognition elements. Various nanomaterials, such as carbon-based materials (carbon nanotubes, graphene, carbon nanoparticles), inorganic and organic nanoparticles (magnetic and metal nanoparticles, nanosized clays), conductive and insulating polymers (nanosized and nanostructured polymers, molecularly imprinted polymers), and hybrid materials, etc., have been successfully applied for the enhancement of the electroanalytical performance of biosensors and for the immobilization of biorecognition elements. Among these, due to their unique physiochemical features, carbon-based materials, such as carbon nanotubes and graphenes, have received special attention in recent years, and examples of surface functionalization using various types of nanoparticles are presented. The future trends in sensor research activities and areas of development that are expected to have an impact in biosensor performance, like immobilization techniques, nanotechnology, miniaturization and multisensor array determinations, are also examined.

Keywords: Carbon-based nanomaterials, metal nanoparticles, magnetic nanoparticles, nanostructure, molecularly imprinted polymers

1. INTRODUCTION

The development of electrochemical sensors has attracted a great deal of interest due to their high sensitivity and selectivity, and they are being increasingly used in many fields, such as analytical chemistry, industrial process monitoring and control, clinical diagnostics, environmental monitoring and security, and food safety. By adding an enzyme at the electrode surface, they can also gain high specificity: thus the biosensor research area was born.

Biosensors (analytical devices coupling a transducer with a biorecognition element) are attractive for pharmaceutical and biomedical analysis due to their sensitivity (ng/mL or less) and high selectivity, and sometimes their specificity, high benefit/cost ratio, simple use and rapidity of data collection [1].

The major advantages of biosensors over traditional analytical methods, which will certainly lead in the near future to their even more pronounced use in the biomedical field, are: the fact that analyte detection can very often be made without prior separation; the short response times that make possible the real-time monitoring of biological and manufacturing processes; their ease of use, allowing in-field or point-of-care measurements; the flexibility and simplicity of preparation; the possibility of mass production and low production costs; and the possibility of miniaturization and automatization. Miniaturization is of great importance because biological samples are available in small amounts, and tissue damage must be minimized in cases of in-vivo monitoring. Therefore, the use of biosensors as components of modern medical devices has improved their portability, functionality and reliability for point-of-care analysis and real-time diagnosis [2].

However, many amperometric biosensors described in the recent literature still display a few drawbacks compared to other analytical methods. The most difficult problems to overcome for biosensors with biomedical applications are: the reduced stability, the electrochemical interferences, and the lack of or low response reproducibility. Removed from their natural environment, most biocomponents tend to rapidly lose their activity and thus limit the lifetime of the sensor. Modifying electrode surfaces in a way that exclusively favours one single electrochemical process is a difficult task and sample matrices in the biomedical field are very complex. Last but not least, in the case of in-vivo measurements biocompatibility and biofouling are critical issues [3].

The selectivity of the biosensor for the target analyte is mainly determined by the biorecognition element, whilst the sensitivity of the biosensor is greatly influenced by the transducer.

New sensing elements like genetically transformed cells and genetically engineered receptor molecules can improve affinity and specificity; therefore, genetic engineering and mass production of the molecular recognition may ultimately dictate the success of biosensor technologies. Enzyme variants that are specific for individual analytes have already been obtained by genetic modification, and novel gene fusions will lead to more sensitive biosensors [4].

As for the electrochemical transducer, important advances have been recently made thanks to the introduction of new platforms for biosensor design, such as nanotechnological materials and nanostructured architectures (i.e., nanoparticles, carbon nanotubes and nanofibres, graphenes, nanostructured surfaces, etc.), which have improved the sensitivity of the assembly [5].

2. STRATEGIES OF ELECTROCHEMICAL BIOSENSOR TECHNOLOGY

The biosensor development and construction strategy includes five features: 1) the detected or measured parameter and the matrix, 2) the working principle of the transducer, 3) the chemical/biochemical model, 4) the field of application and 5) the technology and materials for sensor fabrication [6].

Materials generally used for electrochemical sensors are classified as: (1) materials for the electrode and supporting substrate, (2) materials for improving electroanalytical performances, (3) materials for the immobilization of biological recognition elements and (4) biological elements; the last two are applicable for electrochemical biosensors [5].

Materials used for the electrode and supporting substrate are usually conductive materials exhibiting low currents in an electrolyte solution, free of any electroactive species, over a relatively wide potential window. Among the most frequently used materials for the electrode and supporting substrate, the following may be mentioned: metals (mercury, platinum, gold, silver and stainless steel), metal oxides (indium tin oxide, ITO) carbon-based materials (glassy carbon, graphite, carbon black and carbon fibre), new hybrid materials, and organic electroconductive polymers or salts. Boron-doped diamond is a special case of material, because bare diamond is a non-conducting, allotropic form of carbon. In the last decade, in order to avoid mercury-based electrodes (which are highly toxic), different metallic films (Au, Hg, Bi, Ag, Pb, Sn, Sb, Co, Ga, Se) that are relatively simple to achieve and easy to reproduce were developed. Some strategies to develop two- and three-dimensional nanostructured electrodes with larger surfaces (mesoporous silicates, metal oxides, polymers or carbons) were also recently reported [7]. The major challenge in the field of electrochemical sensors, consisting in the improvement of their electroanalytical performances, mainly in terms of sensitivity and selectivity, was addressed by the development of nanotechnology (especially nanoparticles, carbon nanotubes and graphenes) and nanostructured architectures, and this will be detailed in section 3.

The stable immobilization of a bioelement on an electrode surface, with complete retention of its biological activity and good diffusion properties for substrates, is a crucial problem for the commercial development of biosensors. A successful matrix should stably immobilize or integrate biomolecules at the transducer's surface, it should efficiently maintain their functionality, while providing accessibility for the target analyte, and it should also ensure an intimate contact with the transducer surface [8]. The immobilization matrix may function purely as a support or may also be involved in the mechanism of signal transduction mediation. The selection of an appropriate immobilization method depends on the nature of the biological element, the type of transducer used, the physicochemical properties of the analyte and the biosensor's operating conditions [9]. The choice of materials for the immobilization of biological recognition elements strongly depends on the employed procedure. The procedures that may be employed are as follows: 1) Entrapment behind a permeable membrane as a thin film covering the electrochemical detector (dialysis membrane) [10]; 2) Entrapment within a polymeric matrix (polyacrylonitrile, agar gel, polyurethane or poly(vinyl-alcohol) membranes, sol gels or redox hydrogels with redox centres such as $[Os(bpy)_2Cl]^+$, conducting polymers [11]; 3) Entrapment within self-assembled monolayers (SAMs) [12] or bilayer lipid membranes (BLMs) [13]; 4) Covalent bonding on membranes or surfaces activated by means of bifunctional groups or spacers (glutaraldehyde, carbodiimide, SAMs or multilayers, avidin-biotin, silanization) [14]; 5) Bulk modification of the electrode material (graphite epoxy resin or carbon paste modified with polyoxometallates, clays and double layered hydroxides, zeolites, functionalized silica, sol-gel-derived inorganic and hybrid materials, nanomaterials, sparingly soluble or insoluble inorganic salts, molecular and macrocyclic ligands, complex compounds, organic polymers, surfactants and lipids) [7].

The **biorecognition elements** used for biosensor development are classified into biological and artificial (biomimetic) receptors. They are selected depending on the application and the desired performance criteria, based on the following interaction principles: enzyme/substrate, antigen/antibody, mRNA/DNA, PNA/DNA or RNA, microorganism/substrate or toxic molecules, cell receptor/hormone and (semi)synthetic receptor/target molecule [15]. Initially, three types of biorecognition elements were identified [16]: 1) Enzyme (mono or multienzyme), the most common and best developed recognition system, including antibodies, natural receptors and nucleic acids; 2) Whole cells (microorganisms, such as bacteria, fungi, eukaryotic cells or yeast), or cell organelles or particles (mitochondria, cell walls); 3) Tissue (plant or animal tissue slice). More recently, the range of biochemical receptors was extended with biomimetic recognition elements (aptamers, ribosymes and molecularly imprinted polymers) [17]. The biological recognition elements are based either on catalytic or bioaffinity features, whilst biomimetic ones are based on biocomplexing and bioaffinity.

The **biocatalytic recognition sensors** are based on a reaction catalysed by macromolecules, either in their original biological environment, in their isolated form, or as a product of semisynthesis. Thus, a continuous consumption of substrate(s) is achieved by the immobilized biocatalyst incorporated into the sensor, and transient or steady-state responses are monitored by the integrated detector. In the construction of biosensors, **enzymes** are the most commonly used biological elements, despite their rather high extraction, isolation and purification costs, as they rapidly and cleanly form selective bonds with the substrate. Among the enzymes commercially available the most commonly used are the oxidases, such as glucose oxidase (GOx) and horseradish peroxidase (HRP), beta-lactamase, urease, tyrosinase, and acetylcholine esterase or choline oxidase.

The **biocomplexing or bioaffinity sensors** are based on **antibodies, natural biological chemoreceptors, nucleic acids, aptamers or molecularly imprinted polymers (MIPs)**. These biosensors provide selective interactions of the analyte with a given ligand to form a thermodynamically stable complex. The analyte interacts with macromolecules or organized molecular assemblies that have either been isolated from their original biological environment or engineered. Thus, equilibrium is usually reached and there is no further consumption of the analyte by the immobilized biocomplexing agent. These equilibrium responses are either monitored by the integrated detector or by using a complementary biocatalytic reaction. The most developed examples of biosensors using biocomplexing receptors are based on immunochemical reactions, i.e., the binding of an antigen (Ag) to a specific antibody (Ab), and will be discussed in detail in the chapter concerning immunosensors.

The *biological chemoreceptors* [18] are used for drug and neurotransmitter detection, and they are based on an affinity reaction. Using *nucleic acids* as a biocomponent, one can detect the interaction between the synthetic drug and the genetic material and also the hybridization processes by working with single-strand DNA [19, 20].

The use of *microorganisms and of whole vegetal or animal tissues* is complex and very often their selectivity is limited [21].

3. NANOMATERIALS INVOLVED IN BIOSENSOR CONSTRUCTION

Advances in nanotechnology have led to the recent development of many nanomaterials, including carbon nanomaterials, magnetic and metallic nanoparticles [22]. Nanomaterials, generally defined as materials with feature sizes smaller than 100 nm, have a remarkable impact on various fields of application due to their properties (enhanced electrical conductivity, tensile strength and chemical reactivity), which are imparted by their increased surface area per unit weight. Nanomaterials have already been applied in electronics, foods, cosmetics, electronics, drug development and sensing devices. The use of nanomaterials, especially nanoparticles and nanostructured films, offers advantageous properties that can be exploited in order to maximize the interactions with specific bioelements, to maintain their activity, minimize structural changes, and enhance the catalytic step. In the biosensor field, the analytical exploitation of such protein-nanomaterial interactions are an emerging trend that spans many disciplines [23]. Nanomaterials based on metals, semiconductors and organic compounds contribute to the enhancement of optical, electrical, chemical and magnetic properties that are relevant in the case of sensing devices, and due to these facts they have been frequently used in this research field [24, 25].

Nanomaterials are mainly used *for electrode construction or modification* and *as biomolecule tracers*. Nanoparticles (NPs) are very stable (compared to enzyme labels), offering high sensitivity (thousands of atoms can be released from one nanoparticle) and a wide variety are available on the market. Nanoparticles are used nowadays as electrochemical labels or as vehicles containing several hundreds or thousands of electroactive labels, pushing detection limits down to several hundreds of biomolecules [26].

Various nanomaterials have been successfully applied for the immobilization of bioelements, such as metallic nanoparticles, carbon or metallic nanotubes, nanosilver-coated magnetic beads, magnetic particles, functionalized conductive polymers, etc.

3.1. Carbon-Based Materials

Different types of carbon nanostructures are becoming more and more popular due to their specific structures, properties and the possibility of being used for many applications. In fact, a wide variety of carbon-based materials are available, such as nanoparticles, nanodiamonds, nano-onions, peapods, nanofibers, nanorings, fullerenes and nanotubes, which have been extensively used in analytical applications. The basic structure of fullerenes and nanotubes consists of a layer of sp^2-bonded carbon atoms. This configuration, which resembles that of graphene, is responsible for their good electrical conductivity and their ability to form charge-transfer complexes when in contact with electron donor groups [27]. This configuration is responsible for the development of strong van der Waals forces that significantly hamper the solubility and dispersion of carbon-based nanoparticles. In order to avoid these problems, different pretreatment methods have been proposed [28, 29, 30], such as the addition of polar groups (oxygen, hydroxyl, phenyl and polyvinylpyrrolidone). The surface defects could also affect the stability, the mechanical and electrical properties of carbon nanostructures [31, 32, 33, 34].

The carbon nanomaterials (Figure 1) cover a broad range of structures, beginning with zero-dimensional structures (fullerenes, diamond clusters), continuing with one-dimensional (nanotubes), two-dimensional (graphene), and three-dimensional structures (nanocrystalline diamond, fullerite).

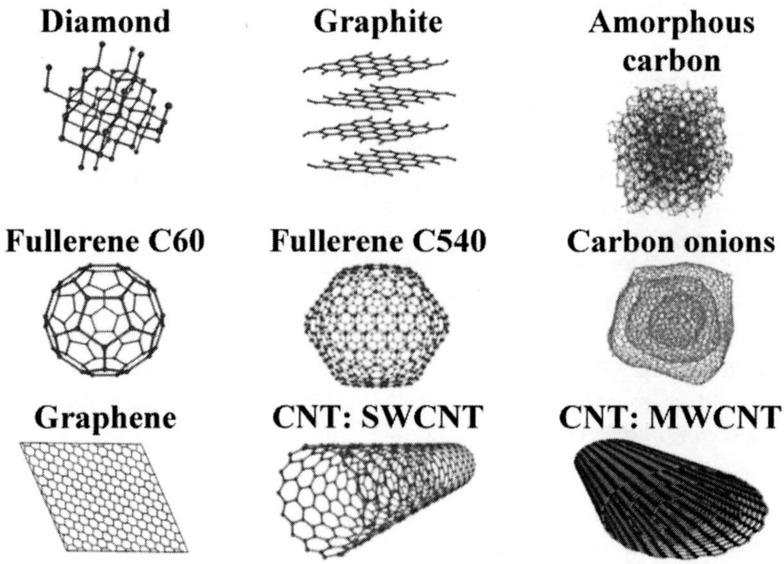

Figure 1. Main carbon entities at nanoscale level

Carbon nanomaterials, including fullerenes, graphene and carbon nanotubes, have many technological advantages such as facile modification by functional groups, high carrier capacity, incorporating both hydrophilic and hydrophobic substances, and high chemical stability.

3.1.1. Carbon nanotubes

Within the family of nanomaterials, carbon nanotubes (CNTs), described for the first time by Iijima [35], are arousing growing interest in relation to several modern devices, such as transducers, for the improvement of biosensing systems, mainly due to their unique structural, electronic and chemical properties, which include large surface area, high electrical conductivity, and excellent properties of electron transfer reactions. Furthermore, CNTs have a unique tubular structure, good biocompatibility and modifiable sidewall, making them ideal candidates for the construction of sensors with high performances [36].

Electrochemical nanobiosensors have been increasingly used in various types of nanomaterials, mainly CNTs, as electrochemical transducers, in order to improve them from the perspectives of automation, miniaturization and multiplexed analysis, but also to improve the analytical performances of such devices [37, 38]. For example, the detection of diseases at an early stage is one of the goals in developing and improving CNT biosensors, focusing on their sensitivity, fast response and small sample volume, in order to provide new strategies for the detection of specific biomarkers at low concentrations in complex sample media (e.g., serum) [39].

Many properties and the quality of CNTs are directly influenced by the way in which the graphene sheets are wrapped around [40], and they are affected by the operating conditions in their fabrication process. The production methods generally require transition metal nanoparticles (i.e., Fe, Co and Ni) to function as growth catalysts. When the catalysts are in contact with a gaseous carbon or a hydrocarbon source, the resulting carbon is deposited on the particle surface and the CNT grows rapidly at the surface of the catalyst. The diameter of the CNTs, their quantity and quality can be controlled by varying some reaction parameters (e.g., temperature, metal concentration, gas type and pressure). The diameter of the NPs defines the nature of the CNTs, whether single-walled (SWCNTs) with diameters in the range 0.4-3 nm, or a group of concentric CNTs sharing a common axis constituting double- and multiwalled (MWCNTs) with diameters in the range 1.4-100 nm [36].

In the case of CNT structures, the reactivity of atoms situated in the plane is different from those at the edges. For a number of biologically important compounds, the electrochemistry of atoms located at the edge-plane sites of the CNT is comparable to that of different planes of graphite [41], and the metal and non-metal impurities contained determine some of the electrochemical catalytic properties and limitations of CNTs [42, 43, 44]. It is also worth mentioning that the electrode surface modification with CNTs leads to a changed mass transport regime from semi-infinite diffusion to effective thin-layer diffusion in a porous layer, and the increased current may frequently be erroneously assigned to electrocatalysis [45, 46, 47].

Currently, SWCNTs and MWCNTs are synthesized by high-temperature processes (e.g., electric arc discharge and laser ablation, which are physical methods to convert carbon into NTs) and chemical vapour deposition (CVD), which is performed at lower temperatures and atmospheric pressure.

Usually, the nanostructured materials obtained using CNTs were morphologically characterized using transmission electron microscopy (TEM) and scanning electron microscopy (SEM). Figure 2 presents the SEM images obtained for SWCNT [48] and MWCNT [49].

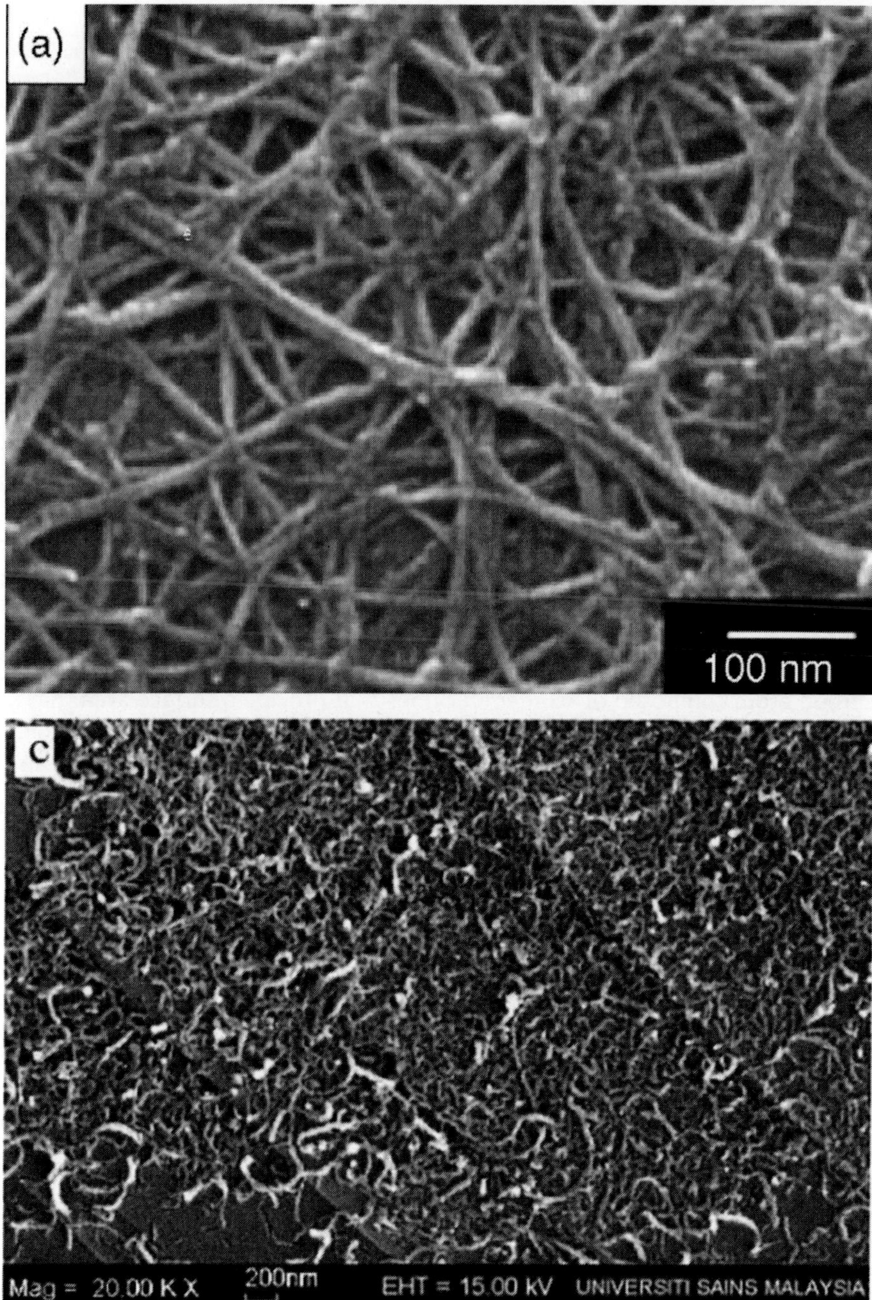

Figure 2. (A) SEM image of SWCNT composite film [48]; (B) FE-SEM images of MWCNT in high magnification [49]

Many papers published in recent years report CNT biosensors based on different oxidases applied for the detection of glucose [50, 51, 52, 53, 54], H_2O_2 [55, 56, 57], *L*-lactate [58, 59, 60], acetaminophen [61, 62, 63, 64, 65], etc. The construction and optimization of new biosensors with HRP immobilized onto the transducer surface, either screen-printed electrode (SPE) or glassy carbon electrode (GCE), by entrapment into polymer films (polyethylenimine (PEI) and polypyrrole (Ppy)) doped with SWCNT and MWCNT, was described [61, 66].

The obtained configurations were used to monitor the signal produced by the electrochemical reduction of the enzymatically generated electroactive *N*-acetylbenzoquinoneimine (the oxidized form of acetaminophen) in the presence of hydrogen peroxide using amperometry, cyclic voltammetry (CV) and differential pulse voltammetry (DPV) in synthetic and real samples, with the best limit of detection (LOD) obtained for acetaminophen of 1.36 μM [61].

3.1.2. Graphene

Graphene is another material that offers great prospects for the future of analytical chemistry, consisting of a one-atom-thick sheet of sp^2 hybridized carbon atom composed of six member rings that provide a surface area that is nearly twice as large as that of SWCNT [43]. This material presents high mechanical strength, high elasticity and high thermal conductivity [43], but also presents a major disadvantage: its poor dispersion in aqueous medium. Stable dispersions of graphene sheets have been achieved using amphiphilic polymers, alkylamines and hydrophilic carboxyl groups, among others, as dispersing agents [67]. The functionalization of graphene is considered an important route for improving its dispersion. Therefore, different authors have tried producing graphene oxide, a graphite derivative with hydroxyl, epoxy and carboxyl groups covalently attached to its layers, with a better dispersion in some solvents being obtained.

Graphene can be interlinked with CNT for the fabrication of high-performance transparent electrodes, and the resulting films present features comparable to indium tin oxide [68]. The electrochemical activity of crystalline graphene is different from that of reduced graphene oxide flakes [69]. A wide variety of methods are available for graphene fabrication, the first being published in 2004 [70], and consist in the mechanical exfoliation (repeated peeling) of small patches of highly ordered pyrolytic graphite. Many other fabrication methods have been developed, including unzipping MWCNT to form graphene ribbons [71, 72], substrate-independent methods using micromoulding inside a capillary [73], and spray deposition of graphene oxide-hydrazine dispersions [74].

SEM images of a graphene film deposited on GCE revealed the typical crumpled and wrinkled graphene sheet structure on the rough surface of the film (Figure 3 A). The TEM image presented inFigure 3B shows the wrinkled graphene sheet with no aggregation, indicating that the functionalized graphene sheets were well dispersed in ethanol, forming stable suspensions [75].

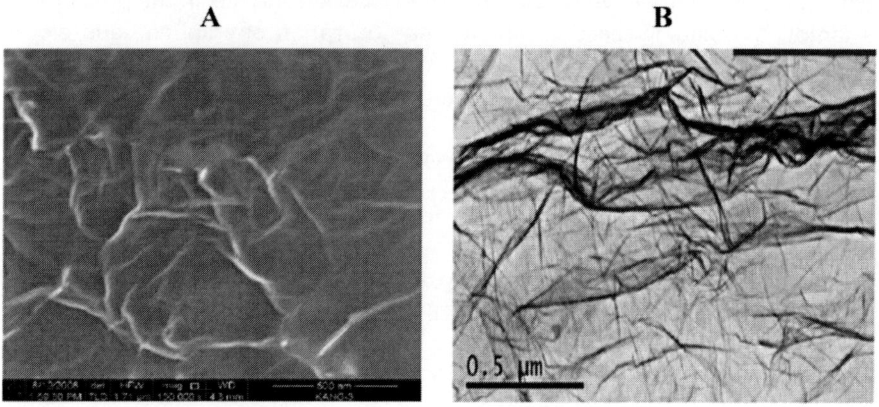

Figure 3. (A) SEM image of graphene-film-modified GCE. (B) TEM image of graphene in ethanol [75]

With regard to aptasensor development, graphene looks like it could have the necessary requirements to implement next-generation and high-performance aptasensors. Graphene naturally adsorbs the unfolded aptamer through the π–π stacking interaction between the purine and pyrimidine bases and the graphene plane, and hinders the adsorption of the folded aptamer [76, 77].

The currently developed strategies for graphene functionalization with aptamers include covalent and non-covalent approaches. The covalent approach often relies on the oxidation or the reduction of graphene oxide that presents carbonyl and carboxyl functional groups [78]. The density of carbonyl and carboxyl functional groups can be increased through chemical oxidation, which is widely used for the chemical modification of the inert graphitic structure [79]. Because this procedure affects the properties of graphene, the covalent modification of graphitic surface through a free radical addition reaction is used, without a preoxidation step and with less impact on the electronic properties [80]. Besides the immobilization methods mentioned above, amide [81] and phosphoramidate reactions [82] were employed towards the fabrication of graphene-based biosensors. Some graphene derivatives, such as graphene oxide, were also used as transducers in electrochemical biosensor development [83].

Graphene can also mediate electron transfer in view of the high-density edge-plane-like defects present within its nanostructure. In the absence of a target, aptamers immobilized on the gold electrode adsorb graphene nanosheets due to the strong π–π interaction that accelerates the electron transfer ratio between the electroactive species and the electrode surface. In the presence of a target, the binding reaction inhibits the adsorption of graphene and blocks the electron transfer. A label-free electrochemical aptasensor was constructed by taking advantage of the ultra-fast electron transfer ratio of graphene [84]. This detection strategy was further improved by nuclease for interferon gamma, yielding an LOD of 0.065 pM [85]. Other authors [86] employed graphene oxide nanoplatelets (GONPs) as electroactive labels for thrombin detection.

A simple label-free electrochemical immunosensor was constructed by modifying a graphite-based SPE with graphene oxide after functionalization with *N*-hydroxysuccinimide in the presence of 1-ethyl-3-(3-dimethyl aminopropyl) carbodiimide hydrochloride, Figure 4.

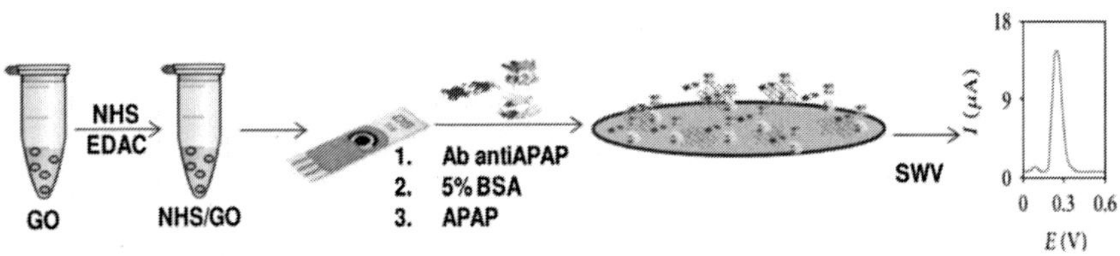

Figure 4. Steps involved in the antiacetaminophen Ab-based immunosensor development; GO=Graphene oxide; NHS=*N*-hydroxysuccinimide; EDAC=1-ethyl-3-(3-dimethyl aminopropyl) carbodiimide hydrochloride; Ab anti APAP=antibody antiacetaminophen; APAP=acetaminophen; SWV=square wave voltammetry [87]

The above-described immunosensor was applied with good results for the determination of acetaminophen in synthetic and real samples by using square wave voltammetry (SWV) with an LOD of 0.17 μM [87].

3.1.3. Carbon Nanoparticle-Based Electrochemical Sensor

A sensitive and selective electrochemical sensor fabricated via the drop-casting of a suspension of carbon nanoparticles (CNPs) onto a GCE was investigated in simultaneous determination of acetaminophen and tramadol in pharmaceuticals and human plasma. CV and DPV studies of acetaminophen and tramadol were carried out at the modified and bare GCE. The results of the electrochemical investigations showed that CNPs enhanced the electroactive surface area and caused a remarkable increase in the peak currents, due to the diffusion within the porous layer. Enhanced sensitivity and a considerable decrease in the anodic overpotential leading to negative shifts in peak potentials was obtained with the modified GCE. By using DPV, the sensor showed good sensitivity and selectivity: LODs for acetaminophen and tramadol were 0.05 and 1 μM, respectively [88].

3.2. Inorganic and Organic Nanoparticles

Nanoparticles of major environmental, pharmaceutical and biomedical importance include magnetic nanoparticles, quantum dots, metal nanoparticles, silica nanoparticles and polymeric types with intrinsic properties contributing to their use in specific applications. Typically, nanoparticles are classified into: *magnetic nanoparticles* (possessing a core of Fe_3O_4) frequently used in drug delivery and lately in immunosensing; *semiconductor/QDs* (based on CdS, CdSe) used in bio- and immunosensing; *metal nanoparticles* (Ag, Au, Pd) used in biosensing and drug delivery; *polymeric nanoparticles* (polystyrene) mainly used in drug and gene delivery; *hybrid nanoparticles* (carbon or Fe_3O_4 covered with different materials like metal oxides, polymers, amino acids, etc.) [89].

3.2.1. Magnetic Nanoparticles

In recent years, considerable efforts were made to develop magnetic nanoparticles (MNPs), due to their inherent advantages (magnetism, nanosize) and low cost of production. MNPs with 10-20 nm diameters exhibit their best performance, due to supermagnetism, which makes them especially suitable for a fast response. Due to their inherent dimensions, they possess a large surface area and high mass transfer capacity [90].

MNPs can be integrated into the transducer materials and/or be dispersed in the sample followed by their attraction by an external magnetic field onto the active detection surface of the (bio)sensor. Biorecognition molecules like enzymes, antibodies or oligonucleotides immobilized onto magnetic particles can be easily trapped by magnets and retained close to or on an electrode surface [90, 91]. Since the properties of MNPs depend strongly on their dimensions, their synthesis and preparation have to be designed in order to obtain particles with adequate size-dependent physicochemical properties. MNPs possessing adequate physicochemistry and tailored surface properties have been synthesized under precise conditions and applied for sensors, biosensors and other detection systems [92, 93].

Magnetic particles consist of a magnetic core surrounded by a non-magnetic shell. Iron oxides, such as magnetite (Fe_3O_4) or maghemite (γ-Fe_2O_3), are often used as core materials instead of iron due to their higher stability [94]. For biological applications, magnetic particles generally consist of nanosized superparamagnetic iron oxide distributed through a matrix, mainly composed of silica [95] or an organic polymer [96]. Silica-based magnetic beads can be prepared in different shapes and sizes with different degrees of porosity, which are much more attractive for the immobilization of biomolecules due to their chemical and mechanical stability and their resistance to bacterial attack [97]. Many types of magnetic particles, functionalized with different groups like carboxyl, amino, hydroxyl or epoxy, are currently available for different applications.

The majority of MNP-based systems use inorganic nanocrystals as magnetic cores. The composition of these inorganic nanocrystal ranges from metals and alloys to metal oxides. Among the metal oxide compounds, superparamagnetic iron oxide nanoparticles, including magnetite (Fe_3O_4) and maghemite (γ-Fe_2O_3), are the most popular [98]. The surfaces of the particles can be modified with organic polymer or inorganic compounds (e.g., gold, silica, alumina), suitable for further functionalization by the attachment of various bioactive molecules [99]. The functionalization of MNPs with bioelements exhibits specificity, due to both the bioreceptor and the magnetism.

MNPs amplify the electrochemical signal, improving the sensitivity of electrochemical devices [90], through their contact with the electrode surface, transport of a redox-active species to the electrode surface, and formation of a thin film on the electrode surface. The detection for MNP-based electrochemical biosensors can be performed by potentiometry [100, 101], amperometry [102, 103], electrochemiluminescence [104, 105], electrochemical impedance spectroscopy [106, 107] and especially voltammetry [108, 109, 110, 111, 112, 113, 114, 115].

Protein G-MNP-based label-free electrochemical immunosensor was elaborated for the ultrasensitive and specific detection of acetaminophen using a carbon-based SPE as immobilization platform (Figure 5). An LOD of 1.76 μM was calculated based on the calibration data. The performances of the optimized immunoassay were tested using two pharmaceutical products containing acetaminophen with excellent recoveries [116].

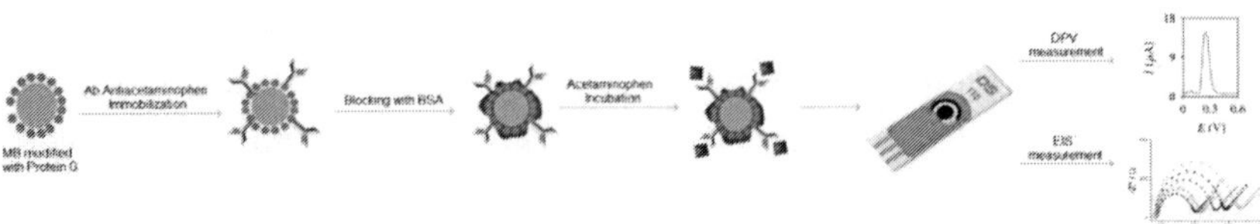

Figure 5. Protein G-MNP-based label-free electrochemical immunosensor development [116]

Protein G-functionalized magnetic beads and graphite-based SPEs were also used to develop an HRP-labelled sandwich type immunosensor for Mucin 1 (MUC1) electrochemical detection with LOD of 1.34 ppb [117].

Fe_3O_4 magnetic dipolar attraction and its large surface area/volume ratio may lead to aggregation in clusters when exposed to biological solutions. Functionalization can overcome this problem and also enhance biocompatibility. Many strategies were used for the functionalization of MNPs, such as core-shell Au-Fe_3O_4 [109], core-shell Au-Fe_3O_4@SiO_2 [102], core-shell Fe_3O_4@SiO_2 [112], Au-Fe_3O_4composite NPs [108], Fe_3O_4@SiO_2/MWCNTs [103], Fe_3O_4 anchored on reduced graphene oxide [113] and Fe_3O_4@ Au-MWCNT-chitosan [114]. Among all these, core-shell Fe_3O_4@SiO_2 contributes to the stabilization of MNPs in solution and enhances the binding of ligands at the surface of MNPs. Electrode surface modified with core-shell Fe_3O_4@SiO_2 shows good electrical conductivity and more electroactive interaction sites, providing enhanced mass transport and easier accessibility to the active sites, thus increasing the analytical signal and sensitivity [90]. A comparison between the analytical performances of some recently developed electrochemical sensors and biosensors based on MNPs is presented in Table 1.

Table 1. Selected examples of sensors and biosensors based on magnetic nanoparticles

Detection method	MNPs type	Analyte	LOD	Reference
Voltammetry	Au-Fe_3O_4	Organochloride pesticides	56 pg/mL	[108]
Voltammetry	Core-shell Au-Fe_3O_4	Carcinoembryonic Ag	10 ng/mL	[109]
Voltammetry	Fe_3O_4 Au NP	Clenbuterol	220 ng/mL	[110]
Voltammetry	Fe_3O_4 Au NP	H_2O_2	20 μM	[111]
Voltammetry	Core-shell Fe_3O_4@SiO_2	Metronidazole	18 nM	[112]
Voltammetry	Fe_3O_4-RGO	Cr(III)	-	[113]
Voltammetry	Fe_3O_4@Au-MWCNT-chitosan	Streptomycin	1.5 nM	[114]
Voltammetry	Core-shell Fe_3O_4@SiO_2/ MWCNT	Uric acid	0.13 μM	[115]
Potentiometry	MaMB Dynabeads Protein G	Zearalenone	7 ng/mL	[100]
Potentiometry	Core-shell-Fe_3O_4	Glucose	0.5 μM	[101]
Amperommetry	Core-shell Au-Fe_3O_4@SiO_2	Glucose	0.01 mM	[102]
Amperommetry	Fe_3O_4@SiO_2-MWCNT	Glucose	800 nM	[118]
EIS	Fe_3O_4-COOH/MNP	OchratoxinA	10 pg/mL	[106]
EIS	Fe@Au NP-2-aminoethanethiol - graphene NP	DNA	2 fM	[107]
Detection method	MNPs type	Analyte	LOD	Reference

3.2.2. Metal Nanoparticles

Metal nanoparticles and CNTs, have been employed to immobilize biomolecules for the construction of sol-gel biosensors, due to their large surface area, high catalytic activity and good biocompatibility. Moreover, the excellent conductivity of gold nanoparticles (AuNPs) provides the potential to construct amperometric biosensors based on redox enzymes (as sensing elements) and nanoparticle arrays (as conductive matrices where enzyme molecules are immobilized) [118, 119]. Metal nanoparticles, such as Pd and Pt, were deposited on SWCNTs and MWCNTs in order to increase the electrocatalytic activity of CNTs. An improved conductivity and electrocatalytic behaviour was obtained by combining sol-gel materials with CNTs, the electron transfer between the redox enzyme and the electrode being facilitated [120].
Gold nanomaterials have unusual optical and electronic properties, high stability and biological compatibility. Aptamer-conjugated gold nanomaterials provide powerful platforms to facilitate targeted recognition, detection and therapy applications, due to their controllable morphology, size dispersion and easy surface functionalization [121].

Many applications of nanoparticles, such as metal NPs (gold, silver) or semiconductor dots in biosensing, have been recently reported [122, 123]. A DNA array detection system based on oligonucleotide targets labelled with AuNPs was developed [124], and AuNPs or silver-enhanced colloidal gold were used as labels for electrochemical detection of DNA hybridization [125, 126, 127]. These NP-based electrochemical biosensors employed the stripping voltammetric technique as the read-out principle due to the preconcentration step resulting in ultra-trace-level LODs.

AuNPs are irreversibly immobilized on the transducer, the electrode surface being difficult to regenerate. To overcome this problem, Chen *et al.* [128] have developed a novel electrochemical system for the sensitive detection of glucose (LOD of 1 mM) as well as single-nucleotide polymorphism, which has the AuNPs directly in the supporting electrolyte. This one-pot detection system can be operated and regenerated very easily, since all the components are integrated in the electrolyte of the gold colloid and the unmodified electrode can be reused.

The sensing mechanism of potentiometric biosensors using NPs is based on a sandwich assay (Figure 6), the target being captured by the immobilized primary receptor and then attached to the NP-labelled secondary receptor. Potentiometry is finally used to detect the dissolution products of NPs after a release step with H_2O_2 [123].

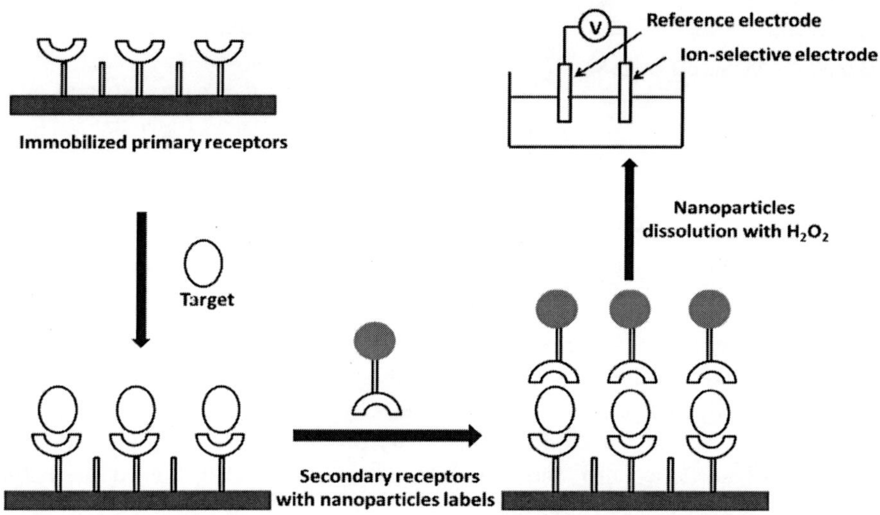

Figure 6. Potentiometric biosensors using nanoparticles as labels [123]

Miniaturized ion-selective electrodes (ISEs) designed to detect ions in microvolume samples, with a concentration of fM, were reported [129, 130], and an LOD of 12.5 pmol/50μL was obtained for IgG using an ISE based on silver-enhanced AuNPs [131]. Metal NPs can also be used as a catalyst for the reduction of metal ion, via the enzymatically generated reducing agent, and the biometallization processes can be monitored by ISEs in real time [132]. Thereby, an ISE based on AuNPs was developed and applied for highly sensitive detection of glucose [132, 133], while the NADH potentiometric detection was successfully obtained by monitoring the NADH-stimulated catalytic reduction of Cu^{2+}in the presence of AuNPs (which have a crucial role in improving the sensitivity) with a Cu^{2+}-ISE [134]. Wang *et al.* [135] reported a highly sensitive potentiometric biosensor for detecting DNA hybridization based on the depletion of Ag^{+} ions induced by the biocatalytic reaction of alkaline phosphatase, recorded in the presence of AuNPs. The sensor has been used to detect DNA and 16S rRNA of *Escherichia coli* with LOD of 50 fM, using only 4 μL of sample solution.

Two simple electrochemical aptasensors were developed for the detection of Mucin 1 (MUC1) tumour marker based on its specific recognition by thiolated aptamers immobilized on Au NP-modified graphite and gold SPEs. The quantitative detection of MUC1 protein was electrochemically achieved by electrochemical impedance spectroscopy (EIS) and DPV. The estimated detection limits for MUC1 were 3.6 ng/mL at AuNP modified graphite SPE by EIS and 0.95 ng/mL at AuNP modified gold SPE by DPV [136].

3.2.3. NANOSIZED CLAYS

Nanosized clays can serve as matrices for electroactive species because they are usually able to incorporate ions by an ion exchange process. Moreover, adsorption of proteins on clay mineral surfaces plays a very important role in the development of biosensors. By modifying the electrode surface with clays, the heterogeneous electron transfer process between the protein and the electrode surface is facilitated.

The entrapment of biomolecules in the clay matrix represents an inexpensive, fast and easy method for the elaboration of enzyme electrodes. The procedure consisting in the adsorption of an enzyme/clay aqueous colloid mixture onto the electrode surface offered the possibility of entrapping a large and known amount of enzyme into the clay film, and could also be adapted for interdigitated microelectrodes [137].

Mesoporous silica nanoparticles can be easily synthesized by using surfactant templating methods [138]. Silica nanoparticles have become an important platform in biomedical applications and bioanalysis because of their high density, which facilitates separation through centrifugation during their synthesis, and their modification and detection steps [139]. The silica surface also provides a biocompatible and versatile substrate for biomolecule immobilization, required for further utilization in biosystems and for drug delivery [140]. Following typical DNA silica surface conjugate chemistry, the aptamer immobilization on silica nanoparticles has been well developed for various applications [141].

Biosensors based on clay materials can play an important role as efficient tools in the field of forensic analysis considering their well-known advantages of sensitivity, selectivity, simple functioning and affordability [142]. Thereby, a bienzyme biosensor based on acid phosphatase (AcP) and polyphenol oxidase (PPO) simultaneously entrapped in anionic clays was constructed and applied for the specific amperometric determination of As(V). The biosensor relied on the successive hydrolysis of phenyl phosphate to phenol catalysed by AcP and the oxidation of phenol to *o*-quinone catalysed by PPO. The electrochemical reduction of the formed *o*-quinone was accomplished at a detection potential of -0.2 V vs Ag/AgCl with a detection limit of 0.15 ng/mL [143]. DNA-functionalized, layer-by-layer assembled SWCNT hybrids with clay nanoparticles were proposed for the detection of As(III) achieving an LOD of 0.05 mg L^{-1} [144]. Another simple amperometric biosensor for cyanide involved the immobilization of PPO into ion exchanger clay based on Zn-Al double-layered hydroxides (LOD of 0.1 nM). The low LOD was a consequence of the anion accumulation in the ion exchanger clay which also contained the immobilized enzyme [145].

3.2.4. Nanosized And Nanostructured Polymeric Films

One of the most important aspects of biosensor development is the bioelement immobilization platform, which needs to ensure high stability and maintain its activity as long as possible. Several types of **conductive** or **insulating polymers** were used to obtain a nano/micropatterned surface with applications in the biosensor field [146]. Electropolymerization in the presence of a template, removed after the polymer formation, is an efficient route for the synthesis of 2D and 3D conductive polymers (CPs). Among the conductive polymers, polypyrrole and polyaniline [147, 148] were extensively studied because of their electrochemical properties, aqueous compatibility and their ability to form nanostructures like nanowires, nanotubes or microcontainers [149].

Hydrogels are cross-linked hydrophilic polymer structures that can hold large amounts of water or biological fluids in their pores. As one of the newest classes of polymer-based systems, target-responsive hydrogels have found numerous biomedical and pharmaceutical applications [121, 150]. Polyacrylamide/acrylate is a copolymer of acrylamide and acrylic acid, and acrydite-modified oligonucleotides can form hydrogels by the introduction of proper complementary DNA as a cross-linker. He *et al.* [151] reported an aptamer-based reversible DNA induced hydrogel system for molecular recognition and separation.

Different types of hydrogel nanoparticles were used together with various biocomponents, such as enzymes, proteins and nucleic acids, during the development of electrochemical biosensing systems, in order to increase their sensing performances [152].

CPs and in particular functionalized CPs are useful for biosensor construction as a platform to immobilize the primary Ab on the electrode surface. Poly *o*-aminobenzoic acid (PABA) can be considered a carboxyl-functionalized aniline CP, capable of self-doping, and is also a soluble derivative of polyaniline. Its carboxyl group can be exploited as a linker, either for PABA immobilization onto the electrode, or for covalent bond formation with biomolecules such as proteins and antibodies [153].

An electrochemical aptamer-based biosensing assay for MUC1 protein detection by the modification of the graphite-based SPE's surface using a functionalized conductive polymer (PABA), and by using methylene blue (MB) as an electrochemical indicator was developed. The recognition of immunoreactions and aptamer binding events was assessed by monitoring the interfacial electron transfer resistance with EIS, while CV and DPV were employed for sensitive indirect quantification of MUC1 via MB with a detection limit of 0.62 ppb [154]. In order to establish the optimal configuration of HRP-based biosensors able to detect acetaminophen in various matrices, another conductive polymer, polypyrrole, was deposited on GCE and SPE by several methods. Chronoamperometric studies performed proved the interaction between the immobilized enzyme and the electroactive species of paracetamol, NAPQI, enzymatically generated in the presence of H_2O_2 [155].

The combination of nanosphere lithography, electrodeposition and click chemistry was applied for the immobilization of ferrocene at GCE. The nanostructuration of the surface was obtained by adsorption of 100 and 900 nm diameter polystyrene nanospheres followed either by electropolymerization of *N*-(10-azidodecyl) pyrrole or by electrografting of 4-azidobenzenediazonium. This approach offered potential applications such as engineering of porous materials and was used for HRP immobilization. The nanostructured biosensor was applied for acetaminophen detection in the presence of hydrogen peroxide, and showed better performances in terms of sensitivity and LOD in comparison with a non-structured electrode [156]. Another template based on nanostructured polymeric film was developed combining latex nanosphere lithography with electropolymerization of *N*-substituted pyrrole monomer in order to immobilize both tyrosinase and GOx and to obtain a biosensor with higher sensing performances [148].

3.2.5. Molecularly Imprinted Polymers

MIPs can be successfully used in applications relying on selective molecular binding as they are fully artificial macromolecular structures with biomimetic properties, imitating receptor-ligand, Ab-Ag, or enzyme-substrate biorecognition [157].

MIPs possess several advantages over their biological counterparts, including low cost, ease of preparation, long term stability and shelf life, inherent reusability, high mechanical strength, and remarkable chemical and thermal stability under various experimental conditions, allowing the very specific binding of the analyte in “harsh” chemical media [158, 159].

The rational design of MIPs is still considered a challenge due to the numerous experimental variables to be addressed, and it has been previously reviewed in detail by several groups of authors [157, 158,160, 161] along with their most important analytical applications.

MIPs prepared by bulk polymerization have proven to be poorly compatible with transducers in terms of efficient immobilization strategies, mass transfer and rebinding kinetics, even if they are ground before use. Therefore, several aspects related to the in-situ polymerization, and more specifically to the electrosynthesis of MIPs and hybrid technologies involving MIPs in the fabrication of biosensors, as well as their most recent advances in addressing two important challenges for MIP-based electrochemical biosensors, the determination of biomacromolecules and enantiospecific analysis, will be very briefly presented.

Electroactive organic functional monomers can polymerize to form conducting, semiconducting or insulating polymers. Polymerization may be performed chemically (using an oxidizer as external reagent) or electrochemically (by electropolymerization). Electropolymerization is the more convenient approach, in terms of simplicity and time-effectiveness, for electrochemical biosensor fabrication as it leads to a tight binding of an MIP film onto the surface of the transducer. This procedure ensures both the synthesis and the surface immobilization of the biomimetic layer, along with the possibility of controlling polymer nucleation and growth, film thickness and morphology by the proper selection of the electropolymerization parameters. The reproducible preparation of thin MIP layers is of the utmost importance in sensing applications, because it reduces response time by significantly shortening diffusional path lengths. Solvent swelling and the inclusion of the supporting electrolyte's ions tunes the rigidity and porosity of the MIP film. Electropolymerization may be performed under galvanostatic, potentiometric or most frequently potentiodynamic conditions, in both non-aqueous and aqueous solutions. The first two techniques are able to generate well-ordered, charged polymeric layers doped with counter ions for charge neutralization, whereas the latter generates a more tangled, charged or neutral polymeric matrix due to the continuous influx and release of solvated ions during its growth upon film charging and discharging [157].

Using electroinsulating MIP films in the fabrication of electrochemical sensors carries a major disadvantage due to the high resistance to charge transfer and the lack of direct path for electron conduction from the recognition sites to the electrode transducer, suffering from an incomplete template removal upon their overoxidation. Using electronically conducting polymers for MIP synthesis avoids these undesirable effects.

Some of the most common electroinsulating layers employed in MIP technology are acrylic- and vinylic-based polymers, but the most frequently reported electrosynthesized insulating MIPs are polyphenylenediamine, polyphenol, polyaminophenol and polythiophenol. Due to their low conductivity, sensors based on such MIPs usually use optical, piezoelectric or impedimetric signal transduction. However, as long as they are electrodeposited in a thickness that does not insulate the underlying conductive electrode surface (<10 nm) [162], a successful voltamperometric signal transduction may also be performed [163].

In the case of electrosynthesis of conducting MIP layers for biosensing, electroactive functional monomers such as pyrrole, thiophene, aniline, 3-aminophenylboronic acid and porphyrin derivatives are most often reported. Analyte determination using such conducting layers may be performed by any available signal transduction procedure, such as optical, piezoelectric, voltammetric, potentiometric, chronoamperometric, impedimetric or conductometric procedures. Poly(pyrrole), poly(thiophene), poly(aniline) and their derivatives are amongst the most widely studied conducting polymers [157,160], due to their high conductivities in their oxidized form and their ability to reversibly switch between conducting and insulating states by doping and undoping [160]. As mentioned earlier, in the non-covalent molecular imprinting, a wide variety of templates are amenable to be used with elevated imprinting efficiencies, ranging from small analytes to macromolecules (proteins, nucleotides). Additional promising approaches include implementing surface MIP for cells (yeast), bacteria, virus or blood [164, 165, 166]. In the case of using small organic molecules as templates, MIP synthesis is well established, however the imprinting of much larger molecules is still considered to be a challenge. Larger templates are less rigid, which does not ensure the formation of well-defined binding cavities during macromolecular imprinting, and, furthermore, the secondary and tertiary structures of biomacromolecules (i.e., proteins) may be also affected under the imprinting conditions. The large size hinders the protein both in reaching and leaving the artificial binding sites, and their molecular complexity increases the chances of non-specific binding with the MIP, leading to poor selectivity and cross-reactivity. More importantly, proteins are often incompatible with the organic solvents used in MIPs synthesis. The use of an aqueous solution for macromolecular imprinting greatly restricts the choice of reactive functional monomers and cross-linkers. Additionally, water could compete for and potentially disrupt any hydrogen bonds between the template and the functional monomers. The protein as template in macromolecular imprinting may be introduced by three different approaches: it may be dissolved into the prepolymerization mixture (bulk imprinting), grafted to the surface of the transducer followed by a partial inclusion into the bulk polymer with the resulting binding sites situated on the MIP surface (surface imprinting), or only a small epitope part of the protein (epitope imprinting) may be used as a template, resulting in an MIP that is able to recognize the whole target protein. Surface and epitope imprinting enhance the imprinting effect for biomolecules, leading to higher affinity and capacity for the template, as well as to reduced non-specific binding; they improve protein binding kinetics, overcoming difficulties to mass transfer and protein removal in imprinted matrices. While surface imprinting is an efficient technique, the conditions used to accomplish the imprinting process must be carefully considered. The conformational stability of proteins is sensitive to their surrounding conditions, including temperature, pH, ion concentration, surfactant and its concentration, the ratio of template to functional monomers, the types of functional monomers used, and initiator type [167]. MIPs generated by epitope imprinting manifest minimal non-specific binding and improved affinity. Furthermore, it is less costly working with short peptides as templates, and more importantly, because of the template's higher stability, organic solvents may also be used during the imprinting process. The detection of the bound non-electroactive target macromolecule is performed indirectly, in the presence or absence of a redox probe, by various signal transduction principles, such as voltammetry, piezomicrogravimetry, impedance spectroscopy, etc.

The high selectivity of such biomimetic polymers is also demonstrated by their efficient use for the molecular imprinting of optically active templates. This special application of MIP-based chiral electrochemical sensing has not been fully exploited, even though there is a great necessity in the pharmaceutical industry for developing fast and cost-effective methods of chiral analysis applicable from the early stages of drug development. The vast majority of MIP-based electrochemical sensors were developed for non-chiral analysis. Only around 20 articles report MIP-based sensors designed for the chiral analysis of different small-molecules, but not pharmaceutically active compounds, like amino acids [168, 169, 170] or monosaccharides [171]. Moreover, enantiospecific signal transduction was only reported in the case of amino acids. Nevertheless, very recently an MIP-based electrochemical sensor has been reported for the simultaneous enantiospecific recognition of severalβ-blocker enantiomers [163].

In spite of the electroinsulating properties of the R(+)-atenolol imprinted acrylate-based MIP, voltamperometric signal transduction was successfully performed, due to the resulting very thin (less than 4 nm) and porous MIP layer. The sensor exhibited distinctive enantiospecific oxidation peaks towards the R-antipodes of four β-blocker representatives and additional oxidation peaks common to both enantiomers of each studied β-blocker, thus allowing the simultaneous analysis of all of their enantiomers in a single determination.

3.3. Hybrid Materials

Metal NPs decorating CNTs or graphene sheets, carbon-coated magnetic NPs, nanoparticles included in carbon paste electrodes, MWCNT-alumina-coated silica, and insulating MIPs combined with various nanostructures are examples of some hybrid nanomaterials used for the enhancement of biosensor sensitivity.

3.3.1. Cnt-based hybrid materials

Electrodes modified by hybrid materials, consisting of CNTs and metal nanoparticles, have been developed for use as fuel cell catalysts and biosensors [172]. For example, platinum nanoparticles (PtNPs) catalyse the electrochemical oxidation of H_2O_2, which is generated by the enzymatic reaction, while MWCNTs can be employed to modify the working electrode for the enlargement of the electroactive surface area [173]. A highly sensitive Pt-CNT-glucose biosensor was developed using the incorporation of GOx on a Pt-CNT electrode but with poor enzyme stability after storage [174]. Electrodes modified by CNTs/PtNPs hybrids protected by dendrimers were first developed for methanol electrochemical oxidation [175], and then adapted for the detection of inorganic phosphorous pesticide through the inhibition of the acetylcholinesterase [172]. These electrochemical biosensors are based on a hybrid material, consisting of MWCNTs, PtNPs and poly(amidoamine) dendrimer (DEN) as a binder for enzyme immobilization (Figure 7).

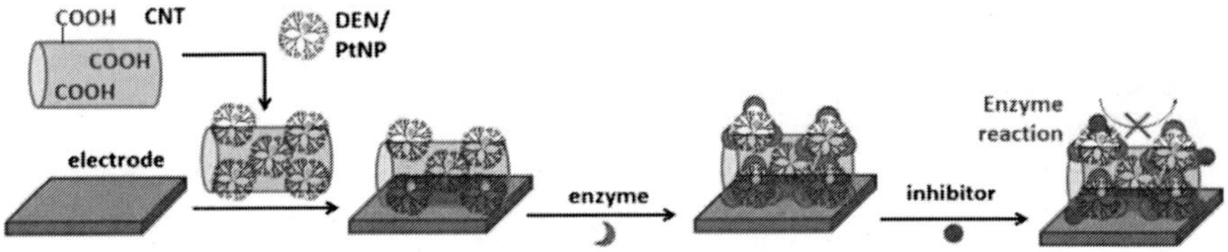

Figure 7. Biosensor preparation by loading MWCNT, PtNPs and enzyme through dendrimer binders [172]

MWCNTs increase the surface area and promote the electron transfer properties, PtNPs intensify the electron transferability in addition to catalytic performance, and dendrimers play a role in the uptake of the redox agent as well as the binder functionality.

Two new xanthine biosensors, based on GC modified with chitosan (CS), Co_3O_4NPs and MWCNTs were developed [176]. Two different enzyme systems were investigated involving monoenzyme xanthine oxidase (XO) immobilization and bienzyme xanthine oxidase (XO)/HRP co-immobilization. Xanthine amperometric determination was achieved by both electrooxidation and electroreduction of the enzymatically generated hydrogen peroxide, with LODs of 0.2 μM and 20 μM with anodic and cathodic processes, respectively. The proposed biosensors were applied for detecting xanthine in fish with average recoveries of 96.2±3.5 % with XO/Co_3O_4MWCNTs/CS/GCE and 95.4±2.1 % with XO/HRP/Co_3O_4MWCNTs/CS/GCE, respectively.

3.3.2. Graphene-Based Hybrid Materials

Graphene can be decorated with AuNPs to enhance the signal transduction or to increase the grafting area for biocomponent immobilization. AuNPs are the most commonly used metal nanoparticles in graphene-based biosensing systems, mainly due to their ability to form strong covalent bonds with the thiol groups [26].

An electrochemical aptasensor using graphene-AuNPs composite obtained by the reduction of tetrachloroauric acid with sodium citrate in a graphene water suspension was reported [109]. Direct electrodeposition technique of AuNPs was also applied to graphene-based electrochemical transducers yielding a high density of nanoparticles that provide an ultra-large specific surface area available for aptamer immobilization [177].

A graphene-CNTs-incorporating zinc oxide NPs hybrid composite (GR-CNT-ZnO) was synthesized and applied to fabricate an enzyme-based glucose biosensor (Figure 8).

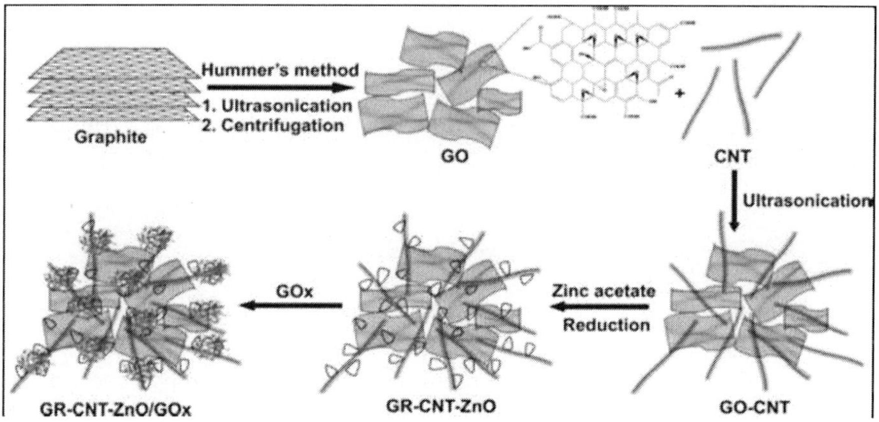

Figure 8. The preparation of GR-CNT-ZnO composite [178]

By using CV, authors have concluded that the modified film in the presence of oxygen has high electrocatalytic ability in glucose detection with a wide linear detection range and a LOD of 4.5 μM of glucose. The sensor was successfully applied for human serum samples, and the results were in good agreement with those determined using a standard photometric method [178].

3.3.3. Molecularly imprinted polymer-based hybrid materials

Hybrid materials combining insulating MIPs with electronically conducting polymers tend to correct the drawbacks of the former and feature networks of molecular wires connecting recognitions sites to the electrode surface [179]. Molecularly imprinted nanoparticles, nanospheres, nanoshells or nanofibres obtained by various synthesis procedures (precipitation or emulsion/suspension polymerization, mini/emulsion precipitation, etc.), along with various organic/inorganic hybrid materials [165, 180, 181], metal nanoparticles, such as Ag [182] and Au [183] or carbon nanomaterials [173], may be combined with electrosynthesized MIP films, leading to composite MIP/nanomaterial biosensing systems [160, 161] with larger specific surface area, higher sensibility and specificity, and higher biocompatibility for trace electrochemical analysis of small analytes and biomacromolecules.

4. PERSPECTIVES AND FUTURE DEVELOPMENT

The selection of materials and fabrication techniques is crucial for adequate sensor function and the performance of a biosensor often ultimately depends upon these, rather than upon other factors. Consequently, future developments in biosensor design will inevitably focus upon the technology of new materials, especially the new copolymers that promise to solve the biocompatibility problem and offer the prospect of more widespread use of biosensors in clinical monitoring. The future trends in sensor development that are expected to have an impact on biosensor performances mainly concern immobilization techniques, nanotechnology, miniaturization and design of multisensor arrays.

REFERENCES

1. Cristea C, Hârceagă V, Săndulescu R. Electrochemical sensor and biosensors. In: Moretto LM, Kalcher K, editors. Environmental Analysis by Electrochemical Sensors and Biosensors. Vol. 1 Fundamentals. 1st ed. New York: Springer; 2014. p. 155-165.

2. Morrison DWG, Dokmeci MR, Demirci U, Khademhosseini A. Clinical applications of micro- and nanoscale biosensors. In: Gonsalves KE, Halberstadt CR, Laurencin CT, Nair LS, editors. Biomedical Nanostructures. 1st ed. New York: John Wiley & Sons Inc.; 2008. p. 1-10.

3. Castillo J, Gaspar S, Leth S, Niculescu M, Mortari A, Botidean I et al. Biosensors for life quality design, development and applications. Sensor. Actuat. B-Chem. 2004;102(2):179-194.

4. Campas M, Prieto-Simon B, Marty JL. A review of the use of genetically engineered enzymes in electrochemical biosensors. Semin. Cell. Dev. Biol. 2009;20(1):3-9.

5. Farré M, Kantiani L, Pérez S, Barceló D. Sensors and biosensors in support of EU directives. TrAC Trends Anal. Chem. 2009;28(2):170-185.

6. Zhang S, Wright G, Yang Y. Materials and techniques for electrochemical biosensor design and construction. Biosens. Bioelectron. 2000;15(5-6):273-282.

7. Walcarius A, Etienne M, Herzog G, Urbanova V, Vila N. Electrode materials (bulk materials and modification). In: Moretto LM, Kalcher K, editors. Environmental Analysis by Electrochemical Sensors and Biosensors. Vol. 1 Fundamentals. 1st ed. New York: Springer; 2014. p. 403-495.

8. Xu Z, Chen X, Dong S. Electrochemical biosensors based on advanced bioimmobilization matrices. TrAC Trends Anal. Chem. 2006;25(9):899-908.

9. Mello LD, Kubota LT. Review of the use of biosensors as analytical tools in the food and drink industries. Food Chem. 2002;77(2):237-256.

10. Lojou E, Bianco P. Membrane electrodes for protein and enzyme electrochemistry. Electroanal. 2004;16(13-14):1113-1121.

11. Teles FRR, Fonseca LP. Applications of polymers for biomolecule immobilization in electrochemical biosensors. Mat. Sci. Eng. C Biomin. 2008;28(8):1530-1543.

12. Samanta D, Sarkar A. Immobilization of bio-macromolecules on self-assembled monolayers: Methods and sensor application. Chem. Soc. Rev. 2011;40(5):2567-2592.

13. Tien HT. Bilayer lipid membrane-based electrochemical biosensors. J. Clin. Lab. Anal. 2005;2(4):256-264.

14. Walcarius A. Electroanalytical applications of microporous zeolites and mesoporous(organo) silicas: Recent trends. Electroanal. 2008;20(7):711-738.

15. Chambers JP, Arulanandam BP, Matta LL, Weis A, Valdes JJ. Biosensor recognition elements. Curr. Issues. Mol. Biol. 2008;10(1-2):1-12.

16. Thevenot DR, Toth K, Durst RA, Wilson GS. Electrochemical biosensors: Recommended definitions and classification. Biosens. Bioelectron. 16;1-2(121-131)

17. Bard AJ, Inzelt G, Scholz F, editors. Electrochemical Dictionary. 2nd ed. Berlin: Springer; 2012.

18. Mascini M, Minunni M, Guilbault GG, Carter R. Methods in biotechnology. In: Rogers KR, Mulchandani A, editors. Affinity Biosensors: Techniques and Protocols. 1st ed. Totowa, NJ: Humana Press Inc.; 1998.

19. Palecek E, Fojta M. DNA hybridization and damage. Anal. Chem. 2001;73(75A-83A).

20. Rauf S, Gooding JJ, Akhtar K, Ghauri MA, Rahman M, Anwar MA, et al. Electrochemical approach of anticancer drugs-DNA interaction. J. Pharm. Biomed. Anal. 2005;37(2):205-217.

21. Mulchandani A, Rogers K. Enzyme and microbial biosensors: Techniques and protocol. Totowa, NJ. 1st ed. Totowa, NJ: Humama Press; 1998.

22. Yamashita T, Yamashita K, Naeshi H, Yoshikawa T, Yoshioka Y, Tsunoda S et al. Carbon nanomaterials: Efficacy and safety for nanomedicine. Materials. 2012;5:350-363.

23. Bhakta SA, Evans E, Benavidez TE, Garcia CD. Protein adsorption onto nanomaterials for the development of biosensors and analytical devices: A review. Anal. Chim. Acta. Forthcoming.

24. Chan CPY, Mak WC, Cheung KY, Sin KK, Yu CM, Rainer TH et al. Evidence-based point-of-care diagnostics: Current status and emerging technologies. Annu. Rev. Anal. Chem. 2013;6:191-211.

25. Carregal-Romero S, Caballero-Díaz E, Beqa L, Abdelmonem AM, Ochs M, Hühn D et al. Multiplexed sensing and imaging with colloidal nano- and microparticles. Annu. Rev. Anal. Chem. 2013;6:53-81.

26. Pumera M, Sanchez S, Ichinose I, Tang J. Electrochemical nanobiosensors. Sensor. Actuat. B-Chem. 2007;123(2):1195-1205.

27. Barnes TM, van de Lagemaat J, Levi D, Rumbles G, Coutts TG, Weeks CL et al. Optical characterization of highly conductive single-wall carbon-nanotube transparent electrodes. Phys. Rev. B. 2007;75(23):23541001-22354110.

28. Zhao J, Chen X, Xie JRH. Optical properties and photonic devices of doped carbon nanotubes. Anal. Chim. Acta. 2006;568:161-170.

29. Wooten M, Gorski W. Facilitation of NADH electrooxidation at treated carbon nanotubes. Anal. Chem. 2010;82(4):1299-1304.

30. Bose S, Khare RA, Moldenaers P. Assessing the strengths and weaknesses of various types of pretreatment of carbon nanotubes on the properties of polymer carbon nanotubes composites: A critical review. Polymer. 2010;51:975-993.

31. Wang HZ, Huang ZP, Cai QJ, Kulkarni K, Chen CL, Carnahan D et al. Reversible transformation of hydrophobicity and hydrophilicity of aligned carbon nanotube arrays and buckypapers by dry processes. Carbon. 2010;48(3):868-875.

32. Soetedjo H, Mora MF, Garcia CD. Optical properties of single-wall carbon nanotube films deposited on Si/SiO2 wafers. Thin Solid Films. 2010;518(14):3954-3959.

33. Qureshi A, Kang WP, Davidson JL, Gurbuz Y. Review on carbon-derived, solid-state, micro and nano sensors for electrochemical sensing applications. Diam. Relat. Mater. 2009;18(12):1401-1420.

34. Masheter AT, Abiman P, Wildgoose GG, Wong E, Xiao L, Rees NV et al. Investigating the reactive sites and the anomalously large changes in

surface pKa values of chemically modified carbon nanotubes of different morphologies. Mater. Chem. 2007;17(25):2616-2626.

35. Iijima S. Helical microtubules of graphitic carbon. Nature. 1991;354:56-58.

36. Hu C, Hu S. Carbon nanotube-based electrochemical sensors: Principles and applications in biomedical systems. Journal of Sensors. 2009;2009:187615.

37. Sadik OA, Mwilu SK, Aluoch A. Smart electrochemical biosensors: From advanced materials to ultrasensitive devices. Electrochim. Acta. 2010;55(14):4287-4296.

38. Erdem A. Nanomaterials based sensor development towards electrochemical sensing of bioreactions. In: Vaseashta A, Braman E, Susmann P, editors. Technological innovation in sensing and detection of chemical, biological, radiological, nuclear threats and ecological terrorism. NATO Science for Peace and Security Series A: Chemistry and Biology. 1st ed. New York: Springer; 2012. p. 165-169.

39. Giljohann DA, Mirkin CA. Drivers of biodiagnostic development. Nature. 2009;462:461-464.

40. Belin T, Epron F. Characterization methods of carbon nanotubes: A review. Mat. Sci. Eng. B-Solid. 2005;119:105-118.

41. Banks CE, Davies TJ, Wildgoose GG, Compton RG. Electrocatalysis at graphite and carbon nanotube modified electrodes: Edge-plane sites and tube ends are the reactive sites. Chem. Commun. 2005;7(21):829-841.

42. Banks CE, Crossley A, Salter C, Wilkins SJ, Compton RG. Carbon nanotubes contain metal impurities which are responsible for the "electrocatalysis" seen at some nanotube-modified electrodes. Angew. Chem. Int. Ed. 2006;45(16):2533-2537.

43. Pumera M, Ambrosi A, Bonanni A, Chng ELK, Poh HL. Graphene for electrochemical sensing and biosensing. TrAC Trends Anal. Chem. 2010;29(9):954-965.

44. Ambrosi A, Pumera M. Nanographite impurities dominate electrochemistry of carbon nanotubes. Chem.-A Eur. J. 2010;16(36):10946-10949.

45. Streeter I, Wildgoose GG, Shao LD, Compton RG. Cyclic voltammetry on electrode surfaces covered with porous layers: An analysis of electron transfer kinetics at single-walled carbon nanotube modified electrodes. Sensor. Actuat. B-Chem. 2008;133:462-466.

46. Sims MJ, Rees NV, Dickinson EJF, Compton RG. Effects of thin-layer diffusion in the electrochemical detection of nicotine on basal plane pyrolytic graphite (BPPG) electrodes modified with layers of multi-walled carbon nanotubes (MWCNT-BPPG). Sensor. Actuat. B-Chem. 2010;144(1):153-158.

47. Henstridge MG, Dickinson EJF, Aslanoglu M, Batchelor-McAuley C, Compton RG. Voltammetric selectivity conferred by the modification of electrodes using conductive porous layers or films: The oxidation of dopamine on glassy carbon electrodes modified with multiwalled carbon nanotubes. Sensor. Actuat. B-Chem. 2010;145(1):417-427.

48. Li W, Jung H, Hoa ND, Kim D, Hong S-K, Kim H. Nanocomposite of cobalt oxide nanocrystals and single-walled carbon nanotubes for a gas sensor application. Sensor. Actuat. B-Chem. 2010;150(1):160-166.

49. Ghadimi H, Tehrani RMA, Mohamed Ali AS, Mohamed N, Ab Ghani S. Sensitive voltammetric determination of paracetamol by poly (4-vinylpyridine)/multiwalled carbon nanotubes modified glassy carbon electrode. Anal. Chim. Acta. 2013;765:70-76.

50. Gao Q, Guo Y, Liu J, Yuan X, Qi H, Zhang C. A biosensor prepared by co-entrapment of a glucose oxidase and a carbon nanotube within an electrochemically deposited redox polymer multilayer. Bioelectrochemistry. 2011;81(2):109-113.

51. Salimi A, Compton RG, Hallaj R. Glucose biosensor prepared by glucose oxidase encapsulated sol–gel and carbon nanotube modified basal plane pyrolytic graphite electrode. Anal. Biochem. 2004;333(1):49-56.

52. Sato N, Okuma H. Development of single-wall carbon nanotubes modified screen-printed electrode using a ferrocene-modified cationic surfactant for amperometric glucose biosensor applications. Sensor. Actuat. B-Chem. 2008;129(1):188-194.

53. Rahman MM, Umar A, Sawada K. Development of amperometric glucose biosensor based on glucose oxidase co-immobilized with multi-walled carbon nanotubes at low potential. Sensor. Actuat. B-Chem. 2009;137(1):327-333.

54. Deng C, Chen J, Chen X, Xiao C, Nie L, Yao S. Direct electrochemistry of glucose oxidase and biosensing for glucose based on boron-doped carbon nanotubes modified electrode. Biosens. Bioelectron. 2008;23(8):1272-1277.

55. Li S, Zhu X, Zhang W, Xie G, Feng W. Hydrogen peroxide biosensor based on gold nanoparticles/thionine/gold nanoparticles/multi-walled carbon nanotubes – chitosan composite film-modified electrode. Appl. Surf. Sci. 2012;258(7):2802-2807.

56. Liu Y, Yuan R, Chai Y, Tang D, Dai J, Zhong X. Direct electrochemistry of horseradish peroxidase immobilized on gold colloid/cysteine/Nafion-modified platinum disk electrode. Sensor. Actuat. B-Chem. 2006;115(1):109-115.

57. Zhang Y, Yuan R, Chai Y, Xiang Y, Hong C, Ran X. An amperometric hydrogen peroxide biosensor based on the immobilization of HRP on multi-walled carbon nanotubes/electro-copolymerized nano-Pt-poly (neutral red) composite membrane. Biochem. Eng. J. 2010;51(3):102-109.

58. Huang J, Li J, Yang Y, Wang X, Wu B, Anzai J-I et al. Development of an amperometric L-lactate biosensor based on L-lactate oxidase immobilized through silica sol-gel film on multi-walled carbon nanotubes/platinum nanoparticle modified glassy carbon electrode. Mater. Sci. Eng. C. 2008;28(7):1070-1075.

59. Wang YT, Yu L, Wang J, Lou L, Du WJ, Zhu ZQ et al. A novel L-lactate sensor based on enzyme electrode modified with ZnO nanoparticles and multiwall carbon nanotubes. J. Electroanal. Chem. 2011;661(1):8-12.

60. Goran JM, Lyon JL, Stevenson KJ. Amperometric detection of L-lactate using nitrogen-doped carbon nanotubes modified with lactate oxidase. Anal. Chem. 2011;83(21):8123-8129.

61. Tertiş M, Florea A, Săndulescu R, Cristea C. Carbon based electrodes modified with horseradish peroxidase immobilized in conducting polymers for acetaminophen analysis. Sensors Basel. 2013;13:4841-4854.

62. Baghayeri M, Namadchian M. Fabrication of a nanostructured luteolin biosensor for simultaneous determination of levodopa in the presence of acetaminophen and tyramine: Application to the analysis of some real samples. Electrochim. Acta. 2013;108(1):22-31.

63. Amiri-Aref M, Raoof JB, Ojani R. A highly sensitive electrochemical sensor for simultaneous voltammetric determination of noradrenaline, acetaminophen, xanthine and caffeine based on a flavonoid nanostructured modified glassy carbon electrode. Sensor. Actuat. B-Chem. 2014;192:634-641.

64. Babaei A, Afrasiabi M, Babazadeh M. A glassy carbon electrode modified with multiwalled carbon nanotube/chitosan composite as a new sensor for simultaneous determination of acetaminophen and mefenamic acid in pharmaceutical preparations and biological samples. Electroanal. 2010;22(15):1743-1749.

65. Shahrokhian S, Asadian E. Simultaneous voltammetric determination of ascorbic acid, acetaminophen and isoniazid using thionine immobilized multi-walled carbon nanotube modified carbon paste electrode. Electrochim. Acta. 2010;55(3):666-672.

66. Tertiş M, Florea A, Feier B, Marian IO, Silaghi-Dumitrescu L, Cristea A et al. Electrochemical impedance studies on single and multi-wall carbon nanotubes-polymer nanocomposites for biosensors development. J. Nanosci. Nanotechnolog. 2015;15(5):3383-3393.

67. Bustos-Ramírez K, Martínez-Hernández AL, Martínez-Barrera G, de Icaza M, Castaño VM, Velasco-Santos C. Covalently bonded chitosan on graphene oxide via redox reaction. Materials. 2013;6:911-926.

68. Guoqing X, Hwang W, Kim N, Cho SM, Chae H. A graphene sheet exfoliated with microwave irradiation and interlinked by carbon nanotubes for high-performance transparent flexible electrodes. Nanotechnol. 2010;21:405201.

69. Lim CX, Hoh HY, Ang PK, Loh KP. Direct voltammetric detection of DNA and pH sensing on epitaxial graphene: An insight into the role of oxygenated defects. Anal. Chem. 2010;82(17):7387-7393.

70. Novoselov KS, Geim AK, Morozov SV, Jiang D, Zhang Y, Dubonos SV et al. Electric field effect in atomically thin carbon films. Science. 2004;306(5696):666-669.

71. Bai J, Huang Y. Fabrication and electrical properties of graphene nanoribbons. Mat. Sci. Eng. R. 2010;70(3-6):341-353.

72. Kim WS, Moon SY, Bang SY, Choi BG, Ham H, Sekino T et al. Fabrication of graphene layers from multiwalled carbon nanotubes using high DC pulse. Appl. Phys. Lett. 2009;95:083103-183103.

73. He Q, Sudibya HG, Yin Z, Wu S, Li H, Boey F et al. Centimeter-long and large-scale micropatterns of reduced graphene oxide films: Fabrication and sensing applications. ACS Nano. 2010;4(6):3201-3208.

74. Pham VH, Cuong TV, Hur SH, Shin EW, Kim JS, Chung JS et al. Fast and simple fabrication of a large transparent chemically-converted graphene film by spray-coating. Carbon. 2010;48:1945-1951.

75. Kang X, Wang J, Wu H, Liu J, Aksay IA, Lin Y. A graphene-based electrochemical sensor for sensitive detection of paracetamol. Talanta. 2010;81:754-759.

76. Liu YX, Dong XC, Chen P. Biological and chemical sensors based on graphene materials. Chem. Soc. Rev. 2012;41(6):2283-2307.

77. Park JW, Lee SJ, Choi EJ, Kim J, Song JY, Gu MB. An ultra-sensitive detection of a whole virus using dual aptamers developed by immobilization-free screening. Biosens. Bioelectron. 2014;51:324-329.

78. Biju V. Chemical modifications and bioconjugate reactions of nanomaterials for sensing, imaging, drug delivery and therapy. Chem. Soc. Rev. 2014;43:744-764.

79. Maiti UN, Lee WJ, Lee JM, Oh Y, Kim JY, Kim E et al. 25th anniversary article: Chemically modified/doped carbon nanotubes & graphene for optimized nanostructures & nanodevices. Adv. Mater. 2014;26(1):40-67.

80. Bekyarova E, Itkis ME, Ramesh P, Berger C, Sprinkle M, de Heer WA et al. Chemical modification of epitaxial graphene: Spontaneous grafting of aryl groups. J. Am. Chem. Soc. 2009;131(4):1336-1337.

81. Liu S, Xing XR, Yu JH, Lian WJ, Li J, Cui M et al. A novel label-free electrochemical aptasensor based on graphene–polyaniline composite film for dopamine determination. Biosens. Bioelectron. 2012;36(1):186-191.

82. Yan M, Sun GQ, Liu F, Lu JJ, Yu JH, Song XR. An aptasensor for sensitive detection of human breast cancer cells by using porous GO/Au composites and porous PtFe alloy as effective sensing platform and signal amplification labels. Anal. Chim. Acta. 2013;798:33-39.

83. Zhao J, Chen GF, Zhu L, Li GX. Graphene quantum dots-based platform for the fabrication of electrochemical biosensors. Electrochem. Commun. 2011;13(1):31-33.

84. Wang L, Xu M, Han L, Zhou M, Zhu CZ, Dong SJ. Graphene enhanced electron transfer at aptamer modified electrode and its application in biosensing. Anal. Chem. 2012;84(17):7301-7307.

85. Yan GP, Wang YH, He XX, Wang KM, Liu JQ, Du YD. A highly sensitive label-free electrochemical aptasensor for interferon-gamma detection based on graphene controlled assembly and nuclease cleavage-assisted target recycling amplification. Biosens. Bioelectron. 2013;44:57-63.

86. Loo AH, Bonanni A, Pumera M. Thrombin aptasensing with inherently electroactive graphene oxide nanoplatelets as labels. Nanoscale. 2013;5:4758-4762.

87. Tertis M, Hosu O, Fritea L, Farcau C, Cernat A, Sandulescu R et al. A novel label-free immunosensor based on activated graphene oxide for acetaminophen detection. Electroanal. 2015; 27(3):638-647.

88. Ghorbani-Bidkorbeh F, Shahrokhian S, Mohammadi A, Dinarvand R. Simultaneous voltammetric determination of tramadol and acetaminophen using carbon nanoparticles modified glassy carbon electrode. Electrochim. Acta. 2010;55(8):2752-2759.

89. Weingart J, Vabbilisetty P, Sun XL. Membrane mimetic surface functionalization of nanoparticles: Methods and applications. Adv. Colloid. Interfac. Sci. 2013;197-198:68-84.

90. Rocha-Santos TAP. Sensors and biosensors based on magnetic nanoparticles. TrAC Trends Anal. Chem. 2014;62:28-36.

91. Rusling JF, Sotzing G, Papadimitrakopoulos F. Designing nanomaterials-enhanced electrochemical immunosensors for cancer biomarker proteins. Bioelectrochemistry. 2009;76(1-2):189-194.

92. Aguilar-Arteaga K, Rodriguez JA, Barrado E. Magnetic solids in analytical chemistry: A review. Anal. Chim. Acta. 2010;674(2):157-165.

93. Xu Y, Wang E. Electrochemical biosensors based on magnetic micro/nanoparticles. Electrochim. Acta. 2012;84:62-73.

94. Nomura A, Shin S, Mehdi OO, Kauffmann JM. Preparation, characterization, and application of an enzyme-immobilized magnetic microreactor for flow injection Analysis. Anal. Chem. 2004;76(18):5498-5502.

95. Zhang XF, Dong XL, Huang H, Lv B, Zhu KG, Lei JP et al. Synthesis, structure and magnetic properties of SiO2-coated Fe nanocapsules. Mater. Sci. Eng. A. 2007;454-455:211-215.

96. Gruttner C, Rudershausen S, Teller J.J. Improved properties of magnetic particles by combination of different polymer materials as particle matrix. J. Magn. Magn. Mater. 2001;225(1-2):1-7.

97. Deere J, Magner E, Wall JG, Hodnett BK. Adsorption and activity of cytochrome c on mesoporous silicates. Chem. Commun. 2001;5:465-466.

98. Yang L, Zhang X, Ye M, Jiang J, Yang R, Fu T. Aptamer-conjugated nanomaterials and their applications. Adv. Drug Deliv. Rev. 2011;63(14-15):1361-1370.

99. Berry CC, Curtis ASG. Topical review: Functionalisation of magnetic nanoparticles for applications in biomedicine. J. Phys. D: Appl. Phys. 2003;36(13):R198–R206.

100. Hervas M, Lopez MA, Escarpa A. Simplified calibration and analysis on screen-printed disposable platforms for electrochemical magnetic bead-based inmunosensing of zearalenone in baby food samples. Biosens. Bioelectron. 2010;25(7):1755-1760.

101. Yang Z, Zhang C, Zhang J, Bai W. Potentiometric glucose biosensor based core-shell Fe3O4-enzyme-polypyrrole nanoparticles. Biosens. Bioelectron. 2014;51:268-273.

102. Chen X, Zhu J, Chen Z, Xu C, Wang Y, Yao C. A novel bienzyme glucose biosensor based on three layers Au-Fe3O4@SiO2 magnetic nanocomposite. Sensor. Actuat. B-Chem. 2011;159(1):220-228.

103. Baby TT, Ramaprabhu R. SiO2 coated Fe3O4 magnetic nanoparticle dispersed multiwalled carbon nanotubes based amperometric glucose biosensor. Talanta. 2010;80(5):2016-2022.

104. Zhou H, Gan N, Li T, Cao Y, Zeng S, Zheng L et al. The sandwich-type electroluminescence immunosensor for α-fetoprotein based on enrichment by Fe3O4-Au magnetic nano probes and signal amplification by CdS-Au composite nanoparticles labeled anti-AFP. Anal. Chim. Acta. 2012;746:107-113.

105. Li J, Xu Q, Wei X, Hao Z. Electrogenerated chemiluminescence immunosensor for Bacillus thuringiensis Cry1Ac based on Fe3O4@Au nanoparticles. J. Agric. Food Chem. 2013;61(7):1435-1440.

106. Zamfir L-G, Geana I, Bourigua S, Rotariu L, Bala C, Errachid A, et al. Highly sensitive label-free immunosensor for ochratoxin A based on functionalized magnetic nanoparticles and EIS/SPR detection. Sensor. Actuat. B-Chem. 2011;159(1):178-184.

107. Yola ML, Eren T, Atar N. A novel and sensitive electrochemical DNA biosensor based on Fe@Au nanoparticles decorated graphene oxide. Electrochim. Acta. 2014;125:38-47.

108. Gan N, Yang X, Xie D, Wu Y, Wen W. A disposable organophosphorus pesticides enzyme biosensor based on magnetic composite nano-particles modified screen printed carbon electrode. Sensors. 2010;10(1):625-638.

109. Li J, Gao H, Chen Z, Wei X, Yang CF. An electrochemical immunosensor for carcinoembryonic antigen enhanced by self-assembled nanogold coatings on magnetic particles. Anal. Chim. Acta. 2010;665(1):98-104.

110. Yang X, Wu F, Chen D-Z, Lin H-W. An electrochemical immunosensor for rapid determination of clenbuterol by using magnetic nanocomposites to modify screen printed carbon electrode based on competitive immunoassay mode. Sensor. Actuat. B-Chem. 2014;192:529-535.

111. Xin Y, Fu-Bing X, Hong-Wei L, Feng W, Di-Zhao C, Zhao-Yang W. A novel H2O2 biosensor based on Fe3O4-Au magnetic nanoparticles coated horseradish peroxidase and grapheme sheets-Nafion film modified

screen-printed carbon electrode. Electrochim. Acta. 2013;109:750-755.

112. Chen D, Deng J, Liang J, Xie J, Hue C, Huang K. A core-shell molecularly imprinted polymer grafted onto a magnetic glassy carbon electrode as a selective sensor for the determination of metronidazole. Sensor. Actuat. B-Chem. 2013;183:594-600.

113. Prakash A, Chandra S, Bahadur D. Structural, magnetic, and textural properties of iron oxide-reduced graphene oxide hybrids and their use for the electrochemical detection of chromium. Carbon. 2012;50(11):4209-4212.

114. Hu Y, Zang Z, Zhang H, Luo L, Yao S. Selective and sensitive molecularly imprinted sol-gel film-based electrochemical sensor combining mecaptoacetic acid modified PbS nanoparticles with Fe3O4@Au-multi-walled carbon nanotubes-chitosan. J. Solid State Electrochem. 2012;16(3):857–867.

115. Arvand M, Hassannezhad M. Magnetic core-shell Fe3O4@SiO2/MWCNT nanocomposite modified carbon paste electrode for amplified electrochemical sensing of uric acid. Mater. Sci. Eng. C. 2014;36:161-167.

116. Hosu O, Tertis M, Cristea C, Sandulescu R. Protein G magnetic beads based immunosensor for sensitive detection of acetaminophen. Farmacia. 2015;63(1):143-146.

117. Taleat Z, Cristea C, Marrazza G, Sandulescu R. Electrochemical sandwich immunoassay for the ultrasensitive detection of human MUC1 cancer biomarker. Int. J. Electrochem. 2013; Article ID 740265,

118. Jia J, Wang B, Wu A, Cheng G, Li Z, Dong S. A method to construct a third-generation horseradish peroxidase biosensor: Self-assembling gold nanoparticles to three-dimensional sol-gel network. Anal. Chem. 2002;74:2217-2223.

119. Di J, Shen C, Peng S, Tu Y, Li S. A one-step method to construct a third-generation biosensor based on horseradish peroxidase and gold nanoparticles embedded in silica sol–gel network on gold modified electrode. Anal. Chim. Acta. 2005;553:196-200.

120. Xu Z, Chen X, Dong S. Electrochemical biosensors based on advanced bioimmobilization matrices. TrAC Trends Anal. Chem. 2006;25(9):899-908.

121. Yang L, Zhang X, Ye M, Jiang L, Yang R, Fu T et al. Aptamer-conjugated nanomaterials and their applications. Adv. Drug Deliv. Rev. 2011;63(14-15):1361-1370.

122. Merkoçi A. Nanoparticles-based strategies for DNA, protein and cell sensors. Biosens. Bioelectron. 2010;26(4):1164-1177.

123. Yin T, Qin W. Applications of nanomaterials in potentiometric sensors. TrAC Trends Anal. Chem. 2013;51:79-86.

124. Taton TA, Mirkin CA, Letsinger RL. Scanometric DNA array detection with nanoparticle probes. Science. 2000;289(5485):1757-1760.

125. Wang J, Polsky R, Xu D. Silver-enhanced colloidal gold electrochemical stripping detection of DNA hybridization. Langmuir. 2001;17(19):5739-5741.

126. Wang J, Xu D, Kawde A-N, Polsky R. Metal nanoparticle-based electrochemical stripping potentiometric detection of DNA hybridization. Anal. Chem. 2001;73(22):5576-5581.

127. Wang J, Liu G, Merkoçi A. Electrochemical coding technology for simultaneous detection of multiple DNA targets. J. Am. Chem. Soc. 2003;125(11):3214-3215.

128. Chen G, Tong H, Gao T, Chen Y, Li G. Direct application of gold nanoparticles to one-pot electrochemical biosensors. Anal. Chim. Acta. 2014;849:1-6.

129. Malon A, Vigassy T, Bakker E, Pretsch E. Potentiometry at trace levels in confined samples: Ion-selective electrodes with subfemtomole detection limits. J. Am. Chem. Soc. 2006;128(25):8154-8155.

130. Rubinova N, Chumbimuni-Torres K, Bakker E. Solid-contact potentiometric polymer membrane microelectrodes for the detection of silver ions at the femtomole level. Sensor. Actuat. B-Chem. 2007;121(1):135-141.

131. Chumbimuni-Torres K, Dai Z, Rubinova N, Xiang Y, Pretsch E, Wang J et al. Potentiometric biosensing of proteins with ultrasensitive ion-selective microelectrodes and nanoparticle labels. J. Am. Chem. Soc. 2006;128(42):13676-13677.

132. Willner I, Baron R, Willner B. Growing metal nanoparticles by enzymes. Adv. Mater. 2006;18:1109-1120.

133. Ngeontae W, Janrungroatsakul W, Maneewattanapinyo P, Ekgasit S, Aeungmaitrepirom W, Tuntulani TW. Novel potentiometric approach in glucose biosensor using silver nanoparticles as redox marker. Sensor. Actuat. B-Chem. 2009;137(1):320-326.

134. Chumbimuni-Torres KY, Wang J. Nanoparticle-induced potentiometric biosensing of NADH at copper ion-selective electrodes. Analyst. 2009;134(8):1614-1617.

135. Wu J, Chumbimuni-Torres KY, Galik M, Thammakhet C, Haake DA, Wang J. Potentiometric detection of DNA hybridization using enzyme-induced metallization and a silver ion selective electrode. Anal. Chem. 2009;81(24):10007-10012.

136. Florea A, Taleat Z, Cristea C, Mazloum-Ardakani M, Săndulescu R. Label free MUC1 aptasensors based on electrodeposition of gold nanoparticles on screen printed electrodes. Electrochem. Comm. 2013;33:127-130.

137. Mousty C. Sensors and biosensors based on clay-modified electrodes-new trends. Appl. Clay Sci. 2004;27:159-177.

138. Direnzo F, Cambon H, Dutartre R. A 28-year-old synthesis of micelle-templated mesoporous silica. Micropor. Mater. 1997;10:283-286.

139. Qhobosheane M, Santra S, Zhang P, Tan WH. Biochemically functionalized silica nanoparticles. Analyst. 2001;126:1274-1278.

140. Yan L, Chen X. Nanomaterials for drug delivery. In: Tjong SC, editor. Nanocrystalline Materials. 2nd ed. London: Elsevier; 2014. p. 221-268.

141. Cheng SH, Lee CH, Yang CS, Tseng FG, Mou CY, Lo LW. Mesoporous silica nanoparticles functionalized with an oxygen-sensing probe for cell photodynamic therapy: Potential cancer theranostics. J. Mater. Chem. 2009;19(9):1252-1257.

142. Yáñez-Sedeño P, Agüí L, Villalonga R, Pingarrón JM. Biosensors in forensic analysis. A review. Anal. Chim. Acta. 2014;823:1-19.

143. Cosnier S, Mousty C, Cui XQ, Yang X, Dong S. Specific determination of As(V) by an acid phosphatase polyphenol oxidase biosensor. Anal. Chem. 2006;78(14):4985-4989.

144. Liu YX, Wei WZ. Layer-by-layer assembled DNA functionalized single-walled carbon nanotube hybrids for arsenic(III) detection. Electrochem. Comm. 2008;10(6):872-875.

145. Shan D, Mousty C, Cosnier S. Subnanomolar cyanide detection at polyphenol oxidase/clay biosensors. Anal. Chem. 2004;76(1):178-183.

146. Santos L, Martin P, Ghilane L, Lacaze P-C, Randriamahazaka H, Abrantes LM et al. Electrosynthesis of well-organized nanoporous poly(3,4-ethylenedioxythiophene) by nanosphere lithography. Electrochem. Commun. 2010;12(7):872-875.

147. Cernat A, Griveau S, Richard C, Bedioui, Săndulescu R. Horseradish peroxidase nanopatterned electrodes by click chemistry: Application to the electrochemical detection of paracetamol. Electroanal. 2013;25(6):1369-1372.

148. Cernat A, Le Goff A, Holzinger M, Sandulescu R, Cosnier S. Micro- to nanostructured poly(pyrrole-nitrilotriacetic acid) films via nanosphere templates: Applications to 3D enzyme attachment by affinity interactions. Anal. Bioanal. Chem. 2014;406(4):1141-1147.

149. Stetter JR, Penrose WR, Yao S. Sensors, chemical sensors, electrochemical sensors, and ECSJ. Electrochem. Soc. 2003;150(2):S11-S16.

150. Ravi Kumar MNV, Kumar N, Kashyap N. Hydrogels for pharmaceutical and biomedical applications. Crit. Rev. Ther. Drug Carrier Syst. 2005;22(2):107-150.

151. He X, Wei B, Mi Y. Aptamer based reversible DNA induced hydrogel system for molecular recognition and separation. Chem. Commun. 2010;46(34):6308-6310.

152. Hasanzadeh M, Shadjou N, de la Guardia M. Electrochemical biosensing using hydrogel nanoparticles. TrAC Trends Anal. Chem. 2014;62:11-19.

153. Sriwichai S, Baba A, Phanichphant S, Shinbo K, Kato K, Kaneko F. Electrochemically controlled surface plasmon resonance immunosensor for the detection of human immunoglobulin G on poly(3-aminobenzoic acid) ultrathin films. Sensor. Actuator. B-Chem. 2010;147(1):322-329.

154. Taleat Z, Cristea C, Marrazza G, Mazloum-Ardakani M, Săndulescu R. Electrochemical immunoassay based on aptamer–protein interaction and functionalized polymer for cancer biomarker detection. J. Electroanal. Chem. 2014;717-718:119-124.

155. Cernat A, Tertis M, Griveau S, Bedioui F, Săndulescu R. New modified electrodes with HRP immobilized in polymeric films for paracetamol analysis. Farmacia. 2012;1:1-12.

156. Cernat A, Griveau S, Martin P. Lacroix JC, Farcau C, Săndulescu R et al. Electrografted nanostructured platforms for click chemistry. Electrochem. Comm. 2012;23:141-144.

157. Sharma PS, Pietrzyk-Le A, D'Souza F, Kutner W. Electrochemically synthesized polymers in molecular imprinting for chemical sensing. Anal. Bioanal. Chem. 2012;402:3177-3204.

158. Yan H, Row KH. Characteristic and synthetic approach of molecularly imprinted polymer. Int. J. Mol. Sci. 2006;7:155-178.

159. Moreira FTC, Dutra RAF, Noronha JPC, Sales MGF. Electrochemical biosensor based on biomimetic material for myoglobin detection. Electrochim. Acta. 2013;107:481-487.

160. Malitesta C, Mazzotta E, Picca RA, Poma A, Chianella I, Piletsky SA. MIP sensors – the electrochemical approach. Anal. Bioanal. Chem. 2012;402:1827-1846.

161. Irshad M, Iqbal N, Mujahid A, Afzal A, Hussain T, Sharif A et al. Molecularly imprinted nanomaterials for sensor applications. Nanomaterials. 2013;3:615-637.

162. Herlem G, Gharbi T. Electrodeposition of insulating thin film polymers from aliphatic monomers as transducers for biosensor applications. In: Serra PA, editor. Biosensors – Emerging Materials and Applications. 1st edition. InTech; 2011.

163. Iacob BC, Bodoki E, Florea A, Bodoki AE, Oprean R. Simultaneous enantiospecific recognition of several β-blocker enantiomers using molecularly imprinted polymer-based electrochemical sensor. Anal. Chem. Forthcoming.

164. Ge Y, Turner APF. Too large to fit? Recent developments in macromolecular imprinting. Trends Biotechnol. 2008;26:218-224.

165. Takeuchi T, Hishiya T. Molecular imprinting of proteins emerging as a tool for protein recognition. Org. Biomol. Chem. 2008;6:2459-2467.

166. Hussain M, Wackerlig J, Lieberzeit PA. Biomimetic strategies for sensing biological species. Biosensors. 2013;3:89-107.

167. Ding XC, Heiden PA. Recent developments in molecularly imprinted nanoparticles by surface imprinting techniques. Macromol. Mat. Engineer. 2014;299:268-289.

168. Chen Y, Chen L, Bi R, Xu L, Liu Y. A potentiometric chiral sensor for L-phenylalanine based on crosslinked polymethylacrylic acid-polycarbazole hybrid molecularly imprinted polymer. Anal. Chim. Acta. 2012;754:83-90.

169. Prasad BB, Pandey I, Srivastava A, Kumar D, Tiwari MP. Multiwalled carbon nanotubes-based pencil graphite electrode modified with an electrosynthesized molecularly imprinted nanofilm for electrochemical sensing of methionine enantiomers. Sensor. Actuat. B-Chem. 2013;176:863-874.

170. Prasad BB, Pandey I. Metal incorporated molecularly imprinted polymer-based electrochemical sensor for enantio-selective analysis of pyroglutamic acid isomers. Sensor. Actuat. B-Chem. 2013;186:407-416.

171. Granot E, Tel-Vered R, Lioubashevski O, Willner I. Stereoselective and enantioselective electrochemical sensing of monosaccharides using

imprinted boronic acid-functionalized polyphenol films. Adv. Funct. Mater. 2008;18:478-484.

172. Siriviriyanun A, Imae T, Nagatani N. Electrochemical biosensors for biocontaminant detection consisting of carbon nanotubes, platinum nanoparticles, dendrimers, and enzymes. Anal. Biochem. 2013;443(2):169-171.

173. Ahmed MU, Hossain MM, Tamiya E. Electrochemical biosensors for medical and food applications. Electroanal. 2008;20:616-626.

174. Liu G, Riechers SL, Mellen MC, Lin Y. Sensitive electrochemical detection of enzymatically generated thiocholine at carbon nanotube modified glassy carbon electrode. Electrochem. Commun. 2005;7(11):1163-1169.

175. Siriviriyanun A, Imae T. Advantages of electrodes with dendrimer-protected platinum nanoparticles and carbon nanotubes for electrochemical methanol oxidation. Phys. Chem. Chem. Phys. 2013;15:4921-4929.

176. Dalkiran B, Kacar C, Erden PE, Kilic E. Amperometric xanthine biosensors based onchitosan-Co3O4-multiwall carbon nanotube modified glassy carbon electrode. Sensor. Actuat. B-Chem. 2014;200:83-91.

177. Jiang BY, Wang M, Chen Y, Xie JQ, Xiang Y. Highly sensitive electrochemical detection of cocaine on graphene/AuNP modified electrode via catalytic redox-recycling amplification. Biosens. Bioelectron. 2012;32(1):305-308.

178. Hwa K-Y, Subramani B. Synthesis of zinc oxide nanoparticles on graphene-carbon nanotube hybrid for glucose biosensor applications. Biosens. Bioelectron. 2014;62:127-133.

179. Lakshmi D, Bossi A, Whitcombe MJ, Chianella I, Fowler SA, Subrahmanyam S et al. Electrochemical sensor for catechol and dopamine based on a catalytic molecularly imprinted polymer-conducting polymer hybrid recognition element. Anal. Chem. 2009;81:3576-3584.

180. Lei R, Guo C, Xiong H, Dong C, Zhang X, Wang S. A novel electrochemical sensor for β2-agonists with high sensitivity and selectivity based on surface molecularly imprinted sol-gel doped with antimony-doped tin oxide. Electroanal. 2014;26:1004-1012.

181. Fatoni A, Numnuam A, Kanatharana P, Limbut W, Thavarungkul P. A novel molecularly imprinted chitosan–acrylamide, graphene, ferrocene composite cryogel biosensor used to detect microalbumin. Analyst. 2014;139:6160-6167.

182. Du D, Chen S, Cai J, Tao Y, Tu H, Zhang A. Recognition of dimethoate carried by bi-layer electrodeposition of silver nanoparticles and imprinted poly-o-phenylenediamine. Electrochim. Acta. 2008;53:6589-6595.

183. Kan X, Liu T, Zhou H, Li C, Fang B. Molecular imprinting polymer electrosensor based on gold nanoparticles for theophylline recognition and determination. Microchim. Acta. 2010;171:423-429.

Chapter 3

Biosensors for Pesticide Detection: New Trends

Audrey Sassolas[1*], Beatriz Prieto-Simón[2], Jean-Louis Marty[1]

[1]*Laboratoire IMAGES EA 4218, Université de Perpignan via Domitia, Perpignan, France*

[2]*Nanobioengineering Group, Institute for Bioengineering of Catalonia, Barcelone, Spain*

ABSTRACT

Due to the large amounts of pesticides commonly used and their impact on health, prompt and accurate pesticide analysis is important. This review gives an overview of recent advances and new trends in biosensors for pesticide detection. Optical, electrochemical and piezoelectric biosensors have been reported based on the detection method. In this review biosensors have been classified according to the immobilized biorecognition element: enzymes, cells, antibodies and, more rarely, DNA. The use of tailor-designed biomolecules, such as aptamers and molecularly imprinted polymers, is reviewed. Artificial Neural Networks, that allow the analysis of pesticide mixtures are also presented. Recent advances in the field of nanomaterials merit special mention. The incorporation of nanomaterials provides highly sensitive sensing devices allowing the efficient detection of pesticides.

Keywords: Biosensors; Pesticides

1. INTRODUCTION

In agriculture, farmers use numerous pesticides to protect crops and seeds before and after harvesting. Pesticide is a term used in broad sense for organic toxic compounds used to control insects, bacteria, weeds, nematodes, rodents and other pests. The pesticide residues may enter into the food chain through air, water and soil. They affect ecosystems and cause several health problems to animals and humans. Pesticides can be carcinogenic and cytotoxic. They can produce bone marrow and nerve disorders, infertility, and immunological and respiratory diseases.

Detection of pesticides at the levels established by the Environmental Protection Agency (EPA) remains a challenge. Chromatographic methods coupled to selective detectors have been traditionally used for pesticide analysis due to their sensitivity, reliability and efficiency. Nevertheless, they are time-consuming and laborious, and require expensive equipments and highly-trained technicians. Over the past decade, considerable attention has been given to the development of biosensors for the detection of pesticides as a promising alternative. A biosensor is a self-contained device that integrates an immobilized biological element (e.g. enzyme, DNA probe, antibody) that recognizes the analyte (e.g. enzyme substrate, complementary DNA, antigen) and a transduction element used to convert the (bio) chemical signal resulting from the interaction of the analyte with the bioreceptor into an electronic one. According to the signal transduction technique, biosensors are classified into electrochemical, optical, piezoelectric and mechanical biosensors. Electrochemical transducers have been widely used in biosensors for pesticides detection due to their high sensitivity [1-3]. Additionally, their low cost, simple design and small size, make them excellent candidates for the development of portable biosensors [4-8]. According to the biorecognition element, enzymatic, whole cell, immunochemical, and DNA biosensors have been developed for pesticides detection.

This review presents a state-of-the-art update in pesticide biosensors. To clearly report the last advances, biosensors have been classified according to the immobilized recognition element. New trends in the field of pesticide analysis are also reviewed. Aptamers are shown as good candidates to replace the conventional antibodies and, thus, to be the biorecognition elements in more robust and stable biosensors for pesticide detection. Due to exceptional characteristics, molecular imprinted polymers (MIPs) are innovative affinity-based recognition elements that are exploited for the development of environmental sensors. The use of Artificial Neural Networks (ANNs) coupled with a sensor array could substantially improve the selectivity and allow exact identification of pesticides present in a sample. Recent reports on the properties of nanomaterials show nanoparticles and nanotubes as promising tools to improve the efficiency of biosensors for the detection of pesticides.

2. ENZYME BIOSENSORS

Enzyme biosensors for pesticide detection are based on measurements of enzyme inhibition or on direct measurements of compounds involved in the enzymatic reaction.

2.1. Inhibition-Based Biosensors

2.1.1. Cholinesterase-Based Biosensors

Enzymatic detection of pesticides is mainly based on cholinesterase (ChE) inhibition [6-10]. Organophosphate and carbamate insecticides are the main ChE inhibitors (**Table 1**). Other compounds, such as heavy metals, fluoride, nerve gas or nicotine, can also inhibit ChE enzyme. Although this lack of selectivity, ChE-based biosensors are shown as powerful tools when a rapid toxicity screening is required.

2.1.1.1. Mono-Enzymatic Biosensors

Two types of natural ChE enzymes are known: acetylcholinesterase (AChE) and butyrylcholine-sterase (BChE). These enzymes have different substrates: AChE preferentially hydrolyzes acetyl esters, such as acetylcholine (Equation (1)), whereas BChE hydrolyzes butyrylcholine (Equation (2)):

$$\text{Acetylcholine} + H_2O \xrightarrow{\text{AChE}} \text{Choline} + \text{Acetic acid} \quad (1)$$

$$\text{Butyrylcholine} + H_2O \xrightarrow{\text{BChE}} \text{Choline} + \text{Butyric acid} \quad (2)$$

The pH variation produced by the acid formation can be measured using electrochemical methods, such as potentiometry [11]. This pH change can also be measured using pH-sensitive spectrophotometric indicators [12,13] or pH sensitive fluorescence indicators [14].

Artificial substrates, acetylthiocholine for AChE and butyrylthiocholine for BChE, have been also used. The enzymatic hydrolysis of these substrates produces electroactive thiocholine (Equations (3) and (4)).

$$\text{Acetylthiocholine} + H_2O \xrightarrow{\text{AChE}} \text{Thiocholine} + \text{Acetic acid} \quad (3)$$

$$\text{Butyrylthiocholine} + H_2O \xrightarrow{\text{AChE}} \text{Thiocholine} + \text{Butyric acid} \quad (4)$$

$$2\ \text{Thiocholine} + H_2O \xrightarrow[\text{oxidation}]{\text{Anodic}} \text{Dithiobischoline} + 2H^+ + 2e^- \quad (5)$$

This system has two advantages over the bi-enzymatic ChE/ChOD biosensors. First, it has a simpler design. Secondly, the detection potential is lower than the one used for the oxidation of H_2O_2.

La Rosa et al. proposed the use of 4-aminophenyl acetate as alternative ChE substrate [15,16]. They oxidize the enzymatic product 4-aminophenol at +250 mV vs SCE. Electrochemical biosensors for pesticide detection based on the use of this substrate avoid interferences from the oxidation of other electroactive compounds [15- 18]. However, 4-aminophenyl acetate is not commercially available and its use involves a laborious and timeconsuming synthesis. Moreover, this substrate is unstable and requires special storage conditions (nitrogen atmosphere, below 0°C).

2.1.1.2. Bi-Enzymatic Biosensors

In most cases, ChE is coupled to choline oxidase (ChOD) [6]. AChE hydrolyzes its natural substrate to choline and acetic acid (Equation (1)). Since choline is not electrochemically active, ChOD is used to produce H_2O_2, which can be oxidized onto the platinum electrode at around + 0.7 V vs Ag/AgCl (Equations (7) and (8)). However, an over-potential is required, favouring the oxidation of interfering electroactive species present in real samples. To overcome this drawback, different approaches have been proposed, such as the use of nanomaterial-modified electrodes. A biosensor for the detection of pesticides and nerve agents was developed by immobilizing AChE and ChOD onto Au-Pt bimetallic NPs [19]. The synergistic effect of these nanoparticles increased the surface area and facilitated the electron transfer process, reducing the applied potential for the detection of H_2O_2. Alternatively to H_2O_2 oxidation, ChE inhibition can be followed using a Clark electrode able to measure the oxygen consumed by the ChOD catalyzed reaction (Equations (6)-(8)) [20].

$$\text{Acetylcholine} + H_2O \xrightarrow{\text{AChE}} \text{Choline} + \text{Acetic acid} \quad (1)$$

$$\text{Choline} + O_2 \xrightarrow{\text{ChOD}} \text{Betaine} + H_2O \tag{6}$$

$$H_2O \xrightarrow[\text{vs Ag/AgCl}]{+0.7\text{ V}} O_2 + 2H^+ + 2e^- \tag{7}$$

$$O_2 + 4H^+ + 4e^- \xrightarrow[\text{vs Ag/AgCl}]{-0.6\text{ V}} H_2O \tag{8}$$

AChE was also coupled to tyrosinase [18]. In this case, AChE enzymatic hydrolysis of phenyl acetate produces phenol compounds, characterized by a high oxidation potential. For this reason, tyrosinase enzyme was used to convert the phenol to quinone, compound that can be electrochemically reduced to catechol at –150 mV vs Ag/AgCl.

2.1.1.3. Tri-Enzymatic Biosensors

Peroxidase may be added to the bi-enzyme system to develop a tri-enzymatic biosensor. Karousos et al. used a Quartz Crystal Microbalance (QCM) sensor based on three enzymes for the determination of organophosphorus and carbamate pesticides [21]. Acetylcholine was converted to choline by AChE and then, choline was converted to hydrogen peroxide by choline oxidase. In the presence of HRP, H_2O_2 oxidized 3,3'-diaminobenzidine to an insoluble product that precipitated out and adsorbed on the crystal surface causing a decrease in the resonant frequency of the crystal. AChE inhibition caused by pesticides reduced the amount of QCM-detectable precipitate produced. This QCM-enzyme sensor system allowed detecting carbaryl and dicholorvos concentrations down to 1 ppm.

2.1.1.4. ChE Sources

The enzyme source has an important effect on the biosensor performance. Several AChE enzymes are available from different sources, such as Electric eel, Bovine or Human erythrocytes, Horse serum and Human blood. Generally, ChE enzymes isolated from insects are more sensitive than those extracted from other sources. The use of recombinant ChE enzymes also allows improvements on the sensitivity of biosensors [22]. As an example, Valdes-Ramirez and co-workers compared the use of three AChEs in biosensors for the detection of dichlorovos in a sample of apple skin [23]. The use of genetically modified AChE decreased four orders of magnitude the detection limit found for the use of AChE from wild type Drosophila melanogaster and Electriceel.

Table 1. Characteristics of electrochemical cholinesterase-based biosensors for pesticide detection.

Target analyte	Detection technique	Enzyme immobilization technique	Electroactive materials	Linearity range (M)	Detection limit (M)	References
			Organophosphorus insecticides			
Chloropyrifos	CV	Covalent binding	Exfoliated graphite nanoplatelets	ND	1.58×10^{-10}	[170]
Chloropyrifos	SWV	Cross-linking	SWCNT	10^{-11} - 10^{-6}	10^{-12}	[171]
Chloropyrifos	Amperometry	Covalence	ZnS NPs	1.5×10^{-9} - 4×10^{-8}	ND	[172]
Chlorpyrifos oxon	CV	Entrapment	PEDOT:PSS	ND	4×10^{-9}	[173]
Chlorpyrifos oxon	Amperometry	Entrapment	7,7,8,8-tetracyano quinodimethane	6×10^{-9} - 2.4×10^{-9}	6×10^{-9}	[174,175]
Paraoxon	Amperometry	Affinity	MWCNT	3.6×10^{-14} - 3.6×10^{-11}	5×10^{-15}	[176]
Paraoxon	Fluorescence	Adsorption	CdTe QDs			
Paraoxon	Amperometry	Entrapment	-	1.3×10^{-7} - 5×10^{-6} M.	3.5×10^{-2}	[177]
Paraoxon	Amperometry	Adsorption	AuNPs, grapheme oxide nanosheets	ND	10^{-13}	[178]
Paraoxon	Amperometry	Cross-linking	CoPc-Prussian blue	7.3×10^{-9} - 1.8×10^{-8}	7.3×10^{-9}	[179]
Methylparaoxon	Amperometry	Entrapment	CoPc	2×10^{-9} - 4×10^{-6}	2.6×10^{-9}	[180]
Methylparaoxon	Amperometry	Affinity	MWCNT	3.8×10^{-14} - 3.8×10^{-11}	5.3×10^{-15}	[176]
Triazophos	Amperometry	Adsorption	MWCNT	3×10^{-8} - 7.8×10^{-6}	10^{-8}	[181]
Dichlorvos	Amperometry	Adsorption	-	ND	10^{-10}	[182]
Dichlorvos	Amperometry	Entrapment	CoPc	ND	7×10^{-12}	[23]
Dichlorvos	Amperometry	Entrapment	CoPc	2×10^{-10} - 10^{-8}	9.6×10^{-11}	[180]
Dichlorvos	Amperometry	Adsorption	-	Up to 10^{-16}	10^{-17}	[183]
Dichlorvos	Amperometry	Cross-linking	Prussian blue	4.52×10^{-11} - 4.52×10^{-8}	1.13×10^{-11}	[184]
Omethoate				2.34×10^{-10} - 4.69×10^{-8}	7.04×10^{-11}	
Trichlorfon				1.16×10^{-10} - 1.94×10^{-8}	1.94×10^{-11}	
Phoxim				1.68×10^{-10} - 3.35×10^{-8}	3.35×10^{-11}	
Trichlorfon	Amperometry	Adsorption	TiO_2 and PbO_2 particles	10^{-8} - 2×10^{-5}	10^{-10}	[185]
Monocrotophos	Amperometry	Adsorption	AuNPs	4.5×10^{-9} - 4.5×10^{-6}	2.7×10^{-9}	[186]
Monocrotophos	Amperometry	Covalent binding	AuNPs-QDs	4.5×10^{-9} - 4.5×10^{-6}	1.3×10^{-9}	[155]
Acephate	FET	Affinity	CNT	ND	5.45×10^{-14}	[187]
Dimethoate	Amperometry	Adsorption	CNTs, zirconia NPs, Au colloid coated Fe_3O_4 magnetic NPs, prussian blue	4.4×10^{-6} - 4.4×10^{-2}	2.4×10^{-6}	[188]
			Carbamate insecticides			
Aldicarb	Amperometry	Cross-linking	CoPc-Prussian blue	6.3×10^{-8} - 1.6×10^{-7}	1.3×10^{-7}	[179]
Carbaryl	Amperometry	Adsorption	-	2.5×10^{-8} - 5×10^{-7}	1.5×10^{-8}	[189]
Carbaryl	Amperometry	Covalent binding	QDs	5×10^{-9} - 2.5×10^{-7}	3×10^{-9}	[155]
Carbaryl	Amperometry	Cross-linking	CoPc-Prussian blue	1.2×10^{-7} - 4.9×10^{-7}	1.2×10^{-7}	[179]
Carbaryl	Amperometry	Adsorption	MWCNT	5×10^{-13} - 5×10^{-10}	5×10^{-15}	[190]
Carbaryl	Amperometry	Entrapment	CoPc	9×10^{-8} - 4×10^{-6}	1.6×10^{-7}	[180]
Carbofuran	DPV	Adsorption	CNTs-AuNPs	4.8×10^{-9} - 9×10^{-8}	4×10^{-8}	[191]
Carbofuran	Amperometry	Cross-linking	CoPc	10^{-10} - 10^{-7}	4.9×10^{-10}	[192]
Carbofuran	Amperometry	Entrapment	CoPc	4×10^{-9} - 8×10^{-8}	4.5×10^{-9}	[180]

2.1.2. Tyrosinase-Based Biosensors

Tyrosinase oxidizes monophenols in two consecutive steps: first, the enzyme catalyzes the o-hydroxylation of monophenol to o-diphenol (cresolate activity, Equation (9)) which, in a second step, is oxidized to its corresponding o-quinone (catecholase activity, Equation (10)):

$$\text{Monophenol} + O_2 \xrightarrow{\text{Cresolate activity}} \text{Catechol} \tag{9}$$

$$\text{Catechol} + O_2 \underset{-0.2\text{ V vs Ag/AgCl}}{\overset{\text{Catecholase activity}}{\rightleftarrows}} \text{O-quinone} \tag{10}$$

Tyrosinase is inhibited by different compounds, such as carbamate pesticides and atrazine. Numerous electrochemical biosensors based on the inhibition of tyrosinase activity have been reported [24-29] (**Table 2**).

Tyrosinase biosensors suffer from poor specificity since many substrates and inhibitors can interfere. The enzyme is inherently unstable, reducing the lifetime of the tyrosinase-based biosensors. However, tyrosinase can stand high temperatures and the organic solvents used to dissolve the pesticides.

2.1.3. Alkaline Phosphatase (ALP)-Based Biosensors

Alkaline phosphatase catalyses the following reaction:

$$\text{Phosphate monoester} + H_2O \rightarrow \text{alcohol} + \text{phosphate} \tag{11}$$

ALP is inhibited by different compounds. Several ALPbased biosensors for the detection of pesticides have been developed using different enzyme substrates depending on the transduction method.

Ayyagari et al. described a chemiluminescent ALPbased biosensor for the detection of paraoxon [30]. The biosensor was based on the measurement of the intensity of the light generated by ALP-catalyzed dephosphorylation of a chemiluminescent substrate, chloro 3-(4-methoxy spiro [1,2-dioxetane-3-2'-tricyclo-[3.3.1.1]-decan]-4- yl) phenyl phosphate.

Table 2. Characteristics of electrochemical inhibition-based biosensors using tyrosinase for pesticide detection.

Target analyte	Detection technique	Enzyme immobilization technique	Electroactive materials	Linearity range (M)	Detection limit (M)	References
			Organophosphorus insecticides			
Methyl parathion	Amperometry	Cross-linking	CoPc	2.28×10^{-8} - 3.8×10^{-7}	ND	[26]
Diazinon	Amperometry	Cross-linking	CoPc	6.24×10^{-8} - 1.64×10^{-7}	ND	[26]
Dichlorvos	Amperometry	Cross-linking + entrapment	1,2-naphthoquinone-4-sulfonate (NQS)	Up to 8×10^{-6}	6×10^{-8}	[25]
Dimethoate				2×10^{-6} - 2×10^{-4}	10^{-6}	
Pirimicarb	Amperometry	Adsorption	-	2×10^{-5} - 5×10^{-3}	10^{-5}	[29]
Paraoxon				10^{-5} - 10^{-2}	5×10^{-6}	
Malathion				10^{-5} - 10^{-2}	5×10^{-6}	
Paraoxon	Amperometry	Cross-linking	Prussian blue	10^{-7} - 10^{-8}	10^{-7}	[193]
			Carbamate insecticides			
Carbofuran	Amperometry	Cross-linking	CoPc	2.26×10^{-8} - 4.07×10^{-7}	ND	[26]
Carbofuran				10^{-5} - 10^{-2}	5×10^{-6}	[29]
Aldicarb	Amperometry	Adsorption	-	10^{-5} - 10^{-2}	5×10^{-6}	[29]
Carbaryl				10^{-5} - 10^{-2}	5×10^{-6}	[29]
Carbaryl	Amperometry	Cross-linking	CoPc	4.97×10^{-8} - 2.48×10^{-7}	ND	[26]
Thiodicarb	Square Wave Voltammetry (SWV)	Entrapment	-	3.75×10^{-7} - 2.23×10^{-6}	1.58×10^{-7}	[24]

A fluorescent ALP-based biosensor for the detection of organochlorine, pesticides (carbamate and fenitrothion), heavy metals and CN^- was also described [31]. ALP enzyme catalyzed the hydrolysis of 1-naphthyl phosphate to fluorescent 1-naphthol.

Mazzei and co-workers developed electrochemical ALP-based biosensors for the detection of malathion and 2,4-dichlorophenoxyacetic acid (2,4-D) by using 3-indoxyl phosphate, phenyl phosphate or ascorbate-2- phosphate as enzyme substrates [32]. Another electrochemical ALP-based biosensor was also described for the screening of several environmental pollutants. The biosensor was based on the entrapment of ALP in a hybrid sol-gel/chitosan film, deposited on the surface of a screen-printed electrode [33]. The substrate ascorbic acid 2-phosphate was catalyzed by the enyme to produce ascorbic acid, which was monitored by amperometry.

2.1.4. Peroxidase-Based Biosensors

Peroxidase molecules can be first oxidized by H_2O_2 and then reduced by phenolic compounds. This process involves two enzyme intermediates: compounds I and II (**Figure 1**). Phenolic compounds are thus oxidized to quinones or free radical products, able to be electrochemically reduced on the electrode surface. Several organic and inorganic compounds have been reported to inhibit the enzyme activity of peroxidase by coordinating compound I. A biosensor based on the inhibition of peroxidase was described for the detection of thiodicarb, a carbamate pesticide [34]. HRP was covalently bound on a gold electrode. In the presence of hydrogen peroxide, hydroquinone was oxidized by peroxidase to p-benzoquinone which could be electrochemically reduced to hydroquinone at a potential of –0.072 V vs Ag/AgCl. The presence of inhibitor compounds induced a decrease of the biosensor current response.

2.1.5. Acid Phosphatase

Acid phosphatase (AP) is reversibly inhibited by some pesticides. AP has been used with glucose oxidase (GOD) to develop a bienzymatic biosensor for the electrochemical detection of Malathion, methyl parathion and paraoxon [37]. Both enzymes were coupled on a commercial H_2O_2 sensing electrode. This system is based on the following reactions:

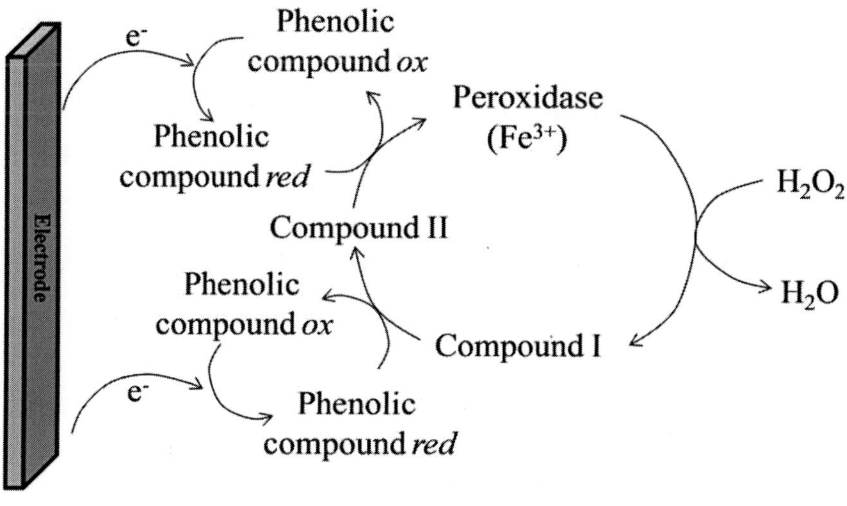

Figure 1. Scheme of the reactions occurring at the surface of a peroxidase-modified electrode. ox: oxidized form, red: reduced form [35,36].

$$\text{Glucose} - 6 - \text{phosphate} + H_2O \xrightarrow{AP} \text{Glucose} + \text{inorganic phosphate}$$

$$\text{Glucose} + O_2 \xrightarrow{\text{GOD}} \text{Gluconolactone} + H_2O$$

2.2. Catalytic Biosensors

2.2.1. Organophosphorus Hydrolase (OPH)

OPH is an enzyme that hydrolyzes organophosphorus pesticides [38], such as parathion, methyl parathion [39] or paraoxon [40,41]. This enzyme hydrolyzes P-O, P-S and P-CN bonds generating two protons, able to be electrochemically detected, and an alcohol, which in many cases is chromophoric and/or electroactive.

However, these biosensors show lower sensitivity values and higher detection limits than cholinesterase-based biosensors. Moreover, they can only detect some organophosphorus (OP) compounds.

Table 3 summarizes the performances of some OPHbased biosensors reported in the literature.

2.2.2. Glutathion-S-Transferase

Glutathion-S-transferase (GST) was used to develop a fiber-optic biosensor for the detection of atrazine [42]. The enzyme was immobilized by cross-linking on a membrane that was supported on an inner glass disk by means of an intermediate binder sol-gel layer. Bromcresol green was incorparated in the sol-gel as pH indicator. GST catalyzed the nucleophile attack of GSH on atrazine, releasing H^+. This pH variation was optically measured by colour changes of bromcresol green.

3. WHOLE CELL BIOSENSORS

3.1. Microbial Biosensors

To develop a microbial biosensor, microorganisms have to be immobilized onto a transducer using different chemical (e.g. cross-linking) or physical techniques (e.g. entrapment) [43]. Microorganisms have several advantages as sensing elements in the development of biosensors. They are able to metabolise a wide range of chemical compounds. The use of whole cells, as a source of intracellular enzymes, avoids expensive protocols of enzyme purification. The enzyme is maintained in its natural environment improving its stability and activity. The main limitation of the use of whole cells is the diffusion of substrate and products through the cell wall resulting in a slow response as compared to enzyme-based biosensors. To overcome this drawback, cells can be permeabilised [44].

Table 3. Characteristics of hydrolase-based biosensors.

Target analyte	Detection technique	Enzyme immobilization technique	Electroactive materials	Linearity range (M)	Detection limit (M)	References
Paraoxon	Amperometry	Covalent binding	SWCNT	5×10^{-7} - 8.5×10^{-6}	10^{-8}	[41]
Paraoxon	Amperometry	Entrapment	Mesoporous carbon	2×10^{-7} - 8×10^{-6}	1.2×10^{-7}	[40]
Paraoxon	Amperometry	Entrapment	MCNT	Up to 4×10^{-6}	15×10^{-8}	[194]
Paraoxon	Amperometry	Cross-linking	MWNT	5×10^{-7} - 2×10^{-6}	0.314×10^{-6}	[195]
Paraoxon	Optic	Cross-linking	-	10^{-5} - 4.8×10^{-4}	2×10^{-6}	[196]
Ethyl Parathion	Amperometry	Covalent binding	-	ND	$<3.4 \times 10^{-9}$	[197]
Methyl parathion	Amperometry	Covalent binding	AuNPs – MWCN-QDs	1.9×10^{-8} - 7.6×10^{-7}	3.8×10^{-9}	[39]
Methyl parathion	Amperometry	Entrapment	MCNT	Up to 2×10^{-6}	8×10^{-7}	[194]
Parathion	Optic	Cross-linking	-	10^{-5} - 2×10^{-4}	2×10^{-6}	[196]
Paraoxon	Fluorescence	Affinity	-	10^{-6} - 8×10^{-4}	ND	[198]
Diisopropyl phosphorofluoridate (DFP)	Fluorescence	Affinity	-	2×10^{-6} - 4×10^{-4}	ND	[198]

3.1.1. Electrochemical Microbial Biosensors

3.1.1.1. Amperometric Detection

Amperometric microbial biosensors have been widely developed for the determination of biochemical oxygen demand (BOD) in order to measure biodegradable organic pollutants in aqueous samples. Most of BOD biosensors consist of a microbial film sandwiched between a porous cellulose membrane and a gas-permeable membrane. Organic substrates, present in wastewater samples, diffuse through the dialysis membrane and are assimilated by the immobilized microbial population, increasing the bacterial respiration rate. Therefore, less dissolved oxygen diffuses through the gas-permeable Teflon membrane to be detected by a Clark oxygen electrode [45]. Different microbial strains were used as biosensing element such as Arxula adeninivorans [46], Bacillus subtilis [47], Serratia marcescens [48] or yeast [49]. Single microorganisms metabolize a limited range of organic pollutants, which may result in an inaccurate estimation of BOD values. To overcome this problem, mixed cultures (e.g. Bacillus subtilis and Trichosporon cutaneum [50]) or activated sludges [51] were used.

Mulchandani's group developed amperometric microbial biosensors for the determination of organophosphate pesticides with p-nitrophenyl substituent (e.g. paraoxon, methyl parathion, parathion, fenitrothion and ethyl pnitrophenol thiobenzenephosphonate (EPN)) [52]. These biosensors were based on the co-immobilization of microorganisms and OPH (free or expressed on the cell surface of other microorganisms). OPH hydrolyzes the pesticide and releases p-nitrophenol. Released p-nitrophenol can be oxidized by some microorganisms, such as or Pseudomonas putida JS444. Two detection strategies were used:

- OPH hydrolyzes the organophosphorus compounds to produce p-nitrophenol. Released p-nitrophenol was degraded by some bacteria, such as Pseudomonas putida JS444. This degradation resulted in electroactive compounds, amperometrically detected [53,54].
- The degradation of p-nitrophenol by some microbes, such as Arthrobacter sp. JS443, consumes oxygen. A Clark oxygen electrode was used to measure oxygen concentration changes [55-57].

3.1.1.2. Potentiometric Detection

Conventional potentiometric microbial biosensors have been developed using ion-selective electrodes (e.g. pH, ammonium) or gas sensing electrodes (e.g. pCO_2) coated with an immobilized microbial layer. Assimilation of substrates by microbes causes changes in potential due to ion accumulation or depletion [43].

A potentiometric biosensor for the direct detection of paraoxon was based on the immobilization of recombinant E. Coli on a glass pH electrode. Bacteria was engineered to contain the opd gene that encodes the OPH enzyme [58]. Entrapped OPH-active bacteria hydrolyzed OP compounds producing two protons. The quantity of released H^+ was correlated to the concentration of hydrolyzed paraoxon.

3.1.2. Optical Microbial Biosensors

Optical microbial biosensors allowing the detection of pollutants such as phenols and heavy metals have been developed [52,59]. However, only few optical microbial biosensors allowing the detection of pesticides have been reported. A disposable colorimetric microbial biosensor for the detection of methyl parathion pesticide was described [60]. Whole cells of Flavobacterium sp. were immobilized on a glass fiber filter paper. The OPH activity of Flavobacterium sp. hydrolyzed methyl parathion into p-nitrophenol that can be detected at 410 nm.

3.2. Plant Tissue and Photosynthesis-Based Biosensors

3.2.1. Plant Tissue-Based Biosensors

The use of plant tissue is an attractive alternative to enzymatic biosensors. Tissue that acts as enzyme source presents many advantages [61]:

- High stability and activity resulting from the maintenance of the enzyme in its natural environ-ment;
- Long lifetime of biosensors;
- High reproducibility of the experimental results;
- Availability and low price of a wide range of plant tissues;
- Avoidance of tedious and time-consuming enzyme extraction and purification steps;
- Presence of the required cofactors in the used tissue.

Planktonic algae have been widely used to develop biosensors for pollutants present in the aquatic ecosystems. Biosensors based on immobilized Chlorella vulgaris microalgae were reported [62-64]. Those biosensors were based on the inhibition of enzymes located on the external membrane, such as alkaline phosphatases and esterases, by heavy metals and pesticides.

3.2.2. Photosynthesis-Based Biosensors

Different types of photosynthetic materials were used as recognition element for the development of biosensors: whole cells (e.g. microalgae), chloroplasts or thylakoids and photosystem II (PS II) [65]. PS II is a supramolecular pigment-protein complex located in the thylakoid membrane. It catalyzes the light-induced transfer of electrons from water to plastoquinone in a process that evolved oxygen. The activity of PS II can be inhibited by several groups of herbicides and heavy metals [66].

The measurement of oxygen evolution using a Clarktype electrode is a standard procedure for the determination of the photosynthetic activity [61,67,68]. The incorporation of several types of photosystem II specific artificial electron acceptors as electroactive mediators allows to maximize the photosynthetic activity. Other biosensors are not based on the use of a Clark-type electrode. In these cases, alga were immobilized on the surface of ITO electrode [69] or SPE [70].

Optical photosynthesis-based biosensors have also been described based on the fluorescence induced by chlorophyll a. The light absorbed by chlorophyll molecules of PSII may be assimilated into the light reactions of the photosynthesis or may be released as fluorescence or thermal energy. Herbicides inhibit photosynthetic electron flow by blocking the PSII quinone binding site causing an increase in the chlorophyll fluorescence emission. Based on this principle, herbicide biosensors based on the measurement of the algal chlorophyll fluorescence at 682 nm (under 469 nm excitation light) were developed [71-73]. Recently, three microalgae species (Dictyosphaerium chlorelloides, Scenedesmus intermedius and Scenedesmus sp.) were entrapped within a silica matrix and the increase in the amount of chlorophyll fluorescence signal was used to quantify simazine [74].

4. IMMUNOSENSORS

Immunosensors are characterized by the highly selective affinity interactions between immobilized antibodies (Ab) or antigens (Ag), on the transducer surface, and their specific analytes, Ag or Ab respectively [75-77]. Unlike enzyme-based biosensors, able to evaluate total toxicity, immuno-sensors are specific for a molecule.

Several immunosensors for pesticides detection have been described, based on electrochemical, optical, piezoelectric and mechanical transduction methods.

4.1. Electrochemical Immunosensors

Table 4 presents performance characteristics of some electrochemical immunosensors.

4.1.1. Amperometric Detection

Numerous amperometric immunosensors for the detection of pesticides have been reported [78-81]. A simple amperometric immunosensor was developed for the analysis of 2,4-D in the presence of organic solvents, required to solubilize it from soil [81]. An amperometric immunosensor based on a carbon paste SPE incorporateing a conducting polyaniline (PANI)/poly(vinylsulphonic acid) (PVS) copolymer was developed for the detection of atrazine [82]. Free and HRP-labeled atrazine competed for their binding to the Ab previously immobilized onto the PANI-PVS-modified electrode surface. The addition of HRP substrate, hydrogen peroxide, enabled the catalytic reaction inducing a flow of electrons from the electrode surface through the molecular wires of the PANI/PVS copolymer.

Table 4. Characteristics of some electrochemical immunosensors.

Target analyte	Immobilization technique	Electroactive materials	Linearity range (M)	Detection limit (M)	References
			Amperometry		
Phenanthrene (PAH)	Adsorption of BSA-phenanthrene conjugate	-	2.8×10^{-9} - 2.5×10^{-7}	4.5×10^{-9}	[199]
Paraoxon	Adsorption + Nafion film	AuNPs	8.7×10^{-8} - 6.9×10^{-6}	4.4×10^{-8}	[200]
Atrazine	Affinity	-	7×10^{-10} - 1.35×10^{-8}	1.7×10^{-10}	[201]
Atrazine	Adsorption	PANI-PVS copolymer	5.5×10^{-10} - 2.3×10^{-8}	4.6×10^{-10}	[82]
17-β estradiol	Adsorption of BSA-estradiol conjugate	-	2.2×10^{-12} - 3.6×10^{-5}	9.2×10^{-13}	[202]
Pichloram	Adsorption of BSA-pichloram conjugate	Gold nanoclusters	3.6×10^{-9} -3.6×10^{-5}	1.8×10^{-9}	[203]
Diuron	Adsoprtion of hapten-BSA conjugate	Prussian blue, AuNPs	4.3×10^{-12} - 4.3×10^{-5}	4.3×10^{-12}	[156]
Naphthalene	Adsorption	AuNPs	3.9×10^{-9} - 7.8×10^{-7}	6.2×10^{-10}	[204]
			Impedance spectroscopy		
Atrazine	Affinity	-	4.6×10^{-8} - 1.4×10^{-6}	9.3×10^{-8}	[97]
Atrazine	Adsorption of BSA-atrazine conjugate	-	ND	2.7×10^{-8}	[90]
Atrazine	Adsorption of BSA-atrazine conjugate	-	ND	3.9×10^{-8}	[91]
Atrazine	Covalent immobilization of BSA-atrazine conjugate	-	ND	1.9×10^{-10}	[92]
Atrazine	Affinity	-	4.6×10^{-10} - 4.6×10^{-8}	4.6×10^{-11}	[93]
Atrazine	Affinity/entrapment	-	4.6×10^{-10} - 9.3×10^{-7}	4.6×10^{-10}	[94]
2,4-D	Covalent binding	-	2×10^{-10} - 2×10^{-6}	ND	[205]

4.1.2. Conductometric Detection

To our knowledge, only a few conductometric immunosensors for environmental analysis have been described, probably due to their low specificity.

Valera and co-workers used this method for atrazine detection [83-85]. Atrazine was covalently immobilized on interdigitated μ-electrodes (IDμE) [85]. Detection of free atrazin was achieved through a competitive reaction with immobilized atrazine for the antibody added in solution. The detection method was based on the use of antibodies labeled with gold nanoparticles. Their presence amplified the conductive signal. This biosensor is adapted for the detection of atrazine in red wines since none matrix effect related to red wine samples were observed.

4.1.3. Potentiometric Detection

A few potentiometric immunosensors have been described for environmental analysis [86-88]. A potentiometric biosensor was developed for terbuthylazine (TBA), a herbicide widely used in agriculture [88]. Free TBA and immobilized TBA-BSA conjugate competed for their binding to urease-labeled specific Ab. The addition of urease substrate, urea, enabled the potentiometric measurement of the ammonia produced in an inversely proportional amount of the TBA present in the sample.

4.1.4. Electrochemical Impedance Spectroscopy

Several impedimetric immunosensors for the detection of pesticides have been reported [89-97]. The electron transfer resistance at the interface between the electrode and the solution changes slightly when the immobilized biomolecule binds the analyte. Impedance spectroscopy allows a label-free detection with many potential advantages, such as higher signal-to noise ratio, ease of detection, lower assay cost, faster assays and shorter analysis times. However, regeneration of the sensing surface is typically time-consuming and not reproducible [4].

Table 4 presents characteristics of some impedimetric immunosensors for pesticide analysis.

Valera and co-workers developed biosensors for atrazine detection using a competitive immuno-assay. AtrazineBSA conjugate was immobilized on the IDμE surface, either by adsorption [90,91] or by covalent binding [92]. Best performances were obtained when the conjugate was covalently bound to the IDμE area previously activated with (3-glycidoxypropyl)trimethoxysilane. The detection limit was 1.9×10^{-10} M. Recently, an impedimetric immunosensor for atrazine detection was developed by immobilizing anti-atrazine antibody modified with histidine-tag onto a polypyrrole (PPy) film N-substituted by nitrilotriacetic acid (NTA) electrogenerated on a gold electrode [93]. After coordination of Cu^{2+} to poly-NTA, anchoring of histidine-tagged atrazine was achieved by affinity interactions between histidine groups and the chelated Cu^{2+} centers. In the presence of atrazine, the interaction of the analyte with the immobilized antibody triggered an increase of the charge transfer resistance proportional to the pesticide concentration. The detection limit was 4.6×10^{-11} M.

4.2. Optical Immunosensors

Optical immunosensors are based on the measurement of changes in the optical characteristics induced by the formation of Ab-Ag complexes.

4.2.1. SPR

SPR-based biosensors have been reported for the detection of different pesticides (**Table 5**). SPR allows realtime monitoring and does not require labeled molecules [98].

Mauriz and co-workers used a commercial SPR (SENSIA) for the on-line monitoring of pesticides in real water samples [99-104]. Pesticides-BSA conjugates were immobilized through a self-assembled monolayer (SAM) onto a gold electrode to obtain a reusable sensing surface. The same sensing surface was used for more than 200 assays, showing good reproducibility. Several conjugates were immobilized on the sensing surface of one individual flow cell to simultaneously detect DTT, chlorpyrifos and carbaryl.

Miura and co-workers developed a SPR-based immunosensor for the competitive detection of 2,4-D [105- 108]. A 2,4-D-ovalbumine conjugate was immobilized onto the Au surface of the sensor chip to compete with free 2,4-D for their selective binding to monoclonal anti- 2,4-D. Amplification through avidin-biotin interactions was described to enhance the sensitivity of the sensor. The amplification was based on successive incubations with a biotinylated secondary antibody against anti-2,4-D, avidin and finally biotin-BSA molecules. The sensor signal was amplified by a factor of 10 [108].

4.2.2. Fluorescence Polarisation

Immunospecies have to be conjugated to fluorescent labels (e.g. cyanine) to develop immuno-sensors. However, fluorescent organic dyes suffer from a photo-bleaching problem. This drawback can be overcome by the use of nanoparticles as fluorescent reporters. For this purpose, europium chelate-dyed polystyrene NPs have been used to develop immunosensors for the detection of atrazine [109].

4.2.3. Total Internal Reflection Fluorescence (TIRF)

TIRF was used to develop immunosensors for water pollution control [75,110-115]. Barzen et al. described a prototype of a portable TIRF-based immunosensor combined with a flow injection system to monitor surface water quality [111]. Aminodextran-analyte conjugates were adsorbed onto the transducer surface. Specific antibodies labeled with a fluorophore were incubated with the target analytes present in the sample. Then, free labeled-antibodies bound immobilized analyte derivatives. A collimated laser beam (635 nm) coupled to the transducer was guided by total internal reflection causing excitation of the bound fluorescent antibody in the evanescent field (**Figure 2**). This TIRF-based immunosensor, called the RIver ANAlyser (RIANA), was used to monitor the levels of atrazine, simazine and alachlor in the Ebre Delta and the estuarine area of Portugal [77]. Studies were focused on the evaluation of matrix effects, interferences due to the presence of crossreactant substances and on the validation of the sensor. Results show the efficiency of the RIANA immunosensor for monitoring natural waters in compliance with the Drinking Water Directive of the European Union.

Table 5. Characteristics of some SPR-based immunosensors.

Target	Immobilization technique	Linearity range (M)	Detection limit (M)	References
DTT	Covalent binding	1.7×10^{-10} - 5×10^{-9}	5.64×10^{-11}	[104]
2,4-D	Adsorption of 2,4-D—BSA conjugate	2.3×10^{-9} - 4.5×10^{-6}	2.3×10^{-9}	[107]
2,4-D	Adsorption of 2,4-D—ovalbumin conjugate	4.5×10^{-10} - 1.4×10^{-6}	4.5×10^{-10}	[106]
2,4-D	Covalent immobilization of 2,4-D—BSA conjugate	ND	3.6×10^{-11}	[108]
2,4-D	Covalent immobilization of 2,4-D—BSA conjugate	4.5×10^{-10}	4.5×10^{-10}	[105]
Chorpyrifos	Covalent binding	6.6×10^{-10} - 1.4×10^{-7}	1.4×10^{-10}	[104]
Carbaryl	Covalent binding	8.2×10^{-9} - 7.3×10^{-8}	4.5×10^{-9}	[104]
Isoproturon	Covalent binding	6.3×10^{-9} - 7.9×10^{-8}	4.8×10^{-10}	[206]
Atrazine	Covalent immobilization of atrazine—BSA conjugate	1.3×10^{-10} - 3.7×10^{-9}	9.3×10^{-11}	[99]
TNT	Adsorption of TNP—BSA conjugate	2.6×10^{-11} - 1.3×10^{-6}	2.6×10^{-11}	[207]
TNT	Covalent immobilization on AuNPs	ND	4.4×10^{-11}	[208]
TNT	Covalent binding	3.5×10^{-11} - 1.3×10^{-7}	ND	[209]
TNT	Covalent binding	Up to 3.1×10^{-8}	4.8×10^{-10}	[210]
2,4-dinitrotoluene	Covalent binding	4.4×10^{-9} - 4.4×10^{-7}	8.8×10^{-11}	[211]
2,4-dichlorophenol	Affinity (protein G)	ND	1.2×10^{-7}	[212]

Automated Water Analyser Computer Supported System (AWACSS) was developed using the same principle as RIANA immunosensor. However, improvements were made in three critical areas: 1) expanded multi-analyte analysis capability allowing for simultaneous detection of up to 30 analytes; 2) novel design approaches to the optical detection and fluidics including miniaturized integrated optics and micro-fluidics and 3) intelligent remote surveillance and control for unattended continuous monitoring [116].

4.2.4. Polarisation-Modulation InfraRed Reflection-Absorption Spectroscopy (PM-IRRAS)

Reflection-absorption IR spectroscopy combined with polarization modulation was proposed as a novel optical transduction mode. Recently, PM-IRRAS was used to develop immunosensors allowing the detection of environmental pollutants [117-119].

This principle was applied by Pradier and co-workers to determine atrazine using an indirect competitive format [119]. First, sensor chips were coated with ovalbumine-atrazine derivatives. The surfaces were analyzed by PM-IRRAS and the integrated area of the peptide bands was measured. Then, the different steps of the conventional ELISA test were performed. Successive binding of anti-atrazine antibody and secondary anti-rabbit immunoglobulin G antibody resulted in a change of the IR absorption properties of the organic film at the sensor surface. Detection of atrazine was based on the analysis of the amide I and II bands. When the concentration of free pesticide increased, the intensity of amide bands decreased.

4.3. Piezoelectric Immunosensors

Piezoelectric immunosensors are devices based on materials such as quartz crystals with Ab or Ag immobilized on their surface. A QCM sensor is a mass-sensitive device able to measure very small mass changes. Several QCM immunosensors for environmental monitoring have been described [120], but only few of them were developed for the detection of pesticides. A QCM immunosensor was developed for the analysis of carbaryl, and 3,5,6-trichloro-2-pyridinol (TCP), the main metabolite of the insecticide chlorpyrifos and of the herbicide triclopyr [121]. The biosensor was based on the immobilization of hapten conjugates onto the gold electrode via SAM. This covalent immobilization allowed the reusability of the modified electrode surface for at least 150 assays without significant loss of sensitivity.

4.4. Mechanical Immunosensors

Microcantilever sensors are based on a response due to either surface stress variation or mass loading. Interaction between an immobilized ligand (e.g. antibody) and an analyte (e.g. an antigen) causes a surface stress change of the cantilever and can be detected as changes in the cantilever deflection. Microcanlitever sensors offer many advantages: label-free detection, high precision, reliability, reduced size and easy manufacture of multielement sensor arrays [122].

Some microcantilever immunosensors have been developed for the detection of pesticides such as atrazine [123,124], DDT [125] and 2,4-D [123]. Suri et al. developed an ultrasensitive cantilever-based immunosensor for the detection of atrazine. A thiolated atrazine antibody was immobilized on a gold coated cantilever. The binding of atrazine to the immobilized Ab changes the surface stress causing bending of the cantilever. The detection limit was 1 ppt (4.65 pM). Recently, a cantilever-based competitive immunosensor was developed for the detection of 2,6 dichlorobenzamide (BAM), which is the most frequent found pesticide residue in European groundwater [126].

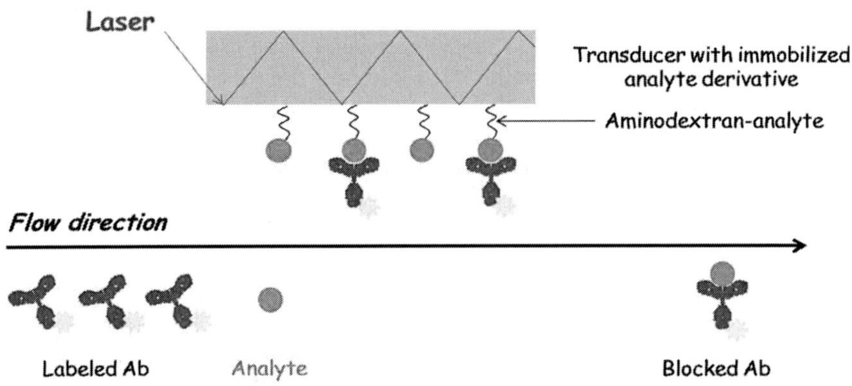

Figure 2. Principle of a TIRF-based immunosensor [111].

5. DNA BIOSENSORS

DNA biosensors exploit the preferential binding of complementary single-stranded nucleic acid sequences. They usually rely on the immobilization of a single-stranded DNA probe onto a surface able to recognize its complementary DNA target sequence by hybridization [122].

Recently, an electrochemical DNA biosensor was developed to study DNA damage caused by several pesticides, such as atrazine, 2,4-D, glufosinate ammonium, carbofuran, paraoxon-ethyl and difluorobenzuron [127]. A biotinylated DNA probe was immobilized on a streptavidin-modified electrode surface. This DNA probe was hybridized with biotinylated complementary DNA target analyte. Streptavidin labeled with ferrocene was further attached to the hybridized biotinylated DNA. The close proximity of ferrocene to the electrode surface induced a current signal. The presence of pesticides caused an unwinding of the DNA and thus a decrease of the ferrocene oxidation current observed in voltammetric experiments. Paraoxon-ethyl and atrazine caused the fastest and most severe damage to DNA.

6. NEW TRENDS

6.1. Aptamers

In 1990, Ellington's group [128], Gold's group and Robertson's group independently reported the development of an in vitro selection technique which allowed the discovery of specific nucleic acid sequences that bind non-nucleic acid targets with high affinity and specificity. The technique was called SELEX (Selection Evolution of Ligands by EXponential enrichment) and the resulting DNA or RNA oligonucleotides are referred to as aptamers [129-131].

Aptamers show high affinity towards a wide range of target analytes, including proteins, metal ions and pathogenic microorganisms. Aptamers possess several competitive advantages over antibodies, such as their accurate and reproducible chemical production [132]. Immunization and animals hosts are not necessary to produce aptamers. The selected nucleic acids bind their targets with affinity and specificity comparable to those of antibodies. Aptamers are more stable than antibodies. They can be selected in extreme conditions whereas antibodies are only stable in physiological conditions. Aptamers can also undergo reversible denaturation and they can be easily modified with new functional groups without affecting their activity. Due to its many advantages, numerous aptamer-based biosensors have been developed for the detection of a wide range of targets [133-137].

To our knowledge, few aptamers for the detection of pesticides have been selected. Recently, a DNA aptamer specific for acetamiprid was described [138]. The potential of aptamers for the pesticide detection has not still been exploited but aptamer-based biosensors could be an alternative to the conventional methods of pesticide analysis.

6.2. Molecularly Imprinted Polymers (MIPs)

Molecular imprinting, which allows the formation of specific recognition sites in polymers, is used to develop MIP-based sensors in the areas of environmental, food and pharmaceutical analysis [139].

The overall principle of molecular imprinting is presented in **Figure 3**. The template interacts with functional monomers either by the formation of covalent bonds or by self-association. Then, these monomers are polymerized around the template with the help of a crosslinker in the presence of a porogenic solvent. Template molecules are removed by extensive washing steps to disrupt the interactions between the template and the monomers. This process allows to obtain synthetic polymers possessing specific cavities complementary to the template in size, shape and position of the functional group [140,141]. The choice of the chemical reagents making up the MIP must be judicious in order to create highly specific cavities designed for the template molecule.

MIPs have been used as artificial recognition elements of biosensors for pesticide detection. An optical sensor for the detection of pesticides (chloropyrifos, diazinon and glyphosate) was developed by forming MIP onto optical fibers [142]. In this case, a luminescent lanthanide (europium), used as spectroscopic probe, was incorporated into the polymer. Detection of the analyte was based upon the changes that occur in the lanthanide spectrum when pesticide was incorporated to Eu^{3+}. The increase of the pesticide concentration induced an increase in the luminescence intensity of the spectra. Several electrochemical MIP-based sensors have been described [143-145]. An electrochemical sensor for 2,4-D was developed by electropolymerization of polypyrrole on a glassy carbon electrode in the presence of template 2,4-D molecules. During the electropolymerization step, 2,4-D molecules were embedded in the imprinted polypyrrole by hydrogen bond and electrostatic interacttions. Then, the template molecules were removed from the polymer by overoxidized process at +1.3 V in 0.2 M Na_2HPO_4 solution for 10 min.

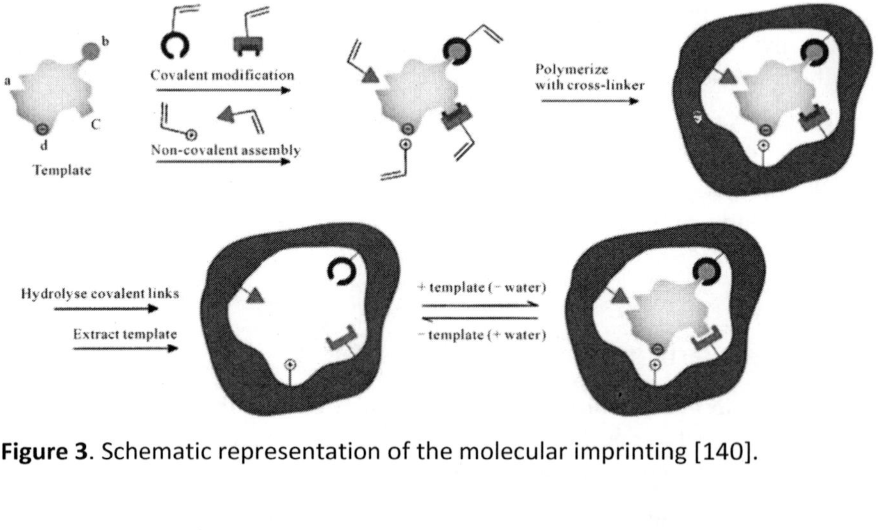

Figure 3. Schematic representation of the molecular imprinting [140].

6.3. Analysis of Pesticide Mixtures: The Artificial Neural Networks

Development of detection systems able to detect several analytes simultaneously represents a promising tool in environmental monitoring and screening. As it has been previously mentioned, numerous organophosphorus and carbamate insecticides can inhibit cholinesterase activity. One limitation of the enzymatic inhibition tests is the difficulty in discriminating between different inhibitors. To solve this problem, a sensor array can be coupled with an Artificial Neural Network (ANN) in order to precisely identify the inhibitors present in the sample. An ANN is a systematic procedure of data processing inspired by the nervous system function in animals. It combines the response of different enzymes to find a pattern that relates inhibitor concentrations with the inhibition percentages observed. Several intelligent biosensors for the analysis of pesticide mixtures have been developed based on the principle of the AChE inhibition and chemometric data analysis using ANNs [146]. Marty's group constructed ANNs to model the combined response of two pesticides (chlorpyrifos oxon and chlorfenvinfos) using sensors incorporating wild-type electric eel AChE and drosophila mutant AChE, associated or not with PTE [147]. These two types of AChEs were selected according to their different sensitivities to OP insecticides. The developed system was applied to the determination of pesticides in real water samples.

6.4. NANOMATERIALS

The emerging synergy between nanotechnology and sensors has been exploited over the past few years [148- 150]. Intensive research efforts have been performed for the design of efficient nanomaterial-based biosensors that exhibit high sensitivity and stability. The immobilization of nanomaterials onto sensing devices generates novel interfaces that enable the sensitive optical or electrochemical detection of analytes. Recently, some nanomaterials have been used for the design of electrochemical enzyme biosensors. Their high conductivity properties have been reported to enhance the electron transfer between the enzyme redox center and the electrode surface [151]. The electrocatalytic action of nanomaterials decreases the overpotential associated to electroactive compounds, minimizing the interferences present in the sample. In few cases, nanomaterials have been used as labels to amplify the signal measured.

6.4.1. Nanoparticle-Based Electrochemical Biosensors

6.4.1.1. Enzyme Biosensors

Various kinds of nanoparticles, such as QDs and AuNPs, have been used for the development of electrochemical enzyme biosensors. An enzyme biosensor was developed for the amperometric detection of trichlorfon using poly (N-vinyl-2-pyrrolidone) (PVP)-capped CdS QDs [152]. The formation of PVP-QD nanostructures on the electrode surface provided a favourable microenvironment and led to a highly sensitive and stable electrochemical detection of the enzymatically generated thiocholine product. The detection limit was 4.8×10^{-8} M. Another kind of nanoparticles are AuNPs. Their unique property to provide a suitable microenvironment for immobilization of biomolecules retaining their bioactivity is a major advantage for the preparation of biosensors. Moreover, AuNPs facilitate direct electron transfer between immobilized redox proteins and the electrode surface [153]. An electrochemical biosensor based on colloidal AuNP modified sol-gel interface was developed for the detection of monocrotophos, carbaryl and methyl parathion [154]. The assembled AuNPs on a sol-gel derived silicate network provided a conductive pathway to electron transfer and favored the interface enzymatic hydrolysis reaction, increasing the sensitivity of the amperometric response. This biosensor presented good stability, retaining 90% of its initial current response after a 30-day storage period. Recently, an efficient biosensor for the detection of monocrotophos was developed by combining the unique properties of AuNPS with those of QDs. This new electrochemical system based on CdTe QDs-AuNPs electrode was more sensitive than those based on QDs or AuNPs alone [155].

6.4.1.2. Immunosensors

Recently, an electrochemical immunosensor was developed for rapid screening of diuron, a substituted phenyl urea herbicide [156]. Low cost ablated electrodes fabriccated on polystyrene substrate were modified with Prussian Blue (PB)-AuNP film. The electrodeposition of PBAuNP film enhanced electron transfer in the vicinity of the gold electrode increasing the sensitivity of the system as compared to unmodified gold electrodes. A conductimetric immunosensor for the detection of atrazine was also developed using antibodies labelled with nanoparticles [85]. The authors showed that AuNPs amplify the conductive signal and hence allow the detection of atrazine by means of DC measurements.

6.4.2. Nanoparticle-Based Optical Biosensors

Nanoparticles have also been used for the development of efficient optical biosensors [157]. QDs are candidates to replace conventional fluorescent markers. These semiconductor particles are more photostable than an organic fluorophore. Moreover, QDs exhibit higher fluorescence quantum yields than conventional organic fluorophores, allowing higher sensitivity. Recently, an optical biosensor was developed for the detection of monocrotophos using CdTe as fluorescence probe [158]. Using positively charged chitosan, CdTe and acetylcholinesterase were assembled onto a quartz surface by a layer-by-layer technique. In the absence of pesticide, acetylcholine was biocatalytically hydrolyzed inducing the production of choline and acetic acid. The released acid resulted in pH decrease that was sensed by the immobilized pH indicator (CdTe). The presence of monocrotophos induced a change of the fluorescence intensity that was related to the pesticide concentration.

Optic properties of AuNPs have been exploited for the development of localized SPR (LSPR) sensor [159,160]. The absorption band of AuNPs results when the incident photon frequency is resonant with the collective oscillation of the conduction electrons and is known as the localized surface plasmon resonance (LSPR). The resonance frequency of the LSPR is highly dependent upon the local environment of the nanoparticle and more specifically upon the binding events that occur to the functionalized NPs. The LSPR was used to develop a biosensor for the detection of paraoxon by immobilizing AChE onto AuNPs layer using a self-assembling technique [157]. In the presence of pesticides, the enzymatic activity was inhibited causing a change of the light attenuation. The detection limit with optimal conditions was 0.2 ppb. The biosensor retained 94% of its original activity after 6 cycles of inhibition with 500 ppb paraoxon followed by reactivation of AChE with 0.5 mM 2- pyriding-adoxime methoiodide. In addition, the sensor retained its activity after 2 months storage in dry state at 4°C.

6.4.3. Nanotube-Based Electrochemical Biosensors

Carbon nanotubes (CNTs) consist of cylindrical graphene sheets with nanometer diameters. They present unique mechanical, physical and chemical properties [148]. CNTs include both single-walled and multiwalled structures. Since their discovery, CNTs have been used in nanoelectronics, biomedical engineering, biosensing and bioanalysis.

Electrochemical biosensors, particularly enzyme electrodes, have benefited from the ability of CNT-modified transducers to promote the electron transfer reactions of enzymatically generated species [161]. Recently, CNTs have been used for the development of biosensors based on the inhibition of AChE activity [162-165]. An amperometric biosensor based on layer-by-layer assembly of single walled CNT-poly (diallyldimethylammonium chloride) and AChE was developed for the analysis of carbaryl [164]. The biosensor showed good sensitivity and stability towards the monitoring of pesticides in water. The detection limit was 4.9×10^{-15} M.

In some cases, the authors developed efficient biosensors for the detection of pesticides by associating the properties of CNTs with those of nanoparticles [39,166].

7. CONCLUSIONS

Biosensors are good candidates for the environmental monitoring. They exploit the remarkable specificity of recognition elements to design efficient analytical tools that can detect the presence of pesticides in complex samples. The biological elements used for environmental sensing are classically enzymes, antibodies and whole cells. Immunosensors, based on the highly selective and sensitive Ab-Ag reaction, allow the identification of a particular pesticide. Their high specificity can be sometimes a disadvantage. Enzyme-based biosensors allow the detection of broad families of pollutants. Thus, they often offer a general toxicity "index" [167], without providing specific information about a particular pesticide. Genetically engineered AChEs, showing better performance than native enzymes, have been widely exploited in enzyme inhibition-based biosensors for the detection of pesticides. Some genetically engineered microorganisms have also been used to develop microbial biosensors for pesticide detection [150]. In the last decade, aptamers have been used as new molecular recognition elements to develop biosensors [133,134]. However, unique properties of aptamers have not been yet exploited for the development of pesticide analysis. We believe that these recognition elements could be used, in the near future, for the development of efficient aptasensors allowing the detection of pesticides. Although biological receptors have specific molecular affinity and have been widely used in biosensing technology, they are often produced via complex protocols with a high cost and require specific handling conditions because of their poor stability [168]. The high specificity and stability of MIPs render them as promising alternatives to enzymes and Abs [169]. Additional advantages include their compatibility with microfabrication technology and their cost-effectiveness compared to conventional biological receptors.

Nanotechnology is playing an important role in the development of efficient biosensors for the pesticide detection [148-150]. Different types of nanomaterials (e.g. nanoparticles and nanotubes) with different properties have been used. They offer exciting new opportunities to improve the performance of biosensors for the detection of pesticides.

The use of biosensors in environmental field is still limited in comparison to medical applications. Most commercial biosensors are for medical applications, whereas only few are adapted for the environmental monitoring. Thus, there is still a challenge to develop improved and more reliable devices allowing the detection of pesticides.

The use of biosensors in environmental field is still limited in comparison to medical applications. Most commercial biosensors are for medical applications, whereas only few are adapted for the environmental monitoring. Thus, there is still a challenge to develop improved and more reliable devices allowing the detection of pesticides.

REFERENCES

1. S. Andreescu and J. L. Marty, “Twenty Years Research in Cholinesterase Biosensors: From Basic Research to Practical Applications,” Biomolecular Engineering, Vol. 23, No. 1, 2006, pp. 1-15.

2. M. D. Luque de Castro and M. C. Herrera, “Enzyme_ Inhibition-Based Biosensors and Biosensing Systems: Questionable Analytical Devices,” Biosensors and Bioelectronics, Vol. 18, No. 2-3, 2003, pp. 279-294.

3. N. Jaffrezic-Renault, “New Trends in Biosensors for Organophosphorus Pesticides,” Sensors, Vol. 1, No. 2, 2001, pp. 60-74.

4. D. Grieshaber, R. MacKenzie, J. Vöros and E. Reimhult, “Electrochemical Biosensors—Sensor Principles and Architectures,” Sensors, Vol. 8, No. 3, 2008, pp. 1400-1458.

5. R. S. Freire, C. A. Pessao, L. D. Mello and L. T. Kubota, “Direct Electron Transfer: An Approach for Electrochemical Biosensors with Higher Selectivity and Sensitivity,” Journal of Brazilian Chemical Society, Vol. 14, No. 2, 2003, pp. 230-243.

6. D. R. Thévenot, K. Toth, R. A. Durst and G. S. Wilson, “Electrochemical Biosensors: Recommended Definitions and Classification,” Pure Applied Chemistry, Vol. 71, No. 12, 1999, pp. 2333-2348.

7. N. J. Ronkainen, H. B. Halsall and W. R. Heineman, “Electrochemical Biosensors,” Chemical Society Reviews, Vol. 39, No. 11, 2010, pp. 1747-1763.

8. U. Yogeswaran and S.-M. Chen, “A Review on the Electrochemical Sensors and Biosensors Composed of Nanowires as Sensing Material,” Sensors, Vol. 8, No. 1, 2008, pp. 290-313.

9. S. Liu, L. Yuan, X. Yue, Z. Zheng and Z. Tang, “Recent Advances in Nanosensors for Organophosphates Pesticide Detection,” Advanced Powder Technology, Vol. 19, No. 5, 2008, pp. 419-441.

10. B. Prieto-Simon, M. Campas, S. Andreescu and J. L. Marty, “Trends in Flow-Based Biosensing for Pesticide Assessment,” Sensors, Vol. 6, No. 10, 2006, pp. 1161- 1186.

11. B. Liu, Y.-H. Yang, Z.-Y. Wu, H. Wang, G.-L. Shen and R.-Q. Yu, “A Potentiometric Acetylcholinesterase Biosensor Based on Plasma-Polymerized Film,” Sensors and Actuators B, Vol. 104, No. 2, 2005, pp. 186-190.

12. V. G. Andreou and Y. D. Clonis, “A Portable Fiber-Optic Pesticide Biosensor Based On Immobilized Cholinesterase and Sol-Gel

Entrapped Bromocresol Purple for In-Field Use," Biosensors and Bioelectronics, Vol. 17, No. 1-2, 2002, pp. 61-69.

13. F. C. Wong, M. Ahmad, L. Y. Heng and L. B. Peng, "An Optical Biosensor for Dichlovos Using Stacked Sol-Gel Films Containing Acetylcholinesterase and a Lipophilic Chromoionophore," Talanta, Vol. 69, No. 4, 2006, pp. 888-893.

14. H. C. Tsai and R. A. Doong, "Simultaneous Determination of Ph, Urea, Acetylcholine and Heavy Metals Using Array-Based Enzymatic Optical Biosensor," Biosensors and Bioelectronics, Vol. 20, No. 9, 2005, pp. 1796-1804.

15. C. La Rosa, F. Pariente, L. Hernandez and E. Lorenzo, "Amperometric Flow-Through Biosensor for the Determination of Pesticides," Analytica Chimica Acta, Vol. 308, No. 1-3, 1995, pp. 129-136.

16. C. La Rosa, F. Pariente, L. Hernandez and E. Lorenzo, "Determination of Organophosphorus and Carbamic Pesticides with an Acetylcholinesterase Amperometric Biosensor Using 4-Aminophenyl Acetate as Substrate," Analytica Chimica Acta, Vol. 295, No. 3, 1994, pp. 273-282.

17. S. Andreescu, T. Noguer, V. Magearu and J. L. Marty, "Screen-Printed Electrode Based on AChE for the Detection of Pesticides in Presence of Organic Solvents," Talanta, Vol. 57, No. 1, 2002, pp. 169-176.

18. S. Andreescu, A. Avramescu, C. Bala, V. Magearu and J. L. Marty, "Detection of Organophosphorus Insecticides with Immobilized Acetylcholinesterase—Comparative Study of Two Enzyme Sensors," Analytical and Bioanalytical Chemistry, Vol. 374, No. 1, 2002, pp. 39-45.

19. S. U padhyay, G. R. Rao, M. K. Sharma, B. K. Bhattacharya, V. K. Rao and R. Vijayaraghavan, "Immobilization of Acetylcholineesterase-Choline Oxidase on a GoldPlatinum Bimetallic Nanoparticles Modified Glassy Carbon Electrode for the Sensitive Detection of Organophosphate Pesticides, Carbamates and Nerves Agents," Biosensors and Bioelectronics, Vol. 25, No. 4, 2009, pp. 832-838.

20. L. Campanella, M. Achilli, M. P. Sammartino and Tomassetti, "Butyrylcholine Enzyme Sensor for Determining Organophosphorus Inhibitors," Journal of Electroanalytical Chemistry, Vol. 321, No. 2, 1991, pp. 237-249.

21. N. G. Karousos, S. Aouabdi, A. S. Way and S. M. Reddy, "Quartz Crystal Microbalance Determination of Organophosphorus and Carbamate Pesticides," Analytica Chimica Acta, Vol. 469, No. 2, 2002, pp. 189-196.

22. M. Campas, B. Prieto-Simon and J. L. Marty, “A Review of the Use of Genetically Engineered Enzymes in Electrochemical Biosensors,” Seminars in Cell and Developmental Biology, Vol. 20, No. 1, 2009, pp. 3-9.

23. G. Valdes-Ramirez, D. Fournier, M. T. Ramirez-Silva and J. L. Marty, “Sensitive Amperometric Biosensor for Dichlorovos Quantification: Application to Detection of Residues on Apple Skin,” Talanta, Vol. 74, No. 4, 2008, pp. 741-746.

24. F. De Lima, B. Lucca, A. M. J. Barbosa, V. S. Ferreira, S. K. Moccelini, A. C. Franzoi and I. C. Vieira, “Biosensor Based on Pequi Polyphenol Oxidase Immobilized on Chitosan Crosslinked with Cyanuric Chloride for Thiodicarb Determination,” Enzyme and Microbial Technology, Vol. 47, No. 4, 2010, pp. 153-158.

25. J. C. Vidal, S. Esteban, J. Gil and J. R. Castillo, “A Comparative Study of Immobilization Methods of a Tyrosinase Enzyme on Electrodes and Their Application to the Detection of Dichlorvos Organophosphorus Insecticide,” Talanta, Vol. 68, No. 3, 2006, pp. 791-799.

26. Y. D. de Albuquerque and L. F. Ferreira, “Amperometric Biosensing of Carbamate and Organophosphate Pesticides Utilizing Screen-Printed Tyrosinase-Modified Electrodes,” Analytica Chimica Acta, Vol. 596, No. 2, 2007, pp. 210-221.

27. G.-Y. Kim, M.-S. Kang and S.-H. Moon, “Sub-strateBound Tyrosinase Electrode Using Gold Nanoparticles Anchored to Pyrroloquinoline Quinone for a Pesticide Biosensor,” Sensors and Actuators B, Vol. 133, No. 1, 2008, pp. 1-4.

28. G. Y. Kim, J. Shim, M. S. Kang and S. H. Moon, “Optimized Coverage of Gold Nanoparticles at Tyrosinase Electrode for Measurement of a Pesticide in Various Water Samples,” Journal of Hazardous Materials, Vol. 156, No. 1-3, 2008, pp. 141-147.

29. L. Campanella, D. Lelo, E. Martini and M. Tomassetti, “Organophosphorus and Carbamate Pesticide Analysis Using an Inhibition Tyrosinase Organic Phase Enzyme Sensor; Comparison by Butyrylcholinesterase + Choline Oxidase Opee and Application to Natural Waters,” Analytica Chimica Acta, Vol. 587, No. 1, 2007, pp. 22-32.

30. M. S. Ayyagari, S. Kamtekar, R. Pande, K. A. Marx, J. Kumar, S. K. Tripathy and D. L. Kaplan, “Biosensors for Pesticide Detection Based on Alkaline Phosphatase-Catalyzed Chemiluminescence,” Materials Science and Engineering: C, Vol. 2, No. 4, 1995, pp. 191-196.

31. F. Garcia Sanchez, A. Navas Diaz, M. C. Ramos Peinado and C. Belledone, "Free and Sol-Gel Immobilized Alkaline Phosphatase-Based Biosensor for the Determination of Pesticides and Inorganic Compounds," Analytica Chimica Acta, Vol. 484, No. 1, 2003, pp. 45-51.

32. F. Mazzei, F. Botrè, S. Montilla, R. Pilloton, E. Podesta and C. Botrè, "Alkaline Phosphatase Inhibition Based Electrochemical Sensors for the Detection of Pesticides," Journal of Electroanalytical Chemistry, Vol. 574, No. 1, 2004, pp. 95-100.

33. L. K. Shyuan, L. Y. Heng, M. Ahmad, S. A. Aziz and Z. Ishak, "Evaluation of Pesticide and Heavy Mtal Toxicity Using Immobilised Enzyme Alkaline Phosphatase with an Electrochemical Biosensor," Asian Journal of Biochemistry, Vol. 3, No. 6, 2008, pp. 359-365.

34. S. K. Moccelini, I. C. Vieira, F. De Lima, B. Lucca, A. M. J. Barbosa and V. S. Ferreira, "Determination of Thiodicarb Using a Biosensor Based on Alfalfa Sprout Peroxidase Immobilized in Self-Assembled Monolayers," Talanta, Vol. 82, No. 1, 2010, pp. 164-170.

35. S. Yang, Y. Li, X. Jiang, Z. Chen and X. Lin, "Horseradish Peroxidase Biosensor Based on Layer-by-Layer Technique for the Determination of Phenolic Compounds," Sensors and Actuators B, Vol. 114, No. 2, 2006, pp. 774-780.

36. S. Yang, Z. Chen, X. Jin and X. Lin, "HRP Biosensor Based on Sugar-Lectin Biospecific Interactions for the Determination of Phenolic Compounds," Electrochimica Acta, Vol. 52, No. 1, 2006, pp. 200-205.

37. F. Mazzei, F. Botrè and C. Botrè, "Acid Phosphatase/Glucose Oxidase-Based Biosensors for the Determination of Pesticides," Analytica Chimica Acta, Vol. 336, No. 1, 1996, pp. 67-75.

38. A. Mulchandani, W. Chen, P. Mulchandani, J. Wang and K. R. Rogers, "Biosensors for Direct Determination of Organophosphate Pesticides," Biosensors and Bioelectronics, Vol. 16, No. 4, 2001, pp. 225-230.

39. D. Du, W. Chen, W. Zhang, D. Liu, H. Li and Y. Lin, "Covalent Coupling of Organophosphorus Hydrolase Loaded Quantum Dots to Carbon Nanotube. Au Nanocomposite for Enhanced Detection of Methyl Parathion," Biosensors and Bioelectronics, Vol. 25, No. 6, 2010, pp. 1370-1375.

40. J. H. Lee, J. Y. Park, K. Min, H. J. Cha, S. S. Choi and Y. J. Yoo, "A Novel Organophosphorus Hydrolase-Based Biosensor Using Mesoporous Carbons and Carbon Black for the Detection of Organophosphate Nerve Agents," Biosensors and Bioelectronics, Vol. 25, No. 7, 2010, pp. 1566-1570.

41. V. A. Pedrosa, S. Paliwal, S. Balasubramanian, D. Nepal, V. Davis, J. Wild, E. Ramanculov and A. Simonian, "Enhanced Stability of Enzyme Organophosphate Hydrolase Interfaced on the Carbon Nanotubes," Colloids and Surfaces B Biointerfaces, Vol. 77, No. 1, 2010, pp. 69-74.

42. V. G. Andreou and Y. D. Clonis, "Novel Fiber-Optic Biosensor Based on Immobilized Glutathione S-Transferase and Sol-Gel Entrapped Bromcresol Green for the Determination of Atrazine," Analytica Chimica Acta, Vol. 460, No. 2, 2002, pp. 151-161.

43. Y. Lei, W. Chen and A. Mulchandani, "Microbial Biosensors," Analytica Chimica Acta, Vol. 568, No. 1-2, 2006, pp. 200-210.

44. S. F. D'Souza, "Microbial Biosensors," Biosensors and Bioelectronics, Vol. 16, 2001, pp. 337-353.

45. J. Liu and B. Mattiasson, "Microbial BOD Sensors for Wastewater Analysis," Water Research, Vol. 36, No. 15, 2002, pp. 3786-3802.

46. C. Chan, M. Lehmann, K. Chan, P. Chan, C. Chan, B. Gruendig, G. Kunze and R. Renneberg, "Designing an Amperometric Thick-Film Microbial BOD Sensor," Biosensors and Bioelectronics, Vol. 15, No. 7-8, 2000, pp. 343-353.

47. K. Riedel, R. Renneberg, M. Kühn and F. Scheller, "A Fast Estimation of Biochemical Oxygen Demand Using Microbial Sensors," Applied Microbiology and Biotechnology, Vol. 28, No. 3, 1988, pp. 316-318.

48. M. N. Kim and H. S. Kwon, "Biochemical Oxygen Demand Sensor Using Serratia marcescens LSY 4," Biosensors and Bioelectronics, Vol. 14, No. 1, 1999, pp. 1-7.

49. M. Hikuma, H. Suzuki, T. Yasuda, I. Karube and S. Suzuki, "Amperometric Estimation of BOD by Using Living Immobilized Yeast," European Journal of Applied Microbiology and Biotechnology, Vol. 8, No. 4, 1979, pp. 289-297.

50. J. Jia, M. Tang, X. Chen, L. Qi and S. Dong, "Co-Immobilized Microbial Biosensor for Bod Estimation Based on Sol-Gel Derived Composite Material," Biosensors and Bioelectronics, Vol. 18, No. 8, 2003, pp. 1023-1029.

51. J. Liu, G. Olsson and B. Mattiasson, "Short-Term BOD (BOD_{st}) as a Parameter for On-Line Monitoring of Biological Treatment Process Part I. A Novel Design of BOD Biosensor for Easy Renewal of Bio-Receptor," Biosensors and Bioelectronics, Vol. 20, No. 3, 2004, pp. 562- 570.

52. L. Su, W. Jia, C. Hou and Y. Lei, "Microbial Biosensors: A Review," Biosensors and Bioelectronics, Vol. 26, No. 5, 2011, pp. 1788-1799.

53. P. Mulchandani, C. M. Hangarter, Y. Lei, W. Chen and A. Mulchandani, "Amperometric Microbial Biosensor for p- Nitrophenol Using Moraxella sp.-Modified Carbon Paste Electrode," Biosensors and Bioelectronics, Vol. 21, No. 12, 2005, pp. 523-527.

54. Y. Lei, P. Mulchandani, W. Chen, J. Wang and A. Mulchandani, "Whole Cell-Enzyme Hybrid Amperometric Biosensor for Direct Determination of Organophosphorous Nerve Agents with p-Nitrophenyl Substituent," Biotechnology and Bioengineering, Vol. 85, No. 7, 2004, pp. 706-712.

55. Y. Lei, P. Mulchandani, W. Chen and A. Mulchandani, "Direct Determination of p-Nitrophenyl Substituent Organophosphorus Nerve Agents Using a Recombinant Pseudomonas putida JS444-Modified Clark Oxygen Electrode," Journal of Agricultural and Food Chemistry, Vol. 53, 2005, pp. 524-527.

56. Y. Lei, P. Mulchandani, W. Chen and A. Mulchandani, "Biosensor for Direct Determination of Fenitrothion and EPN Using Recombinant Pseudomonas putida JS444 with Surface Expressed Organophosphorus Hydrolase. 1. Modified Clark Oxygen Electrode," Sensors, Vol. 6, No. 4, 2006, pp. 466-472.

57. P. Mulchandani, W. Chen and A. Mulchandani, "Microbial Biosensor for Direct Determination of NitrophenylSubstituted Organophosphate Nerve Agents Using Genetically Engineered Moraxella sp.," Analytica Chimica Acta, Vol. 568, No. 1-2, 2006, pp. 217-221.

58. E. I. Rainina, E. N. Efremenco, S. D. Varfolomeyev, A. L. Simonian and J. R. Wild, "The Development Of A New Biosensor Based on Recombinant E. coli for the Direct Detection of Organophosphorus Neurotoxins," Biosensors and Bioelectronics, Vol. 11, No. 10, 1996, pp. 991- 1000.

59. F. Lagarde and N. Jaffrezic-Renault, "Cell-Based Electrochemical Biosensors for Water Quality Assessment," Analytical and Bioanalytical Chemistry, Vol. 400, No. 4, 2011, pp. 947-964.

60. J. Kumar, S. K. Jha and S. F. D'Souza, "Optical Microbial Biosensor for Detection of Methyl Parathion Pesticide Using Flavobacterium sp. Whole Cells Adsorbed on Glass Fiber Filters as Disposable Biocomponent," Biosensors and Bioelectronics, Vol. 21, No. 11, 2006, pp. 2100-2105.

61. M. Campas, R. Carpentier and R. Rouillon, "Plant-Tissue-and Photosynthesis-Based Biosensors," Biotechology Advances, Vol. 26, No. 4, 2008, pp. 370-378.

62. C. Chouteau, S. V. Dzyadevych, J.-M. Chovelon and C. Durrieu, "Development of Novel Conductometric Biosensors Based on

Immobilised Whole Cell Chlorella vulgaris Microalgae," Biosensors and Bioelectronics, Vol. 19, No. 9, 2004, pp. 1089-1096.

63. C. Chouteau, S. V. Dzyadevych, C. Durrieu and J.-M. Chovelon, "A Bi-Enzymatic Whole Cell Conductometric Biosensor for Heavy Metal Ions and Pesticides Detection in Water Samples," Biosensors and Bioelectronics, Vol. 21, No. 2, 2005, pp. 273-281.

64. H. Guedri and C. Durrieu, "A Self-Assembled Monolayers Based Conductimetric Algal Whole Cell Biosensor for Water Monitoring," Microchimica Acta, Vol. 163, No. 3-4, 2008, pp. 179-184.

65. R. Rouillon, S. A. Piletsky, E. V. Piletska, P. Euzet and R. Carpentier, "Comparison of the Immobilization Techniques for Photosystem II," In: M. T. Giardi and E. V. Piletska, Eds., Biotechnological Applications of Photosynthesis Proteins: Biochips, Biosensors and Biodevices, Landes Bioscience, Georgetown, 2006, pp. 73-83.

66. M. T. Giardi and E. Pace, "Photosystem II-Based Biosensors for the Detection of Photosynthetic Herbicides," In: M. Giardi and E. V. Piletska, Eds., Biotechnological Applications of Photosynthetic Proteins; Biochips, Biosensors and Biodevices, Landes Bioscience, Georgetown, 2006, pp. 147-154.

67. R. Rouillon, J.-J. Mestres and J. L. Marty, "Entrapment of Chloroplasts and Thylakoids in Polyvinylalcohol-SbQ. Optimization of Membrane Preparation and Storage Conditions," Analytica Chimica Acta, Vol. 311, No. 3, 1995, pp. 437-442.

68. M. Koblizek, J. Masojidek, J. Komenda, T. Kucera, R. Pilloton, A. K. Mattoo and M. T. Giardi, "A Sensitive Photosystem II-Based Biosensor for Detection of a Class of Herbicides," Biotechnology Bioengeering, Vol. 60, No. 6, 1998, pp. 664-669.

69. I. Shitanda, K. Takada, T. Sakai and T. Tatsuma, "Compact Amperometric Algal Biosensors for the Evaluation of Water Toxicity," Analytica Chimica Acta, Vol. 530, No. 2, 2005, pp. 191-197.

70. I. Shitanda, S. Takamatsu, K. Watanabe and M. Itagaki, "Amperometric Screen-Printed Algal Biosensor with Flow Injection Analysis System for Detection of Environmental Toxic Compounds," Electrochimica Acta, Vol. 54, No. 21, 2009, pp. 4933-4936.

71. H. Nguyen-Ngoc and C. Tran-Minh, "Fluorescent Biosensor Using Whole Cells in an Inorganic Translucent Matrix," Analytica Chimica Acta, Vol. 583, No. 1, 2007, pp. 161-165.

72. C. Védrine, J. C. Leclerc, C. Durrieu and C. Tran-Minh, "Optical Whole-Cell Biosensor Using Chlorella Vulgaris Designed for Monitoring

Herbicides," Biosensors and Bioelectronics, Vol. 18, No. 4, 2003, pp. 457-463.

73. A. Ventrella, L. Catucci and A. Agostiano, "Herbicides Affect Fluorescence and Electron Transfer Activity of Spinach Chloroplasts, Thylakoid Membranes and Isolated Photosystem II," Bioelectrochemistry, Vol. 79, No. 1, 2010, pp. 43-49.

74. E. Pena-Vazquez, E. Maneiro, C. Perez-Conde, M. C. Moreno-Bondi and E. Costas, "Microalgae Fiber Optic Biosensors for Herbicide Monitoring Using Sol-Gel Technology," Biosensors and Bioelectronics, Vol. 24, No.12, 2009, pp. 3538-3543.

75. X. Jiang, D. Li, X. Xu, Y. Ying, Y. Li, Z. Ye and J. Wang, "Immunosensors for Detection of Pesticides Residues," Biosensors and Bioelectronics, Vol. 23, No. 11, 2008, pp. 1577-1587.

76. C.R. Suri, R. Boro, Y. Nangia, S. Gandhi, P. Sharma, N. Wangoo, K. Rajesh and G. S. Shekhawat, "Immunoanalytical Techniques for Analyzing Pesticides in the Environment," Trends in Analytical Chemistry, Vol. 28, No. 1, 2009, pp. 29-39.

77. E. Mallat, D. Barcelo, C. Barzen, G. Gauglitz and R. Abuknesha, "Immunosensors for Pesticide Determination in Natural Waters," Trends in Analytical Chemistry, Vol. 20, No. 3, 2001, pp. 124-132.

78. F. F. Bier, E. Ehrentreich-Förster, R. Dölling, A. V. Eremenko and F. W. Scheller, "A Redox Label Immunosensor on Basis of a Bi-Enzyme Electrode," Analytica Chimica Acta, Vol. 344, No. 1-2, 1997, pp. 119-124.

79. T. Kalab and P. Skadal, "A Disposable Amperometric Immunosensor for 2,4-Dichlorophenoxyacetic Acid," Analytica Chimica Acta, Vol. 304, No. 3, 1995, pp. 361-368.

80. P. Skadal and T. Kalab, "A Multichannel Immunochemical Sensor for the Determination of 2,4-Dichlorophenoxyacetic Acid," Analytica Chimica Acta, Vol. 316, No. 1, 1995, pp. 73-78.

81. S. Kröger, S. J. Setford and A. P. F. Turner, "Immunosensor for 2,4-Dichlorophenoxyacetic Acid in Aqueous/Organic Solvents Soil Extracts," Analytical Chemistry, Vol. 70, No. 23, 1998, pp. 5047-5053.

82. K. Grennan, G. Strachan, A. J. Porter, A. Killard and M. R. Smyth, "Atrazine Analysis Using an Amperometric Immunosensor Based on Single-Chain Antibody Fragments and Regeneration-Free Multi-Calibrant Measurement," Analytica Chimica Acta, Vol. 500, No. 1-2, 2003, pp. 287-298.

83. E. Valera, D. Muniz and A. Rodriguez, "Fabrication of Flexible Interdigitated μ-Electrodes (FIDμEs) for the Development of a Conductimetric Immunosensor for Atrazine Detection Based on Antibodies Labelled with Gold Nanoparticles," Microelectronic Engineering, Vol. 87, No. 2, 2010, pp. 167-173.

84. E. Valera, J. Ramon-Azcon, A. Barranco, B. Alfaro, F. Sanchez-Baeza, M. P. Marco and A. Rodriguez, "Determination of Atrazine Residues in Red Wine Samples. A Conductimetric Solution," Food Chemistry, Vol. 122, No. 3, 2010, pp. 888-894.

85. E. Valera, J. Ramon-Azcon, F. J. Sanchez, M. P. Marco and A. Rodriguez, "Conductimetric Immunosensor for Atrazine Detection Based on Antibodies Labelled with Gold Nanoparticles," Sensors and Actuators B, Vol. 134, No. 1, 2008, pp. 95-103.

86. Y. V. Plekhanova, A. N. Reshetilov, E. V. Yazynina, A. V. Zherdev and B. B. Dzantiev, "A New Assay Format for Electrochemical Immunosensors: PolyelectrolyteBased Separation on Membrane Carriers Combined with Detection of Peroxidase Activity by pH-Sensitive Field-Effect Transistor," Biosensors and Bioelectronics, Vol. 19, No. 2, 2003, pp. 109-114.

87. M. F. Yulaev, R. A. Sidtikov, N. M. Dmitrieva, E. V. Yazynima, A. V. Zherdev and B. B. Dzantiev, "Development of a Potentiometric Immunosensor for Herbicide Simazine and Its Application for Food Testing," Sensors and Actuators B, Vol. 75, No. 1-2, 2001, pp. 129-135.

88. L. Mosiello, C. Laconi, M. Del Gallo, C. Ercole and A. Lepidi, "Development of a Monoclonal Antibody Based Potentiometric Biosensor for Terbuthylazine Detection," Sensors and Actuators B, Vol. 95, No. 1-3, 2003, pp. 315- 320.

89. H. B. Fredj, S. Helali, Z. Sassi, N. Jaffrezic-Renault and A. Abdelghani, "Polyaniline Based Immunosensor for Atrazine Sensing," Sensor Letters, Vol. 7, No. 5, 2009, pp. 661- 666.

90. A. Rodriguez, E. Valera, J. Ramon-Azcon, F. J. Sanchez, M. P. Marco and L. M. Castaner, "Single Frequency Impedimetric Immunosensor for Atrazine Detection," Sensors and Actuators B, Vol. 129, No. 2, 2008, pp. 921-928.

91. E. Valera, J. Ramon-Azcon, A. Rodriguez, L. M. Castaner, F. J. Sanchez and M. P. Marco, "Impedimetric immunosensor for Atrazine Detection Using Interdigitated μ-Electrode (IDμE's)," Sensors and Actuators B, Vol. 125, No. 2, 2007, pp. 526-537.

92. J. Ramon-Azcon, E. Valera, A. Rodriguez, A. Barranco, B. Alfaro, F. Sanchez-Baeza and M. P. Marco, "An Impedimetric Immunosensor

Based on Interdigitated Microelectrodes (IDμE) for the Determination of Atrazine Residues in Food Samples," Biosensors and Bioelectronics, Vol. 23, No. 9, 2008, pp. 1367-1373.

93. R. E. Ionescu, C. Gondran, L. Bouffier, N. Jaffrezic-Renault, C. Martelet and S. Cosnier, "Label-Free Impedimetric Immunosensor for Sensitive Detection of Atrazine," Electrochimica Acta, Vol. 55, No. 21, 2010, pp. 6228-6232.

94. C. Esseghaier, S. Helali, H. Fredj, A. Tlili and A. Abdelghani, "Polypyrrole-Neutravidin Layer for Impedimetric Biosensor," Sensors and Actuators B, Vol. 131, No. 2, 2008, pp. 584-589.

95. S. Helali, C. Martelet, A. Abdelghani, M. A. Maaref and N. Jaffrezic-Renault, "A Disposable Immunomagnetic Electrochemical Sensor Based on Functionalised Magnetic Beads on Gold Surface for the Detection of Atrazine," Electrochimica Acta, Vol. 51, No. 24, 2006, pp. 5182- 5186.

96. B. Corry, J. Uilk and C. Crawley, "Probing Direct Binding Affinity in Electrochemical Antibody-Based Sensors," Analytica Chimica Acta, Vol. 496, No. 1-2, 2003, pp. 103-116.

97. S. Hleli, C. Martelet, A. Abdelghani, N. Burais and N. Jaffrezic-Renault, "Atrazine Analysis Using an Impedimetric Immunosensor Based on Mixed Biotinylated SelfAssembled Monolayer," Sensors and Actuators B, Vol. 113, No. 2, 2006, pp. 711-717.

98. J. Homola, S. S. Yee and G. Gauglitz, "Surface Plasmon Resonance Sensors: Review," Sensors and Actuators B, Vol. 54, No. 1-2, 1999, pp. 3-15.

99. M. Farre, E. Martinez, J. Ramon, A. Navarro, J. Radjenovic, E. Mauriz, L. Lechuga, M. P. Marco and D. Barcelo, "Part per Trillion Determination of Atrazine in Natural Water Samples by a Surface Plasmon Resonance Immunosensor," Analytical and Bioanalytical Chemistry, Vol. 388, No. 1, 2007, pp. 207-214.

100. E. Mauriz, A. Calle, J. J. Manclus, A. Montoya and L. M. Lechuga, "Multi-Analyte SPR Immunoassays for Environmental Biosensing of Pesticides," Analytical and Bioanalytical Chemistry, Vol. 387, No. 4, 2006, pp. 1449- 1458.

101. E. Mauriz, A. Calle, A. Abad, A. Montoya, A. Hildebrandt, D. Barcelo and L. M. Lechuga, "Determination of Car0 baryl in Natural Water Samples by a Surface Plasmon Resonance Flow-Through Immunosensor," Biosensors and Bioelectronics, Vol. 21, No. 11, 2006, pp. 2129-2136.

102. E. Mauriz, A. Calle, J. J. Manclus, A. Montoya, A. M. Escuela, J. R. Sendra and L. M. Lechuga, "Single and Multi-Analyte Surface Plasmon Resonance Assays for Simultaneous Detection Of Cholinesterase Inhibiting Pesticides," Sensors and Actuators B, Vol. 118, No. 1-2, 2006, pp. 399-407.

103. E. Mauriz, A. Calle, L. M. Lechuga, J. Quintana, A. Montoya and J. J. Manclus, "Real-Time Detection of Chlorpyrifos at Part Per Trillion Levels in Ground, Surface and Drinking Water Samples by a Portable Surface Plasmon Resonance Immunosensor," Analytica Chimica Acta, Vol. 561, No. 1-2, 2006, pp. 40-47.

104. E. Mauriz, A. Calle, A. Montoya and L. M. Lechuga, "Determination of Environmental Organic Pollutants with a Portable Optical Immunosensor," Talanta, Vol. 69, No. 2, 2006, pp. 359-364.

105. S. J. Kim, K. V. Gobi, H. Tanaka, Y. Shoyama and N. Miura, "A Simple and Versatile Self-Assembled Monolayer Based Surface Plasmon Resonance Immunosensor for Highly Sensitive Detection of 2,4-D from Natural Water Resources," Sensors and Actuators B, Vol. 130, No. 1, 2008, pp. 281-289.

106. K. V. Gobi, S. J. Kim, H. Tanaka, Y. Shoyama and N. Miura, "Novel Surface Plasmon Resonance (SPR) Immunosensor Based on Monomolecular Layer of Physically-Adsorbed Ovalbumin Conjugate for Detection of 2,4-Dichlorophenoxyacetic Acid and Atomic Force Microscopy Study," Sensors and Actuators B, Vol. 123, No. 1, 2007, pp. 583-593.

107. K. V. Gobi, H. Tanaka, Y. Shoyama and N. Miura, "Highly Sensitive Regenerable Immunosensor for LabelFree Detection of 2,4-Dichlorophenoxyacetic Acid at ppb Levels by Using Surface Plasmon Resonance Imaging," Sensors and Actuators B, Vol. 111-112, 2005, pp. 562- 571.

108. S. J. Kim, K. V. Gobi, H. Iwasaka, H. Tanaka and N. Miura, "Novel Miniature SPR Immunosensor Equipped with All-in-One Multichannel Sensor Chip for Detecting Low-Molecular-Weight Analytes," Biosensors and Bioelectronics, Vol. 23, No. 5, 2007, pp. 701-707.

109. C. M. Cummins, M. E. Koivunen, A. Stephanian, S. J. Gee, B. D. Hammock and I. M. Kennedy, "Application of Europium(III) Chelate-Dyed Nanoparticle Labels in a Competitive Atrazine Fluoroimmunoassay on an ITO Waveguide," Biosensors and Bioelectronics, Vol. 21, No. 7, 2003, pp. 1077-1085.

110. A. Klotz, A. Brecht, C. Barzen, G. Gauglitz, R. D. Harris, G. R. Quigley, J. S. Wilkinson and R. Abuknesha, "Immunofluorescence Sensor for

Water Analysis," Sensors and Actuators B, Vol. 51, No. 1-3, 1998, pp. 181-187.

111. C. Barzen, A. Brecht and G. Gauglitz, "Optical MultipleAnalyte Immunosensor for Water Pollution Control," Biosensors and Bioelectronics, Vol. 17, No. 4, 2002, pp. 289-295.

112. F. Long, H. C. Shi, M. He and A. N. Zhu, "Sensitive and Rapid Detection of 2,4-Dicholorophenoxyacetic Acid in Water Samples by Using Evanescent Wave All-Fiber Immunosensor," Biosensors and Bioelectronics, Vol. 23, No. 9, 2008, pp. 1361-1366.

113. F. Long, M. He, H. C. Shi and A. N. Zhu, "Development of Evanescent Wave All-Fiber Immunosensor for Environmental Water Analysis," Biosensors and Bioelectronics, Vol. 23, No. 7, 2008, pp. 952-958.

114. S. Rodriguez-Mozaz, M. L. de Alda and D. Barcelo, "Analysis of Bisphenol A in Natural Waters by Means of an Optical Immunosensor," Water Research, Vol. 39, No. 20, 2005, pp. 5071-5079.

115. J. Tschmelak, N. Kappel and G. Gauglitz, "TIRF-Based Biosensor for Sensitive Detection of Progesterone in Milk Based on Ultra-Sensitive Progesterone Detection in Water," Analytical and Bioanalytical Chemistry, Vol. 382, No. 8, 2005, pp. 1895-1903.

116. J. Tschmelak, G. Proll, J. Riedt, J. Kaiser, P. Kraemmer, L. Barzaga, J. S. Wilkinson, P. Hua, J. P. Hole, R. Nudd, M. Jackson, R. Abuknesha, D. Barcelo, S. RodriguezMozaz, M. J. de Alda, F. Sacher, J. Stien, J. Slobodnik, P. Oswald, H. Kozmenko, E. Korenkova, L. Tothova, Z. Krascsenits and G. Gauglitz, "Automated Water Analyser Computer Supported System (AWACSS) Part I: Project Objectives, Basic Technology, Immunoassay Development, Software Design and Networking," Biosensors and Bioelectronics, Vol. 20, No. 8, 2005, pp. 1499-1508.

117. S. Boujday, S. Nasri, M. Salmain and C.-M. Pradier, "Surface IR Immunosensors for Label-Free Detection of Benzo[a]pyrene," Biosensors and Bioelectronics, Vol. 26, No. 4, 2010, pp. 1750-1754.

118. S. Boujday, C. Gu, M. Girardot, M. Salmain and C. M. Pradier, "Surface IR Applied to Rapid and Direct Immunosensing of Environmental Pollutants," Talanta, Vol. 78, No. 1, 2009, pp. 165-170.

119. M. Salmain, N. Fischer-Durand and C. M. Pradier, "Infrared optical immunosensors: Application to the MeasUrement of the Herbicide Atrazine," Analytical Biochemistry, Vol. 373, No. 1, 2008, pp. 61-70.

120. S. Kurosawa, J.-W. Park, H. Aiwaza, S.-I. Wakida, H. Tao and K. Ishihara, "Quartz Crystal Microbalance Immunosensor for Environmental

Monitoring," Biosensors and Bioelectronics, Vol. 22, No. 4, 2006, pp. 473-481.

121. C. March, J. J. Manclus, Y. Jimenez, A. Arnau and A. Montoya, "A Piezoelectric Immunosensor for the Determination of Pesticide Residues and Metabolites in Fruit Juices," Talanta, Vol. 78, No. 3, 2009, pp. 827-833.

122. A. Sassolas, B. D. Leca-Bouvier and L. J. Blum, "DNA Biosensors and Microarrays," Chemical Reviews, Vol. 108, No. 1, 2008, pp. 109-139.

123. J. Kaur, K. V. Singh, A. H. Schmid, G. C. Varshney, C. R. Suri and M. Raje, "Atomic Force Spectroscopy-Based Study of Antibody Pesticide Interactions for Characterization of Immunosensor Surface," Biosensors and Bioelectronics, Vol. 20, No. 2, 2004, pp. 284-293.

124. C. R. Suri, J. Kaur, S. Gandhi and G. S. Shekhawat, "Label-free Ultra-Sensitive Detection of Atrazine Based on Nanomechanics," Nanotechnology, Vol. 19, No. 23, 2008, p. 235502.

125. M. Alvarez, A. Calle, J. Tamayo, L. Lechuga, A. Abad and A. Montoya, "Development of Nanomechanical Biosensors for Detection of the Pesticide DDT," Biosensors and Bioelectronics, Vol. 18, No. 5-6, 2003, pp. 649-653.

126. M. Bache, R. Taboryski, S. Schmid, J. Aamand and M. H. Jakobsen, "Investigations on Antibody Binding to a Microcantilever Coated with a BAM Pesticide Residue," Nanoscale Research Letters, Vol. 6, 2011, p. 386.

127. A. M. Nowicka, A. Kowalczyk, Z. Stojek and M. Hepel, "Nanogravimetric and Voltammetric DNA-Hybridization Biosensors for Studies of DNA Damage by Common Toxicants and Pollutants," Biophysical Chemistry, Vol. 146, No. 1, 2010, pp. 42-53.

128. A. D. Ellington and J. W. Szostak, "In Vitro Selection of RNA Molecules that Bind Specific Ligands," Nature, Vol. 346, No. 6287, 1990, pp. 818-822.

129. C. L. A. Hamula, J. W. Guthrie, H. Zhang, X.-F. Li and X. C. Le, "Selection and Analytical Applications of Aptamers," Trends in Analytical Chemistry, Vol. 30, No. 10, 2011, pp. 1587-1597.

130. R. Stoltenburg, C. Reinemann and B. Strehlitz, "SELEX- A(r)evolutionary Method to Generate High-Affinity Nucleic Acid Ligands," Biomolecular Engineering, Vol. 24, No. 4, 2007, pp. 381-403.

131. A. Famulok and J. W. Szostak, "In Vitro Selection of Specific Ligand Binding Nucleic Acids," Angewandte Chemie-International Edition, Vol. 31, No. 8, 1992, pp. 979- 988.

132. S. D. Jayasena, "Aptamers: An Emerging Class of Molecules That Rival Antibodies in Diagnostics," Clinical Chemistry, Vol. 45, No. 9, 1999, pp. 1628-1650.

133. A. Sassolas, L. J. Blum and B. D. Leca-Bouvier, "Optical Detection Systems Using Immobilized Aptamers," Biosensors and Bioelectronics, Vol. 26, No. 9, 2011, pp. 3725- 3736.

134. A. Sassolas, L. J. Blum and B. D. Leca-Bouvier, "Electrochemical Aptasensors," Electroanalysis, Vol. 21, 2009, pp. 1237-1250.

135. T. Yuan, Z.-Y. Liu, L.-Z. Hu and G.-B. Xu, "Electrochemical and Electrochemiluminescent Aptasensors," Chinese Journal of Analytical Chemistry, Vol. 39, No. 7, 2011, pp. 972-977.

136. A.-E. Radi, "Electrochemical Aptamer-Based Biosensors: Recent Advances and Perspectives," International Journal of Electrochemistry, 2011, Article ID 863196.

137. Y. Xu, G. Cheng, P. He and Y. Fang, "A Review: Electrochemical Aptasensors with Various Detection Strategies," Electroanalysis, Vol. 21, No. 11, 2009, pp. 1251- 1259.

138. J. He, Y. Liu, M. Fan and X. Liu, "Isolation and Identification of the Dna Aptamer Target to Acetamiprid," Journal of Agricultural and Food Chemistry, Vol. 59, No. 5, 2011, pp. 1582-1586.

139. A. L. Hillberg, K. R. Brain and C. J. Allender, "Molecular imprinted Polymer Sensors: Implications for Therapeutics," Advanced Drug Delivery Reviews, Vol. 57, No. 12, 2005, pp. 1875-1889.

140. A. G. Mayes and M. J. Whitcombe, "Synthetic Strategies for the Generation of Molecularly Imprinted Organic Polymers," Advanced Drug Delivery Reviews, Vol. 57, No. 12, 2005, pp. 1742-1778.

141. V. Pichon and F. Chapuis-Hugon, "Role of Molecularly Imprinted Polymers for Selective Determination of Environmental Pollutants—A Review," Analytica Chimica Acta, Vol. 622, No. 1-2, 2008, pp. 48-61.

142. A. L. Jenkins, R. Yin and J. L. Jensen, "Molecularly Imprinted Polymer Sensors for Pesticide and Insecticide Detection in Water," Analyst, Vol. 126, 2001, pp. 798- 802.

143. C. Xie, S. Gao, Q. Guo and K. Xu, "Electrochemical Sensor for 2,4-Dichlorophenoxy Acetic Acid Using Molecularly Imprinted Polypyrrole Membrane as Recognition Element," Microchimica Acta, Vol. 169, No. 1, 2010, pp. 145- 152.

144. C. Pellicer, A. Gomez-Caballero, N. Unceta, M. A. Goicolea and R. J. Barrio, "Using a Portable Device Based on Screen-Printed Sensor Modified with a Molecularly Imprinted Polymer for the Determination

of the Insecticide Fenitrothion in Forest Samples," Analytical Methods, Vol. 2, No. 9, 2010, pp. 1280-1285.

145. E. Pardieu, H. Cheap, C. Vedrine, M. Lazerges, Y. Lattach, F. Garnier, S. remira and C. Pernelle, "Molecularly Imprinted Conducting Polymer Based Electrochemical Sensor for Detection of Atrazine," Analytica Chimica Acta, Vol. 649, No. 2, 2009, pp. 236-245.

146. M. Cortina-Puig, G. Istamboulie, T. Noguer and J. L. Marty, "Analysis of Pesticide Mixtures Using Intelligent Biosensors," In: V. S. Somerset, Ed., Intelligent and Biosensors, 2010, pp. 205-216.

147. G. Istamboulie, M. Cortina-Puig, J. L. Marty and T. Noguer, "The Use of Artificial Neural Networks for the Selective Detection of Two Organophosphate Insecticides: Chlorpyrifos and Chlorfenvinfos," Talanta, Vol. 79, No. 2, 2009, pp. 507-511.

148. X. Zhang, Q. Guo and D. Cui, "Recent Advances in Nanotechnology Applied to Biosensors," Sensors, Vol. 9, No. 2, 2009, pp. 1033-1053.

149. R. Singh, "Prospects of Nanobiomaterials for Biosensing," International Journal of Electrochemistry, 2011, Article ID 125487.

150. A. Ballesteros-Gomez and S. Rubio, "Recent Advances in Environmental Analysis," Analytical Chemistry, Vol. 81, No. 12, 2009, pp. 4601-4622.

151. A. Sassolas, L. J. Blum and B. D. Leca-Bouvier, "Immobilization Strategies to Develop Enzymatic Biosensors," Biotechnology Advances, in press,

152. X.-H. Li, Z. Xie, H. Min, C. Li, M. Liu and Y. J. Xian, "Development of Quantum Dots Modified Acetylcholinesterase Biosensor for the Detection of Trichlorfon," Electroanalysis, Vol. 18, No. 22, 2006, pp. 2163-2167.

153. J. Pingarron, P. Yanez-Sedeno and A. Gonzalez-Cortes, "Gold Nanoparticles-Based Electrochemical Biosensors," Electrochimica Acta, Vol. 53, No. 19, 2008, pp. 5848- 5866.

154. D. Du, S. Chen, J. Cai and A. Zhang, "Electrochemical Pesticide Sensitivity Test Using Acetylcholinesterase Biosensor Based on Colloidal Gold Nanoparticle Modified Sol-Gel Interface," Talanta, Vol. 74, No. 4, 2008, pp. 766-772.

155. D. Du, S. Chen, D. Song, H. Li and X. Chen, "Development of Acetylcholinesterase Biosensor Based on CdTe Quantum Dots/Gold Nanoparticles Modified Chitosan Microspheres Interface," Biosensors and Bioelectronics, Vol. 24, No. 3, 2008, pp. 475-479.

156. P. Sharma, K. Sablok, V. Bhalla and C. R. Suri, "A Novel Disposable Electrochemical Immunosensor for Phenyl Urea Herbicide Diuron," Biosensors and Bioelectronics, Vol. 26, No. 10, 2011, pp. 4209-4212.

157. T.-J. Lin, K.-T. Huang and C.-Y. Liu, "Determination of Organophosphorous Pesticides by a Novel Biosensor Based on Localized Surface Plasmon Resonance," Biosensors and Bioelectronics, Vol. 22, No. 4, 2006, pp. 513-518.

158. X. Sun, B. Liu and K. Xia, "A Sensitive and Regenerable Biosensor for Organophosphate Pesticide Based on Self- Assembled Multilayer Film with CdTe as Fluorescence Probe," Luminescence, Vol. 26, No. 6, 2011, pp. 616-621.

159. J. Fu, B. PArk and Y. Zhao, "Limitation of a Localized Surface Plasmon Resonance Sensor for Salmonella Detection," Sensors and Actuators B, Vol. 141, No. 1, 2009, pp. 276-283.

160. L.-K. Chau, Y.-F. Lin, S.-F. Cheng and T.-J. Lin, "Fiber-Optic Chemical and Biochemical Probes Based on Localized Surface Plasmon Resonance," Sensors and Actuators B, Vol. 113, No. 1, 2006, pp. 100-105.

161. J. Wang, "Carbon-Nanotube Based Electrochemical Biosensors: A Review," Electroanalysis, Vol. 17, No. 1, 2005, pp. 7-14.

162. D. Du, X. Huang, J. Cai and A. Zhang, "Comparison of Pesticide Sensitivity by Electrochemical Test Based on Acetylcholinesterase Biosensor," Biosensors and Bioelectronics, Vol. 23, No. 2, 2007, pp. 285-289.

163. A. C. Oliveira and L. H. Mascaro, "Evaluation of Acetylcholinesterase Biosensor Based on Carbon Nanotube Paste in the Determination of Chlorphenvinphos," International Journal of Analytical Chemistry, 2011, Article ID 974216.

164. S. Firdoz, F. Ma, X. Yue, Z. Dai, A. Kumar and B. Jiang, "A Novel Amperometric Biosensor Based on Single Walled Carbon Nanotubes with Acetylcholine Esterase for the Detection of Carbaryl Pestcide in Water," Talanta, Vol. 83, No. 1, 2010, pp. 269-273.

165. Y. Qu, Q. Sun, F. Xiao, G. Shi and L. Jin, "Layer- by-Layer Self-Assembled Acetylcholienesterase/ PAMAMAu on CNTs Modified Electrode for Sensing Pesticides," Bioelectrochemistry, Vol. 77, No. 2, 2010, pp. 139-144.

166. S. Chen, J. Huang, D. Du, J. Li, H. Tu, D. Liu and A. Zhang, "Methyl Parathion Hydrolase Based Nanocomposite Biosensors for Highly Sensitive and Selective Determination of Methyl Parathion," Biosensors and Bioelectronics, Vol. 26, No. 11, 2011, pp. 4320-4325.

167. J. L. Marty, B. Leca and T. Noguer, “Biosensors for the Detection of Pesticides,” Analusis Magazine, Vol. 26, 1998, pp. 144-149.

168. M. Blanco-Lopez, “Electrochemical Sensors Ased on Molecularly Imprinted Polymers,” Trends in Analytical Chemistry, Vol. 23, No. 1, 2004, pp. 36-48.

169. S. A. Piletsky and A. P. F. Turner, “Electrochemical Sensors Based on Molecularly Imprinted Polymers,” Electroanalysis, Vol. 14, No. 5, 2002, pp. 317-323.

170. A. C. Ion, I. Ion, A. Culetu, D. Gherase, C. A. Moldovan, R. Iosub and A. Dinescu, “Acetylcholinesterase Voltammetric Biosensors Based on Carbon Nanostructure-Chitosan Composite Material for Organophosphate Pesticides,” Materials Science and Engineering, Vol. 30, No. 6, 2010, pp. 817-821.

171. S. Viswanathan, H. Radecka and J. Radecki, “Electrochemical Biosensor for Pesticides Based on Acetylcholinesterase Immobilized on Polyaniline Deposited on Vertically Assembled Carbon Nanotubes Wrapped with ssDNA,” Biosensors and Bioelectronics, Vol. 24, No. 9, 2009, pp. 2772-2777.

172. N. Chauhan, J. Narang and C. S. Pundir, “Immobilization of Rat Brain Acetylcholinesterase on ZnS and Poly(Indole-5-carboxylic acid) Modified Au Electrode for Detection of Organophosphorus Insecticides,” Biosensors and Bioelectronics, Vol. 29, No. 1, 2011, pp. 82-88.

173. G. Istamboulie, T. Sikora, E. Jubete, E. Ochoteco, J. L. Marty and T. Noguer, “Screen-Printed Poly(3,4-Ethylenedioxythiophene) (PEDOT): A New Electrochemical Mediator for Acetylcholinesterase-Based Biosensors,” Talanta, Vol. 82, No. 3, 2010, pp. 957-961.

174. A. Hildebrandt, J. Ribas, R. Bragos, J. L. Marty, M. Tresanchez and S. Lacorte, “Development of a Portable Biosensor for Screening Neurotoxic Agents in Water Samples,” Talanta, Vol. 75, No. 5, 2008, pp. 1208-1213.

175. A. Hildebrandt, R. Bragos, S. Lacorte and J. L. Marty, “Performance of a Portable Biosensor for the Analysis of Organophosphorus and Carbamate Insecticides in Water and Food,” Sensors and Actuators B, Vol. 133, No. 1, 2010, pp. 195-201.

176. Y. Ivanov, I. Marinov, K. Gabrovska, N. Dimcheva and T. Godjevargova, “Amperometric Biosensor Based on a SiteSpecific Immobilization of Acetylcholinesterase via Affinity Bonds on a Nanostructured Polymer Membrane With Intergrated Multiwall Carbon Nanotubes,” Journal of Molecular Catalysis B: Enzymatic, Vol. 63, No. 3-4, 2010, pp. 141-148.

177. R. Sinha, M. Ganesana, S. Andreescu and L. Stanciu, "AChE Biosensor Based on Zinc Oxide Sol-Gel for the Detection of Pesticides," Analytica Chimica Acta, Vol. 661, No. 2, 2010, pp. 195-199.

178. Y. Wang, S. Zhang, D. Du, Y. Shao, Z. Li, J. Wang, M. H. Englehard, J. Li and Y. H. Lin, "Self Assembly of Acetylcholinesterase on a Gold Nanoparticles-Graphene Nanosheet Hybrid for Organophosphate Pesticide Detection Using Polyelectrolyte as a Linker," Journal of Materials Chemitsry, Vol. 21, No. 22, 2011, pp. 5319-5325.

179. F. Arduini, F. Ricci, C. S. Tuta, D. Moscone, A. Amine and G. Palleschi, "Detection of Carbamic and Organophosphorous Pesticides in Water Samples Using a Cholinesterase Biosensor Based on Prussian Blue-Modified Screen-Printed Electrode," Analytica Chimica Acta, Vol. 580, No. 2, 2006, pp. 155-162.

180. G. Valdes-Ramirez, M. Cortina, M. T. Ramirez-Silva and J. L. Marty, "Acetylcholinesterase-Based Biosensors for Quantification of Carbofuran, Methylparaoxon, and Dichlorvos in 5% Acetonitrile," Analytical and Bioanalytical Chemistry, Vol. 392, No. 4, 2008, pp. 699-707.

181. D. Du, X. Huang, J. Cai and A. Zhang, "Amperometric Detection of Triazophos Pesticide Using Acetylcholinesterase Biosensor Based on Miltiwall Carbon Nanotube-Chitosan Matrix," Sensors and Actuators B, Vol. 127, No. 2, 2007, pp. 531-535.

182. A. Vakurov, C. E. Simpson, C. L. Daly, T. D. Gibson and P. A. Millner, "Acetylcholinesterase-Based Biosensor Electrodes for Organophosphate Pesticide Detection. I. Modification of Carbon Surface for Immobilization of Acetylcholinesterase," Biosensors and Bioelectronics, Vol. 20, No. 6, 2004, pp. 1118-1125.

183. S. Sotiropoulou, D. Fournier and N. A. Chaniotakis, "Genetically Engineered Acetylcholinesterase-Based Biosensor for Attomolar Detection of Dichlorvos," Biosensors and Bioelectronics, Vol. 20, No. 11, 2005, pp. 2347-2352.

184. X. Sun and X. Wang, "Acetylcholinesterase Biosensor Based on Prussian Blue-Modified Electrode for Detecting Organophosphorous Pesticides," Biosensors and Bioelectronics, Vol. 25, No. 1-2, 2010, pp. 2611-2614.

185. Y. Wei, Y. Li, Y. Qu, F. Xiao, G. Shi and L. Jin, "A Novel Biosensor Based on Photoelectro-Synergistic Catalysis for Flow-Injection Analysis System/Amperometric Detection of Organophosphorous Pesticides," Analytica Chimica Acta, Vol. 643, No. 1-2, 2009, pp. 13-18.

186. D. Du, S. Chen, J. Cai and A. Zhang, "Immobilization of Acetylcholinesterase on Gold Nanoparticles Embedded in Sol-Gel Film for Amperometric Detection of Organophosphorous," Biosensors and Bioelectronics, Vol. 23, No. 1, 2007, pp. 130-134.

187. A. Ishii, S. Takeda, S. Hattori, K. Sueoka and K. Mukasa, "Ultrasensitive Detection of Organophosphate Insecticides by Carbon Field-Effect Transistor," Colloids and Surfaces A, Vol. 313-314, 2008, pp. 456-460.

188. N. Gan, X. Yang, D. Xie, Y. Wu and W. Wen, "A Disposable Organophosphorus Pesticides Enzyme Biosensor Based On Magnetic Nano-Particles Modified Screen Printed Carbon Electrode," Sensors, Vol. 10, No. 1, 2010, pp. 665-638.

189. D. Du, J. Ding, J. Cai and A. Zhang, "Determination of Carbaryl Pesticide Using Amperometric Acetylcholinesterase Sensor Formed by Electrochemically Deposited Chitosan," Colloids and Surfaces B Biointerfaces, Vol. 58, No. 1, 2007, pp. 145-150.

190. S. P. Zhang, L. G. Shan, Z. R. Tian, Y. Zheng, L. Y. Shi and D. S. Zhang, "Study of Enzyme Biosensor Based on Carbon Nanotubes Modified Electrode for Detection of Pesticides Residue," Chinese Chemical Letters, Vol. 19, No. 5, 2008, pp. 592-594.

191. Y. Qu, Q. Sun, F. Xiao, G. Shi and L. Jin, "Layerby-Layer Self-Assembled Acetylcholinesterase/PAMAMAu on CNTs Modified Electrode for Sensing Pesticides," Bioelectrochemistry, Vol. 77, No. 2, 2010, pp. 139-144.

192. S. Laschi, D. Ogonczyk, I. Palchetti and M. Mascini, "Evaluation of Pesticide-Induced Acetylcholinesterase Inhibition by Means of Disposable Carbon-Modiifed Electrochemical Biosensors," Enzyme and Microbial Technology, Vol. 40, No. 3, 2007, pp. 485-489.

193. S. Sajjadi, H. Ghourchian and H. Tavakoli, "Choline Oxidase as a Selective Recognition Element for Determination of Paraoxon," Biosensors and Bioelectronics, Vol. 24, No. 8, 2009, pp. 2509-2514.

194. R. P. Deo, J. Wang, I. Block, A. Mulchandani, K. A. Joshi, M. Trojanowicz, F. Scholz, W. Chen and Y. H. Lin, "Determination of Organophosphate Pesticides at a Carbon Nanotube/Organophosphorus Hydrolase Electrochemical biosensor," Analytica Chimica Acta, Vol. 530, No. 2, 2005, pp. 185-189.

195. T. Laothanachareon, V. Champreda, P. Sritongkham, M. Somasundrum and W. Surareungchai, "Cross-Linked Enzyme Crystals of Organophosphate Hydrolase for Electrochemical Detection of Organophosphorus Compounds," World Journal of Microbiology and Biotechnology, Vol. 24, No. 12, 2008, pp. 3049-3055.

196. A. Mulchandani, S. T. Pan and W. Chen, "Fiber-Optic Enzyme Biosensor for Direct Determination of Organophosphate Nerve Agents," Biotechnology Progress, Vol. 15, No. 1, 1999, pp. 130-134.

197. V. Sacks, I. Eshkenazi, T. Neufeld, C. Dosoretz and J. Rishpon, "Immobilized Parathion Hydrolase: An Amperometric Sensor for Parathion," Analytical Chemistry, Vol. 72, No. 9, 2000, pp. 2055-2058.

198. L. Viveros, S. Paliwal, D. McCrae, J. Wild and A. Simonian, "A Fluorescence-Based Biosensor for the Detection of Organophosphate Pesticides and Chemical Warfare Agents," Sensors and Actuators B, Vol. 115, No. 1, 2006, pp. 150-157.

199. K. A. Fähnrich, M. Pacda and G. G. Guilbault, "Diposable Amperometric Immunosensor for the Detection of Polycyclic Aromatic Hydrocarbons (PAHs) Using ScreenPrinted Electrodes," Biosensors and Bioelectronics, Vol. 18, No. 1, 2003, pp. 73-82.

200. S.-Q. Hu, J.-W. Xie, Q.-H. Xu, K. T. Rong, G.-L. Shen and R.-Q. Yu, "A Label-Free Electrochemical Immunosensor Based on Gold Nanoparticles for Detection of Paraoxon," Talanta, Vol. 61, No. 6, 2003, pp. 769-777.

201. E. Zacco, R. Galve, M. P. Marco, S. Alegret and M. I. Pividori, "Electrochemical Biosensing of Pesticide Residues Based on Affinity Biocomposite Platforms," Biosensors and Bioelectronics, Vol. 22, No. 8, 2007, pp. 1707- 1715.

202. D. Butler and G. G. Guilbault, "Disposable Amperometric Immunosensor for the Detection of 17-β Estradiol Using Screen-Printed Electrodes," Sensors and Actuators B, Vol. 113, No. 2, 2006, pp. 692-699.

203. L. Chen, G. Zeng, Y. Zhang, L. Tang, D. Huang, C. Liu, Y. Pang and J. Luo, "Trace Detection of Pichloram Using an Electrochemical Immunosensor Based on 3D Au Nanoclusters," Analytical Biochemistry, Vol. 407, No. 2, 2010, pp. 172-179.

204. Y. Zhang and H.-S. Zhuang, "Amperometric Immunosensor Based on Layer-by-Layer Assembly of Thiourea and Nano-Gold Particles on Gold Electrode for Determination of Naphthalene," Chinese Journal of Analytical Chemistry, Vol. 38, No. 2, 2010, pp. 153-157.

205. I. Navratilova and P. Skladal, "The Immunosensors for Measurement of 2,4-Dichlorophenoxyacetic Acid Based on Electrochemical Impedance Spectroscopy," Bioelectrochemistry, Vol. 62, No. 1, 2004, pp. 11-18.

206. M. F. Gouzy, M. Kess and P. M. Kramer, "A SPR-Based Immunosensor for the Detection of Isoproturon," Biosensors and Bioelectronics, Vol. 24, No. 6, 2009, pp. 1563-1568.

207. D. R. Shankaran, K. Matsumoto, K. Toko and N. Miura, "Development and Comparison of Two Immunoassays for the Detection of 2,4,6-Trinitrotoluene (TNT) Based on Surface Plasmon Resonance," Sensors and Actuators B, Vol. 114, No. 1, 2006, pp. 71-79.

208. T. Kawaguchi, D. R. Shankaran, G. Y. Kim, K. Matsumoto, K. Toko and N. Miura, "Surface Plasmon Resonance Au Nanoparticle for Detection of TNT," Sensors and Actuators B, Vol. 133, No. 2, 2008, pp. 467-472.

209. T. Kawaguchi, D. R. Shankaran, S. J. Kim, K. V. Gobi, K. Matsumoto, K. Toko and N. Miura, "Fabrication of a Novel Immunosensor Using Functionalized Self-Assembled Monolayer for Trace Lavel Detection of TNT by Surface Plasmon Resonance," Talanta, Vol. 72, No. 2, 2007, pp. 554-560.

210. P. Singh, T. Onodera, Y. Mizuta, K. Matsumoto, N. Miura and K. Toko, "Dendrimer Modified Biochip for Detection of 2,4,6 Trinitrotoluene on SPR Immunosensor: Fabrication and Advantages," Sensors and Actuators B, Vol. 137, No. 2, 2009, pp. 403-409.

211. K. Nagatomo, T. Kawaguchi, N. Miura, K. Toko and K. Matsumoto, "Development of a Sensitive Surface Plasmon Resonance Immunosensor for Detection of 2,4-Dinitrotoluene with a Novel Oligo (Ethylene Glycol)-Based Sensor Surface," Talanta, Vol. 79, No. 4, 2009, pp. 1142- 1148.

212. N. Soh, T. Tokuda, T. Watanabe, K. Mishima, T. Imato, T. Masadome, Y. Asano, S. Okutani, O. Niwa and S. Brown, "A Surface Plasmon Resonance Immunosensor for Detecting a Dioxin Precursor Using a Gold Binding Polypeptide," Talanta, Vol. 60, No. 4, 2003, pp. 733-745.

Chapter 4

A Biosensor for Urea from Succinimide-Modified Acrylic Microspheres Based on Reflectance Transduction

Alizar Ulianas, Lee Yook Heng * and Musa Ahmad

School of Chemical Sciences and Food Technology, Faculty of Science and Technology, University Kebangsaan Malaysia, Bangi, Selangor 43600, Malaysia

ABSTRACT

New acrylic microspheres were synthesised by photopolymerisation where the succinimide functional group was incorporated during the microsphere preparation. An optical biosensor for urea based on reflectance transduction with a large linear response range to urea was successfully developed using this material. The biosensor utilized succinimide-modified acrylic microspheres immobilized with a Nile blue chromoionophore (ETH 5294) for optical detection and urease enzyme was immobilized on the surface of the microspheres via the succinimide groups. No leaching of the enzyme or chromoionophore was observed. Hydrolysis of the urea by urease changes the pH and leads to a color change of the immobilized chromoionophore. When the color change was monitored by reflectance spectrophotometry, the linear response range of the biosensor to urea was from 0.01 to 1,000 mM (R^2 = 0.97) with a limit of detection of 9.97 μM. The biosensor response showed good reproducibility (relative standard deviation = 1.43%, n = 5) with no interference by major cations such as Na^+, K^+, NH_4^+ and Mg^{2+}. The use of reflectance as a transduction method led to a large linear response range that is better than that of many urea biosensors based on other optical transduction methods.

Keywords: urea; reflectance; optical biosensor; succinimide; acrylic microspheres; photopolymerisation

1. INTRODUCTION

Urea biosensors have been reported for urea determinations in biomedical, clinical and food industry applications [1–3], e.g., for monitoring of urea levels in human urine and blood for diagnosis of kidney function and health [4–6], as well as for the determination of urea levels in milk [2]. These numerous applications have motivated the development of urea biosensors. Urea biosensors have been developed using different transducers, including ion-selective field effect transistor electrode [7], ammonium ion-selective electrode [8], CO_2 gas electrode [9], pH electrode [10] and ammonia gas selective electrode [11].

Many urease enzyme immobilization techniques for biosensor construction have involved the incorporation of the enzyme into a polymer membrane or polymer backbone such as tetrafluoroethylene and poly(tetrafluoroethylene) [1], chitosan membranes [12], sol-gel-derived thick-films [3,6] and poly(*N*-vinylcarbazole)/stearic acid [5]. Immobilization of urease in a membrane can lead to a loss of enzyme activity (deactivation and denaturation), as well as a decrease of the lifetime, sensitivity and poor response time of the biosensor. Therefore, it is essential to choose an enzyme immobilization technique that will enhance the operational and storage stability of the enzyme but at the same time avoid any interference of the immobilization matrix on enzyme activity [12–14]. Previously reported urease enzyme immobilization techniques include entrapment and encapsulation, covalent binding, cross-linking and adsorption [15]. Other published reports on the use of nanospheres for biosensors were based on poly-(styrene-*co*-acrylic acid) functionalized with dithioglycol [16,17] and poly(divinylbenzene-*co*-acrylic acid) [18] modified with gold nanoparticles for the immobilization of horseradish peroxidase enzyme. The use of spherical beads for immobilization of urease in biosensor is uncommon and no report in the use of acrylic microspheres for urea biosensor has been published to date. Nile-Blue chromoionophore is commonly used as an indicator in sol-gel film for biosensing of pesticides [19], in sol-gel [6] and in plasticized PVC membrane for biosensing of urea [20,21] with absorption optical detection mode.

In this research, acrylic microspheres modified with succinimide groups at the sphere surface were used to fabricate an optical urea biosensor based on a reflectance method. The microspheres serve two purposes, *i.e.*, first to covalently immobilize urease on the surface of the spheres and secondly to entrap the chromoionophore as an indicator in the spheres. *N*-Acryloxysuccinimide was used to modify acrylic microspheres to produce functional groups as a linker to immobilize enzyme urease via covalent bonds. The efficiency of this bonding has be proven as reported [22,23]. The acrylic microspheres have the advantages of small size and providing a large surface area to volume ratio for enzymatic reactions to occur on the surface, in addition to preventing any barriers to diffusion of reactants and products. Thus this should improve biosensor performance in terms of response time and linear response range. The use of *n*-butyl acrylate in microsphere synthesis is also compatible with the hydrophobicity of the lipophilic chromoionophore ETH5294 where the chromoionophore is used in the reflectance mode instead of the commonly used absorption mode. In addition, microspheres made from *n*-butyl acrylate also possess good adhesion properties that allow these spheres to be coated directly on a plastic substrate for optical biosensor fabrication.

2. EXPERIMENTAL

2.1. Chemicals

2-2-Dimethoxy-2-phenylacetophenone (DMPP) and 1,6-hexanadiol diacrylate (HDDA) were supplied by Aldrich. Sodium dodecyl sulfate (SDS), *N*-acryloxysuccinimide (NAS), urea, urease enzyme (E.C 3.5.1.5; 26.1 units/mg from Jack beans) and phenolphthalein (PP) were supplied by Systerm, Acros, Harnstoff, Sigma-Aldrich and Merck, respectively. Bradford Reagent and bovine serum albumin (BSA) were obtained from Sigma. Chromoionophores (ETH5294), $MgCl_2$ and K_2HPO_4 were supplied by Fluka, while KH_2PO_4 and *n*-butyl acrylate (n-BA) were from Merck. NaCl, KCl and dimethylformamide (DMF) were obtained from Systerm, NH_4Cl from Comak and 4-(*N*-*N*-dimethylamino)-benzaldehyde (DMAB) from Riedel de Haën. All aqueous solutions were prepared using deionized water.

2.2. Synthesis of Acrylic Microspheres

A mixture of 450 μL HDDA, 0.1 g DMPP, 6 mg NAS and 15 mL H_2O with various amounts of n-BA and SDS was sonicated for 10 min, after which it was subjected to photopolymerisation for 600 s with UV light (350 nm) under nitrogen gas flow. The resulting acrylic microspheres were then collected by centrifugation at 4,000 rpm for 30 min. These spheres were washed three times in phosphate buffer at 0.05 M (pH 7.0), then air dried. The size of the acrylic microspheres was determined using scanning electron microscopy (SEM, LEO 1450VP). A Microtrac-X100 particle sizer was used to determine the size distribution of the acrylic microspheres (0.1 mg/mL). FTIR spectra of acrylic microsphere were obtained using a Spectrum FTIR GX infra-red spectrophotometer (Perkin Elmer).

2.3. Immobilization and Activity of Urease Enzyme

The acrylic microspheres were coated onto a transparent plastic sheet by dipping the plastic sheet in the acrylic microsphere suspension just after photocuring and air drying at room temperature. A solution of 2 mg of urease per mL of buffer was prepared in 0.05 M pH 7.0 phosphate buffer solution. The urease was then immobilized onto the acrylic microspheres by immersing the microsphere-coated support in the urease enzyme solution for 24 h at 4 °C. The acrylic microspheres with immobilized urease were then washed and kept in phosphate buffer solution at pH 7.0 until use. The plastic support with no immobilized acrylic microsphere coating was used for the control experiments.

The amount of urease immobilized on these samples was then estimated by checking the enzyme activity. For this measurement, 10 mL of test solution containing urea and phenolphthalein was prepared in deionized water and used for all measurements. Enzyme activity was determined by immersing a plastic support coated with a known amount of enzyme-modified microspheres in the urea solution. Urea hydrolysis by the immobilized urease on the microspheres produced a change in pH, which caused the test solution to turn pink. The enzyme urease catalyzes the urea hydrolysis to change pH according to the following reaction:

$$H_2N{-}CO{-}NH_2 + 3H_2O \xrightarrow{Urease} 2NH_4^+ + HCO_3^- + OH^-$$

The absorbance of the test solution was monitored for 30 min at a wavelength of 553 nm using the spectrophotometer.

2.4. Enzyme Leaching Test and Effects of pH on Enzyme Immobilization

To determine the effects of pH on the immobilization of the enzyme, urease enzyme solutions were prepared using phosphate buffer at pHs of 6.5, 7.0, 7.5 and 8.0. For the leaching test, a fixed amount of acrylic microspheres modified with immobilized urease were immersed in 0.05 M (pH 7.0) phosphate buffer solution, and 0.5 mL phosphate buffer solution was taken in every 10 min before mixing with 2.5 mL of urea solution (1.0 M) and phenolphthalein indicator. The mixture was monitored for 90 min, where the absorbance of that solution was periodically measured at a wavelength of 553 nm using the spectrophotometer.

2.5. Determination of Urease Immobilized

The amount of urease immobilized onto the urea biosensor and the influence of various immobilization times (0.5 to 24 h) were quantified by Bradford reagent method [24] with bovine serum albumin (BSA) as a standard to quantify urease enzyme immobilization. Urease enzyme concentration for immobilisation studies was fixed at 2.0 mg mL^{-1}. The amount of immobilized enzyme was calculated according to the following equation:

$$\% \text{ urease immobilized} = \{(A - B)/A\} \times 100\%$$

where A is the amount of urease used and B is the urease left after immobilisation.

2.6. Construction and Evaluation of Urea Biosensor Performance

A sample of 375.0 mg of dried acrylic microspheres, 300.0 μL chromoionophore solution (0.1 mg mL^{-1} ethanol), 450.0 μL ethanol and 125.0 μL DMF were ultrasonically mixed for 3 min until homogeneous. After that 15.0 μL of the suspensions were drop-coated onto a fixed size plastic support (5.25 ± 0.01 mm diameter and 0.56 ± 0.05 mm thickness) and then dried at room temperature for 1 h. Urease was then immobilized onto these microspheres by dispensing 100 μL urease solution onto the support, then leaving it for 24 h at 4 °C. A urea biosensor in the form of a test-strip was thus fabricated and ready for testing.

To evaluate the response of the urea biosensor to urea, the biosensor was tested with several concentrations of urea solution in the 0.001 to 1,000 mM range in 0.05 M phosphate buffer (pH 6.5). The intensity of the color change of the urea biosensor was measured using a reflectance spectrophotometer (Mikropack DH-2000-Bal). Biosensors with varying amounts of acrylic microspheres (0.64, 6.43 and 643 mg) at different analysis times (5.0 min, 10 min and 20 min) were also evaluated. The linear response range of urea for the biosensor was determined from plots of urea concentration *versus* intensity change as measured by the reflectance spectrophotometer.

Since the enzymatic hydrolysis of the urea and the protonation of chromoionophore can be affected by pH, the working pH range of the optode was also optimized [20]. The effects of pH on the response of the urea biosensor was studied in phosphate buffer at pHs from 5.0 to 8.0 and with urea concentration fixed at 0.05 M.

2.7. Investigation of the Ionic Interference on the Optical Urea Biosensor

To determine the applicability of the optical urea biosensor to real samples, the interference of ions typically found in urine such as NH_4^+, Na^+, K^+ and Mg^{2+} was studied under optimum condition where the urea biosensor was exposed to the interference at various concentrations (0.001 to 100 mM).

2.8. Validation and Recovery Studies of Urea Biosensor

For validation and recovery performance of the urea biosensor, urine samples were used as described previously [6]. A non-enzymatic method using *p*-dimethylaminobenzaldehyde (DMAB) [25] was used as comparison. Briefly, for the chemical analysis of urea, DMAB solution was prepared in ethanol and HCl and the range of the standard urea concentration used was 1.0 to 70.0 mM to construct the calibration curve. For urine samples, urea concentrations of 1.01 to 69.94 mM were spiked into the samples. These spiked urine samples were analyzed for urea with both biosensor and DMAB methods. The percentage of recovery of urea from all urea-spiked urine samples was determined by the urea biosensor method, calculated with the following equation:

$$\% \text{ Recovery} = Cs/C \times 100\%$$

where C is the actual concentration of urea and Cs is the concentration of urea determined by the biosensor in urea-spiked sample.

2.9. Reproducibility and Stability of the Urea Biosensor

The pH of the urea solution was kept constant during evaluation of the reproducibility and stability of the biosensor. Reproducibility of several urea biosensors using reflectance spectrophotometer were tested at a constant concentration of 10.0 mM urea at pH 6.5 (0.05 mM phosphate buffer). The constant reflectance response of each biosensor would indicate good reproducibility. Stability was determined by fabricating several urea biosensors, which were kept at 4 °C before used. Each biosensor was tested for up to 45 days using 1.0 mM and 10.0 mM urea in pH 6.5 (0.05 mM phosphate buffer).

3. RESULTS AND DISCUSSION

3.1. Characterization of Acrylic Microspheres and Urease Immobilization

In the FTIR spectrum of the acrylic microspheres (Figure 1), peaks at 1,188.99 cm^{-1} to 1,066.13 cm^{-1} are associated with the stretching frequencies of the (C–O) groups of the n-BA molecules and the succinimide ester (C–O) of the NAS molecules. The C–N stretch peaks present at 1,256.42 cm^{-1} to 1,165.58 cm^{-1} are from the NAS molecules.

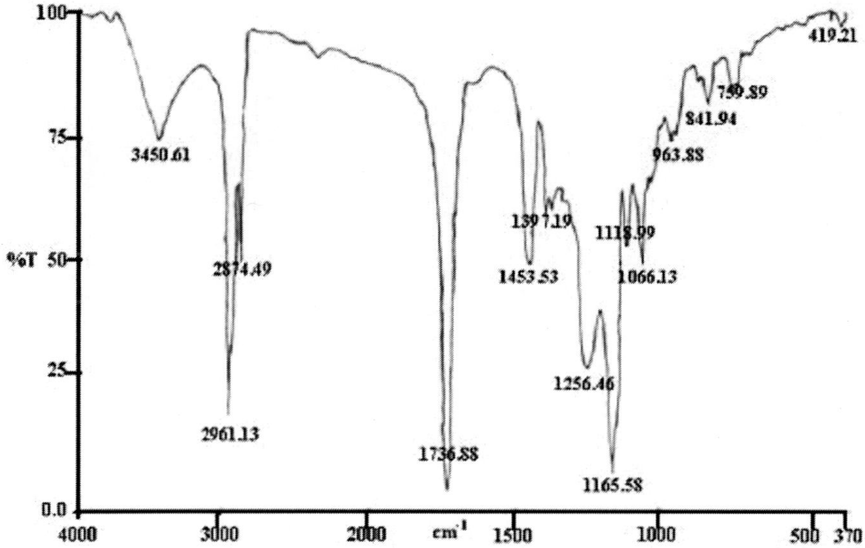

Figure 1. FTIR spectrum of acrylic microspheres.

Urease can be immobilized onto the surface of the acrylic microspheres by covalent binding because the NAS succinimide ester groups are highly reactive towards enzyme amino groups [22] and thus the amino groups of the urease.

These spectra indicated that the acrylic microspheres had been modified with NAS. A typical scanning electron microscope (SEM) image of the acrylic microspheres (Figure 2) demonstrates that the size of the microspheres was approximately less than 5 μm in diameter, depending on the amount of monomer and SDS used. However, when the size of the microspheres was determined by a particle sizer, a size of more than 5 μm was predominant. This is because of aggregation of the acrylic spheres in solution phase during measurement when using a particle sizer.

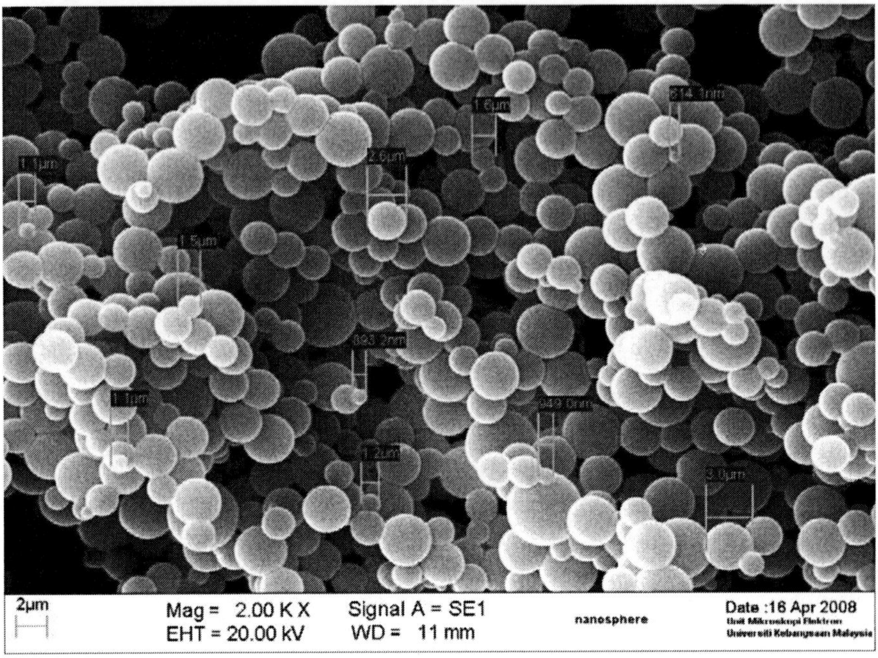

Figure 2. SEM image of acrylic microspheres prepared from photopolymerisation.

Acrylic polymers with higher amounts of n-BA tend to be more sticky and result in microsphere aggregation [26]. The effect of the amount of n-BA monomer used on the size of the microspheres was more pronounced when compared with the variation of the amount of SDS used. Based on the particle sizer determination, with a fixed amount of SDS, an increase in the n-BA from 0.5 to 7 mL led to an increase in the size of the microspheres from 0.32 to 271 µm. This may be explained by the fact that a higher quantity of n-BA leads to the merging of the monomer into larger droplets and hence an increase in the size of the microspheres formed after photopolymerisation. Futhermore, after formation, microspheres with higher n-BA will aggregate more, a phenomenon attributable to the increased sticky properties during particle sizer measurements, where the microspheres appeared to have increased in size. The effect of the quantity of monomer on microsphere size has been reported previously [29] and the synthesis of acrylic microsphere by photopolymerisation in a suspension performed here behaved similarly. Although the SDS used is able to stabilize the monomer droplets to prevent formation of bigger droplets during emulsion polymerisation [27,28], an increase of SDS from 0.01 to 0.05 g produced microspheres of similar particle sizes. The size of the microsphere used for fabricating the biosensor consists is 1–5 µm based on the SEM images. The amount of n-BA monomer used to produce this microsphere composition was able to yield spheres that possessed a good adhesion characteristics, which is essential for coating onto a substrate for optical sensor.

The response of the immobilized urease was indicated by a change of colour of the phenolphthalein indicator to red in the presence of urea. The absorption spectrum of urea solution containing urease modified microspheres is shown in Figure 3. On the other hand, a plastic support coated with acrylic microspheres without immobilized urease showed no change in color when immersed in urea solution. This confirmed that the immobilized urease was still reactive where hydrolysis of urea by the urease producing NH_4^+, HCO_3^-, and OH^-, which changed the pH towards alkaline values. Leaching experiments have shown that no leaching of urease enzyme that was immobilized on the acrylic microspheres occurred because a buffer solution exposed to the immobilised enzyme for 90 min did not change color when urea and the phenolphthalein indicator were added. This suggests that a very strong binding of urease with the succinimide ester groups of NAS on the acrylic microspheres.

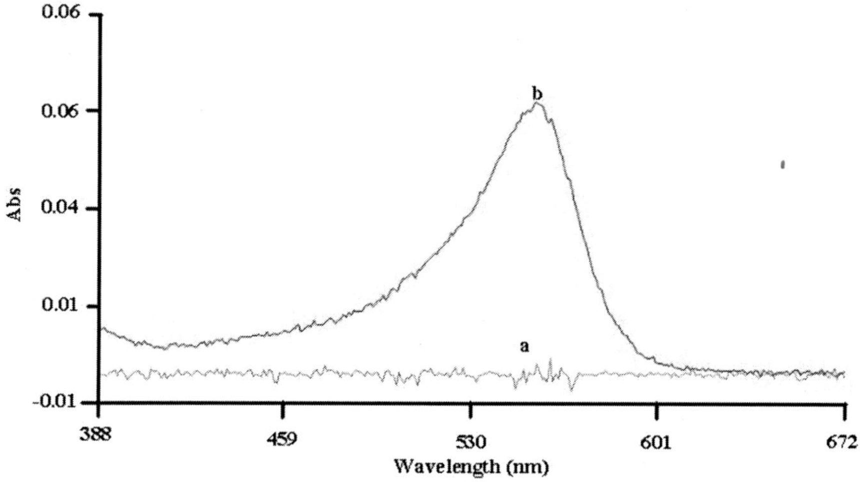

Figure 3. The absorbance response of the phenolphthalein indicator in the presence of acrylic microspheres and urea before immobilization with urease (**a**); and after immobilization (**b**).

3.2. Optimum pH for Enzyme Immobilization

Optimization of pH conditions is essential for successful immobilization of the enzyme since a higher pH value will result in more amine groups being unprotonated and able to attack the succinimide ester group of the NAS moiety, but the ester is also susceptible to hydrolysis at high pH [22]. Figure

4 shows the effects of the use of different pHs for the enzyme immobilization step. The response of the biosensor increased with pH up to 7.0, after which it decreased with further increased in pH, indicating that pH 7.0 was optimum for suitable covalent binding between the amine groups of the enzyme with the succinimide ester of NAS. When the pH was higher than 7.0, which was outside the optimum working range of the enzyme urease, an overall decrease in the response of the measurement was thus observed. This finding is consistent with a previous report [22], which found that the amino group of urease directly reacted with the succinimide ester of NAS under the same pH.

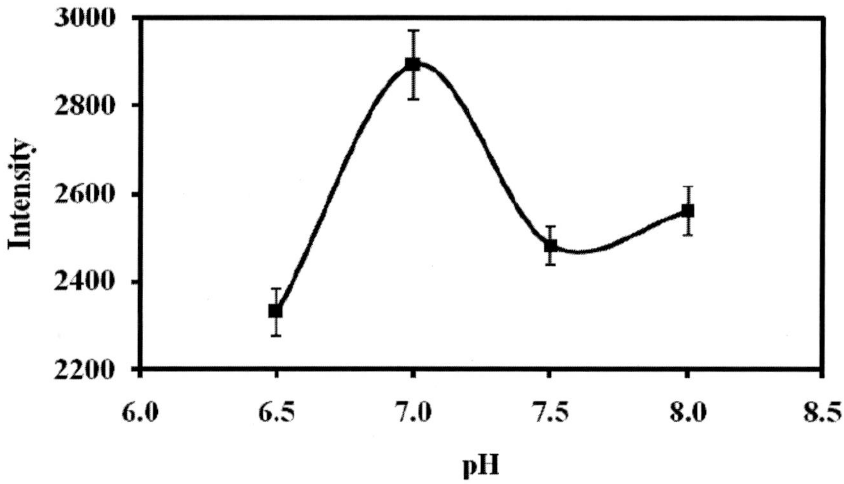

Figure 4. The reflectance intensity of the urease-modified microspheres prepared in the presence of different pH values (at 0.05 M phosphate buffer and 10 mM fixed urea concentration).

3.3. Dependence of Biosensor Response on pH

Both the activity of urease and chromoionophore can be affected by pH of the buffer solution used, making it important to optimize the biosensor response with respect to pH. From the Figure 5, pH 6.5 was found to give the best response in terms of sensitivity and linear range for the urea biosensor (Table 1). This pH yielded a maximum sensitivity from the enzymatic response. Below this pH the histidine group of the urease enzyme is not affected by the phosphate buffer used [30].

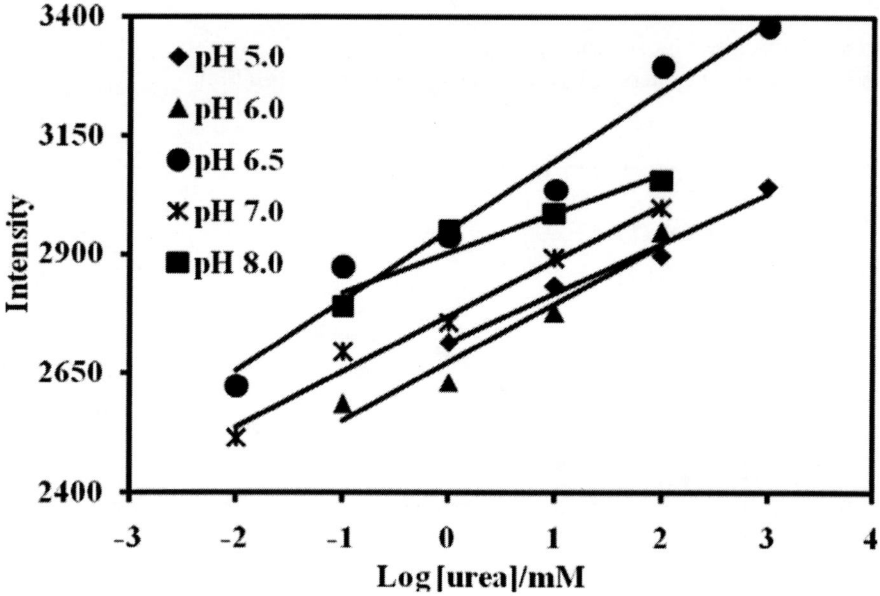

Figure 5. The effect of pH on the urea biosensor sensitivity.

Table 1. The effect of pH on the response of the urea biosensor based on acrylic microspheres and reflectance measurements (in 0.05 M phosphate buffer)

pH	Sensitivity (I/mM)	R^2	Linear range (mM)
5.0	104.80	0.98	1.0–1,000
6.0	122.77	0.95	0.01–100
6.5	146.99	0.97	0.01–1,000
7.0	116.51	0.98	0.01–100
8.0	83.76	0.91	0.1–100

3.4. The Effect of Loading of Acrylic Microspheres on the Response of the Urea Biosensor

Figure 6(a) shows the urea biosensor response with different amounts of acrylic microspheres coated onto a film. The response of the biosensor towards different amounts of microspheres indicated that the use of too few or too many microspheres would lead to poor responses. When too few microspheres were used, the intensity of the colour change will be low and thus the optode is insensitive, but when too many microspheres were loaded onto the film, a tight packing of these microspheres hindered the diffusion of the urea and reaction products. These factors eventually resulted in poor biosensor response. The response of the urea biosensor is best at 6.43 mg of acrylic microspheres for each coated film.

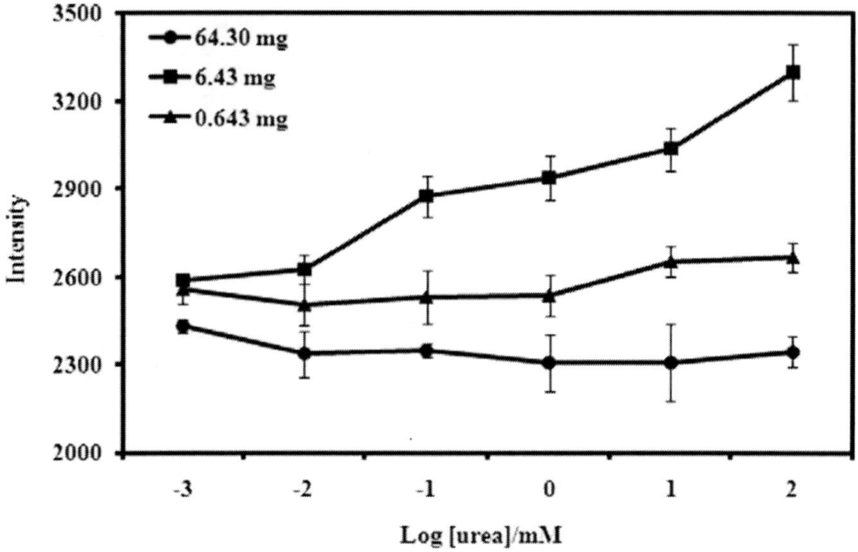

(a)

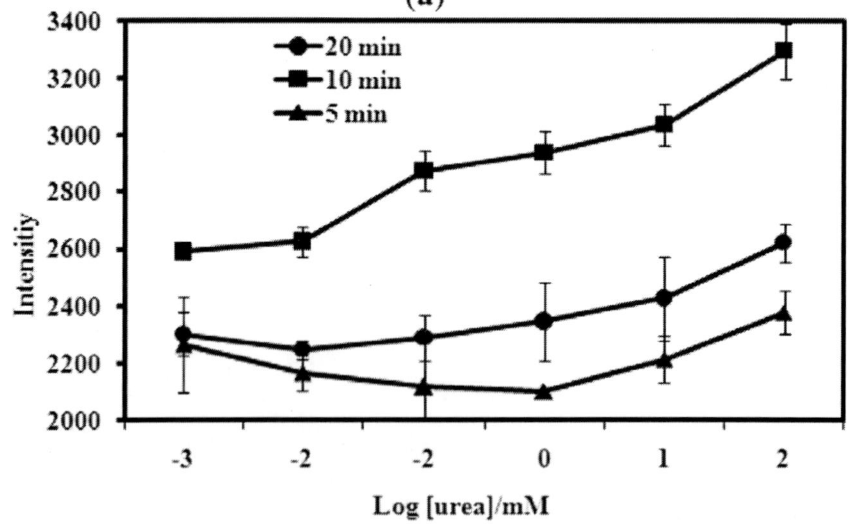

(b)

Figure 6. (**a**) The effects of the amount of acrylic microspheres coated onto a film on the urea biosensor response (measured at fixed time analysis 10 min); and (**b**) The effects of exposure time to urea on the response of the urea biosensor (fixed amount of coated microspheres).

In terms of exposure time, the intensity of the colour change when the biosensor was exposed to urea was recorded after reaction times of 5, 10 and 20 min [Figure 6(b)]. When the intensity was taken at a short time of 5 min, little colour change was observed, indicating that the biosensor had not undergone sufficient reaction to produce products. On the other hand, when the exposure time was too long, the colour change was too dark and this resulted in small differences in reflectance changes. Therefore, exposure time of the biosensor to urea that would yield large changes in reflectance intensity was chosen at 10 min.

3.5. Effects of Ionic Interference on the Optical Urea Biosensor

Acrylic polymer itself is not responsive to ions [26] whilst the chromoionophore ETH 5294 is sensitive only to pH [6,20,31]. Thus, above 0.01 mM urea, there was little response of the urea biosensor to the various ions used in the interference studies when compared to the response of urea (Figure 7). It indicates that these ions do not interfere with the urea response and the biosensor was selective to urea when used in the determination of urea in the urine samples.

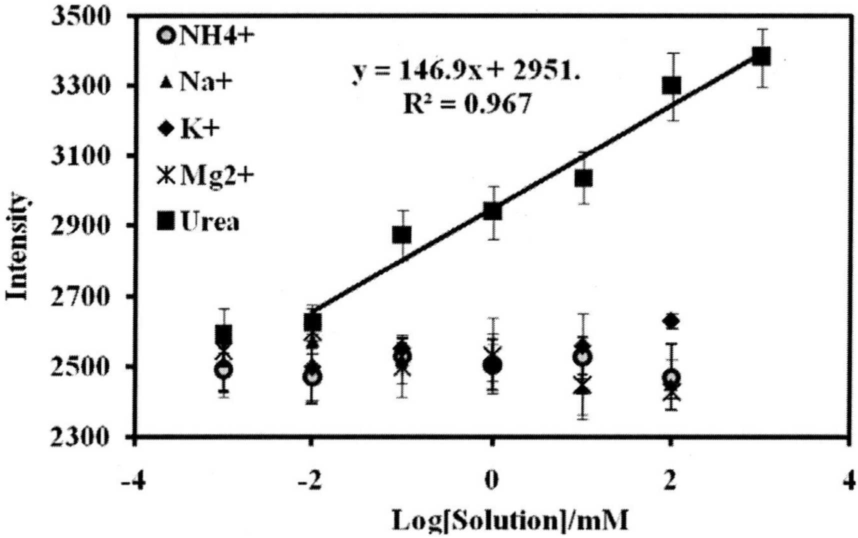

Figure 7. The effect of various ions on the response of the optical urea biosensor based on acrylic microspheres and reflectance transduction.

3.6. Effect of Urease Immobilization Time Biosensor Response

Figure 8 shows the amount of urease immobilized on the urea biosensor with the time employed during the process of urease immobilisation. The length of time for the immobilzation process affected the amount of enzyme immobilized onto the urea biosensor. Increasing in immobilization time obviously increased the amount of enzyme immobilized because binding of more enzyme via the succinimide ester could occur given a longer period of time. Thus, urease immobilisation on microspheres was performed over a 24 h period.

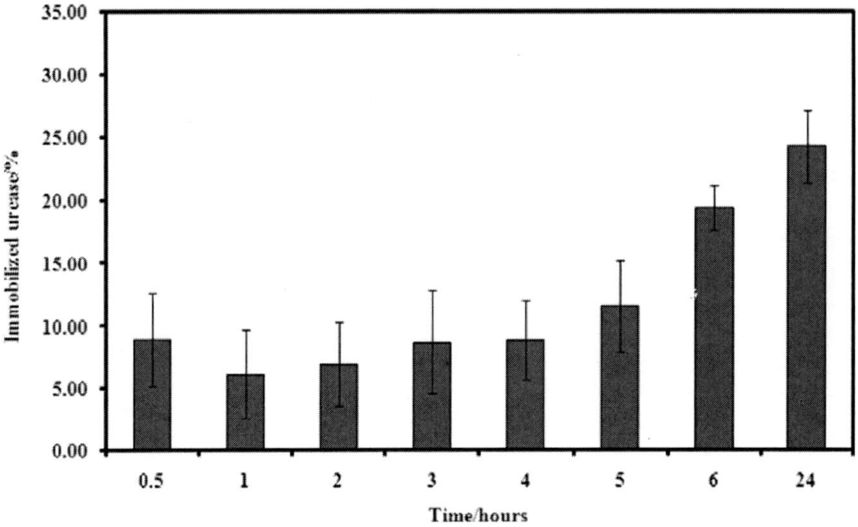

Figure 8. The amount of urease immobilized onto the urea biosensor at various immobilization times.

3.7. Validation and Recovery of Optical Urea Biosensor

Figure 9 correlates the results using the urea biosensor and the DMAB methods for urea determined in the urine sample. A good correlation coefficient (R^2 = 0.997) and slope (1.18) for responses of the urea biosensor and DMAB methods to urea concentration was observed, indicating that the urea biosensor developed in this work can be used for the measurement of urea concentrations in urine samples.

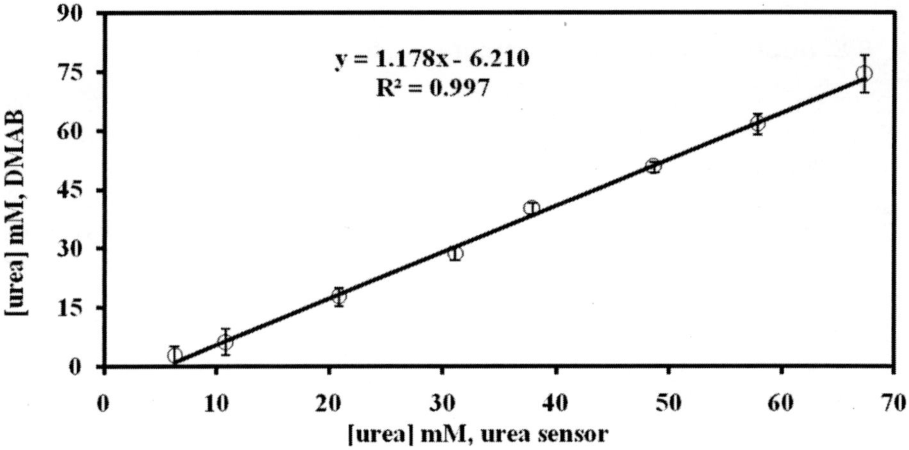

Figure 9. Urea concentrations in urine samples as determined by biosensor and DMAB.

Recovery by the urea biosensor is a very important indication of the accuracy of its detection of urea concentration in real samples of urine. Normally, the kidney excretes urea into the urine at 6 to 90 mg/dL (10.0 to 15.0 mM) [6]. Since the variation in level of urea in urine reflects the health of the kidney, therefore it is very important that the biosensor be able to determine the urea in urine very accurately. Table 2 shows the recovery of the optical urea biosensor containing different spiked urea concentrations in urine. The average recovery percentage of the urea biosensor across the concentrations of urea was 94.91% to 118.73% indicating no clear interference during determination of urea in urine samples by the optical biosensor.

Table 2. Recovery performance of urea biosensor for urine samples spiked with urea at different concentrations (n = 3).

Concentration of spiked urea (mM)	Concentration of urea as determined by urea biosensor (mM)	% of recovery by urea biosensor
0	14.87	-
1.01	1.20	118.73
19.99	20.90	104.58
39.97	37.94	94.91
49.96	48.72	97.51
69.94	67.42	96.40

3.8. Reproducibility and Stability of Urea Biosensor

The reproducibility of the urea biosensor was satisfactory, with a small RSD of 1.43%. In terms of stability, the reflectance intensity of the biosensor was stable and remained at about 100% for the first 15 days and thereafter it was still remained at 87.01% and 82.37% of the initial intensity on day-1 for 1.0 mM and 10.0 mM urea, respectively, until the 45th day (Figure 10). The sensitivity of an optical urea biosensor based on sol-gel entrapment of urease enzyme was observed to decrease to 45% after 28 days storage [20]. The improvement in the long term stability of the optical urea biosensor based on acrylic microspheres was attributed to the covalent immobilization of the enzyme on the microsphere surface.

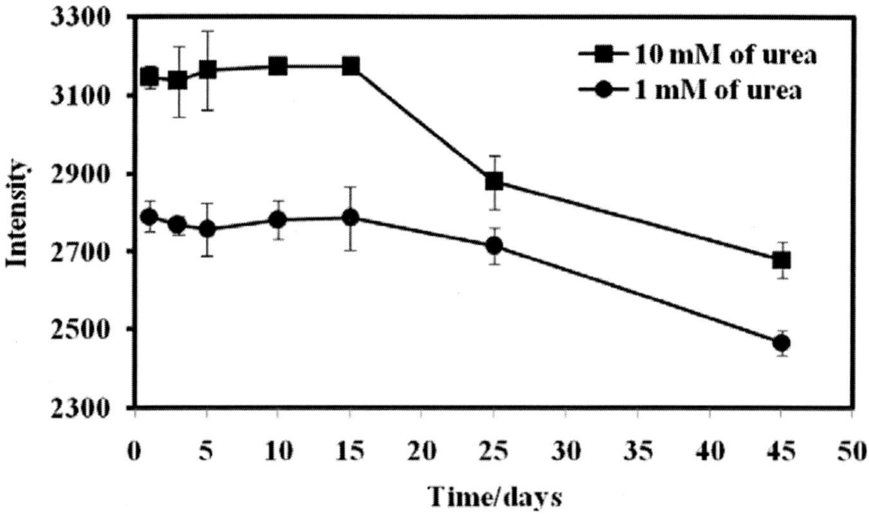

Figure 10. The stability of the urea biosensor response over a period of 45 days (in 0.05 M phosphate buffer solution at pH 7.0 containing 1.0 and 10.0 mM urea).

3.9. Comparison with Other Urea Biosensors

Table 3 presents a detailed comparison of optical urea biosensors reported in the literature with the biosensor reported in this work. By comparison, the urea biosensor presented in this work using acrylic microspheres as the immobilization matrix demonstrated a larger linear response range when compared to other types of optical based urea biosensors using various enzyme immobilization matrices. The acrylic microspheres based urea biosensor using reflectance transduction also compared favorably with other reported optical urea biosensors in terms of detection limit and response time.

Table 3. A comparison of optical urea biosensor with previously reported urea biosensors.

Urease immobilisation	Linear range (mM)	Limit of detection (mM)	Response time(min)	Reference
Acrylic microspheres	0.01–1,000	0.01	10	This work
Sol-gel films	5–100	0.01	10	[6]
Poly(vinyl chloride) film	0.1–1	–	0.33	[20]
Sol-gel encapsulation	0.0025–0.05	0.0025	10	[32]
Alginate microspheres	0.1–60	0.1	8	[33]
Poly(vinyl chloride) membrane	0.3–100	0.3	3	[34]
Polypyrrole-polyvinyl sulphonate film	1–100	–	–	[35]

4. CONCLUSIONS

An optical urea biosensor was successfully developed using acrylic microspheres as a matrix for urease immobilization. Urease enzyme can be immobilized onto the acrylic microspheres because of the introduction of NAS for enzyme binding onto the microsphere surface. In general the use of these microspheres coupled with reflectance transduction yielded a urea biosensor with an extended linear response range when compared with other reported optical urea biosensors, whilst maintaining similar detection limit, response time, selectivity and reproducibility. Besides that the optical urea biosensor is demonstrated to be suitable for measuring the concentration of urea in urine with good recovery when compared to the standard spectrophotometric method DMAB for the analysis of urine.

ACKNOWLEDGMENTS

We would like to thank Universiti Kebangsaan Malaysia for its support of this research through funding under the University Operational Research Grant UKM-OUP-NBT-29-151/2011.

REFERENCES

1. Liu, B; Hu, R; Deng, J. Studies on a potentiometric urea biosensor based on an ammonia electrode and urease, immobilized on a γ-aluminium oxide matrix. *Anal. Chim. Acta* **1997**,*341*, 161–169.

2. Trivedi, UB; Lakshminarayana, D; Kothari, IL; Patel, NG; Kapsed, HN; Makhija, KK; Patel, PB; Panchale, CJ. Potentiometric biosensor for urea determination in milk. *Sens. Actuat. B***2009**, *140*, 260–266.

3. Lee, WY; Kim, SR; Kim, TH; Lee, KS; Shin, MC; Park, JK. Sol-gel-derived thick-film conductometric biosensor for urea determination in serum. *Anal. Chim. Acta* **2000**, *404*, 195–203.

4. Luo, YC; Do, JS. Urea biosensor based on PANi (urease)®/Au composite electrode. *Biosens. Bioelectron* **2004**, *20*, 15–23.

5. Singhal, R; Gambhir, A; Pandey, MK; Annapoorni, S; Malhotra, BD. Immobilization of urease on poly(N-vinyl carbazole)/stearic acid Langmuir-Blodgett films for application to urea biosensor. *Biosens. Bioelectron* **2002**, *17*, 697–703.

6. Muawia, SA; Lee, YH; Musa, A. A urea biosensor from stacked sol-gel films with immobilized Nile blue chromoionophore and urease enzyme. *Sensor* **2007**, *7*, 2251–2262.

7. Senillou, A; Renault, NJ; Martelet, C; Consnier, S. A miniaturized urea sensor based on the integration of both ammonium based urease enzyme field effect transistor and a reference field effect transistor for in a single chip. *Talanta* **1999**, *50*, 219–226.

8. Yashuda, K; Miyagi, H; Hamada, Y; Takata, Y. Determination of urea in whole blood using a urea electrode with immobilized urease membrane. *Analyst* **1984**, *109*, 61–64.

9. Guilbault, GG; Shu, FR. Enzyme electrode based on the use of carbon dioxide sensor. *Anal. Chem* **1972**, *44*, 2161–2165.

10. Ruzicka, J; Hansen, EH; Ghose, K. Enzymatic determination of urea in serum based on pH measurement with flow injection method. *Anal. Chem* **1979**, *51*, 199–206.

11. Guilbault, GG; Trap, MA. Specific enzyme electrode for urea. *Anal. Chim. Acta* **1974**, *73*, 355–365.

12. Magalhães, JM; Machado, AA. Urea potentiometric biosensor based on urease immobilized on chitosan membranes. *Talanta* **1998**, *47*, 183–191.

13. Gouda, MD; Kumar, MA; Thakur, MS; Karanth, NG. Enhancement of operational stability of an enzyme biosensor for glucose and sucrose using protein based stabilizing agents. *Biosens. Bioelectron* **2002**, *17*, 503–507.

14. Lakard, B; Herlem, G; Lakard, S; Antoniou, A; Fahys, B. Urea potentiometric biosensor based on modified electrodes with urease

immobilized on polyethylenimine films. *Biosens. Bioelectron* **2004**, *19*, 1641–1647.

15. Singh, M; Verma, N; Kumar Garg, A; Redhu, N. Urea biosensors. *Sens. Actuat. B* **2008**, *134*, 345–351.

16. Xu, S; Han, X. A novel method to construct a third-generation biosensor: Self-assembling gold nanoparticles on thiol-functionlized poly(styrene-*co*-acrylic acid) nanospheres. *Biosens. Bioelectron* **2004**, *19*, 1117–1120.

17. Xu, S; Tu, G; Peng, B; Han, X. Self-assembling gold nanoparticles on thiol-functionalized poly(styrene-*co*-acrylic acid) nanospheres for fabrication of a mediatorless biosensor. *Anal. Chim. Acta* **2006**, *570*, 151–157.

18. Xu, S; Peng, B; Han, X. A third-generation H_2O_2 biosensor based on horseradish peroxidase-labeled Au nanoparticles self-assembled to hollow porous polymeric nanopheres. *Biosens. Bioelectron* **2007**, *22*, 1807–1810.

19. Wong, FCM; Ahmad, M; Heng, LY; Peng, LB. An optical biosensor for dichlovos using stacked sol-gel containing acetylcholinesterase and a lipophilic chromoionophore. *Talanta***2006**, *69*, 888–893.

20. Kovács, B; Nagy, GR; Dombi, R; Tóth, K. Optical biosensor for urea with improve response time. *Biosens. Bioelectron* **2003**, *18*, 111–118.

21. Wofbies, OS; Li, H. Fluorescence optical urea biosensor with an ammonium optrode as transducer. *Biosens. Bioelectron* **1993**, *8*, 161–166.

22. Chen, JP; Chiu, SH. A poly(N-isopropylacrylamide-co-N-acryloxysuccinimede-co-2-hydroxyethyl methacrylate) composite hydrogel membrane for urease immobilization to enhance urea hydrolysis rate by temperature swing. *Enzym. Micro. Technol* **2000**, *26*, 359–367.

23. Chaix, C; Pacard, E; Elaissari, A; Hilaire, JF; Picot, C. Surface functionalization of oil-in-water nanoemulsion with a reactive copolymer: Colloidal characterization and peptide immobilization. *Colloids Surf. B* **2002**, *29*, 39–52.

24. Zor, T; Selinger, Z. Linearization of the bradford protein assay increases its sensitivity: Theoretical and experimental studies. *Anal. Biochem* **1996**, *36*, 302–308.

25. Williams, S. *Official Methods of Analysis of the Association of Official Analytical Chemists*; Association of Official Analytical Chemists: Colombia, MD, USA, 1984.

26. Lee, YH; Hall, EAH. Methacrylic-acrylic polymers in ion-selective membranes: Achieving theright polymer recipe. *Anal. Chim. Acta* **2000**, *403*, 77–89.

27. Landfester, K; Schork, FJ; Kusuma, VA. Particle size distribution in mini-emulsion polymerization. *Compt. Rendus Chim* **2003**, *6*, 1337–1342.

28. Liu, XQ; Guan, YP; Liu, HZ; Ma, ZY; Yang, Y; Wu, XB. Preparation and characterization of magnetic polymer nanospheres with high protein binding capacity. *J. Magn. Magn. Mater***2005**, *293*, 111–118.

29. Usmani, A; Akmal, N. *Diagnostic Biosensor Polymer*; ACS Symposium Series 556; American Chemical Society: Washington, DC, USA, 1994.

30. Krajewska, B; Zaborska, W. The effect of phosphate buffer in the range of pH 5.80–8.07 on jack bean urease activity. *J. Mol. Catal. B: Enzym* **1999**, *6*, 75–81.

31. Francis, CMW; Musa, A; Lee, YH; Lim, BP. An optical biosensor for dichlovos using stacked sol-gel films containing acetylcholinesterase and a lipophilic chromoionophore. *Talanta***2006**, *69*, 888–893.

32. Tsai, HC; Dong, R. Simultaneous determination of pH, urea, acethylcholine and heavy metals using array-based enzymatic optical biosensor. *Biosens. Bioelectron* **2005**, *20*, 1796–1804.

33. Swati, M; Hase, NK; Srivastava, R. Nanoengineered optical urea biosensor for estimating hemodialysis parameters in spent dialysate. *Anal. Chim. Acta* **2010**, *676*, 68–74.

34. Koncki, R; Mohr, GJ; Wolfe's, OS. Enzyme biosensor for urea based on a novel pH bulk optode membrane. *Biosens. Bioelectron* **1995**, *10*, 653–659.

35. Kharat, HJ; Datta, K; Ghosh, P; Shirsat, MS. Development of an optical urea biosensor using polypyrrole-polyvinyl sulphonate film. *Sens. Trans. J* **2009**, *101*, 112–122.

Chapter 5

Graphene — A Platform for Sensor and Biosensor Applications

Nada F. Atta[1], Ahmed Galal[1] and Ekram H. El-Ads[1]

[1] Departement of Chemistry, Faculty of Science, Cairo University, 12613, Giza, Egypt

ABSTRACT

Graphene, mother of all carbon materials, has opened up new era of exploration due to its unique properties. Graphene, one-atom thick, exhibits a unique chemical structure and outstanding electronic, optical, thermal, and mechanical properties that made it compelling for various engineering applications. Graphene and graphene-based materials are promising candidates for fabricating state-of-the-art nano-scale sensors and biosensors. They featured with good conductivity and large specific surface area thereby; graphene-based sensors/biosensors performed well with good accuracy, rapidness, high sensitivity and selectivity, low detection limits, and long-term stability. They are ideally used as gas sensors, electrochemical sensors for heavy metal ions, immunosensors and dihydronicotinamide dinucleotide NADH, DNA, catecholamine neurotransmitters, paracetamol, glucose, H2O2, hemoglobin, and myoglobin biosensors. This chapter reviews the applications of graphene in nanotechnology since it came to the field particularly in sensing and biosensing applications. It updates the reader with the scientific progress of the current use of graphene as sensors and biosensors. There is still much room for the scientific research and application development of graphene-based theory, materials, and devices. Despite the vast amount of research already conducted on graphene for various applications, the field is still growing and many questions remain to be answered.

Keywords: Graphene, Graphene-based materials, Nano-composites, Sensors, Biosensors

1. INTRODUCTION

Carbon is an abundant element playing a major role in the chemistry of life [1]. Some of the advantageous properties of carbon-based electrodes include wide potential windows, fairly inert electrochemistry, and good electrocatalytic activity for many redox reactions [2]. Various carbon allotropes including diamond (the carbon atoms are bonded together in a tetrahedral lattice arrangement, 3D), graphite (the carbon atoms are bonded together in sheets of a hexagonal lattice, 3D), graphene (2D), carbon nanotubes (single-walled carbon nanotubes (SWCNT), multi-walled carbon nanotubes (MWCNT), 1D), and fullerenes (the carbon atoms are bonded together in spherical, tubular, or ellipsoidal formations, 0D) have very different physical and electrochemical properties from each other (see Table 1 and Figure 1) [2-4]. The advents of sp^2 and/or sp^3 hybridized structures, such as fullerenes, carbon nanotubes, and conducting diamond, provide a route for surface modification and are very beneficial to electrochemical research, especially electrocatalysis [2]. The former 3D allotropes have been known and widely used for centuries whereas fullerenes and nanotubes have been only discovered and studied in the last two decades. With the exception of diamond, it is possible to think of fullerenes, nanotubes, and graphite as different structures built from the same hexagonal array of sp^2carbon atoms, namely, graphene [3].

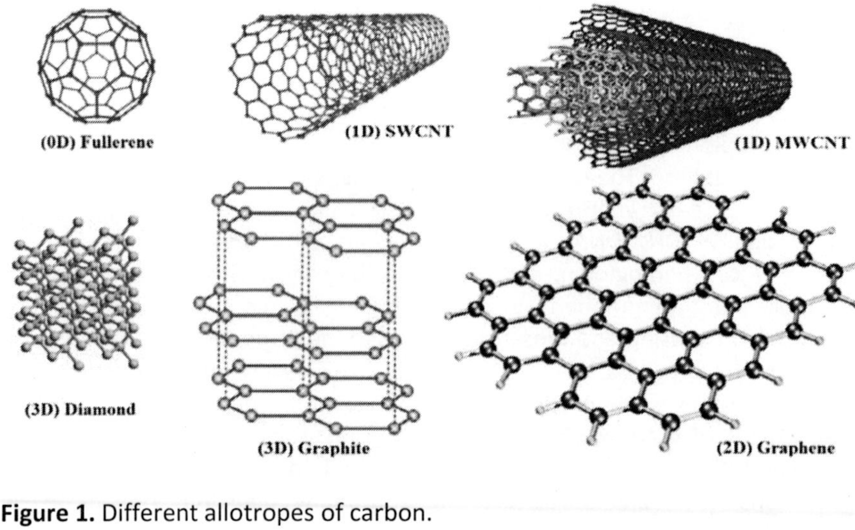

Figure 1. Different allotropes of carbon.

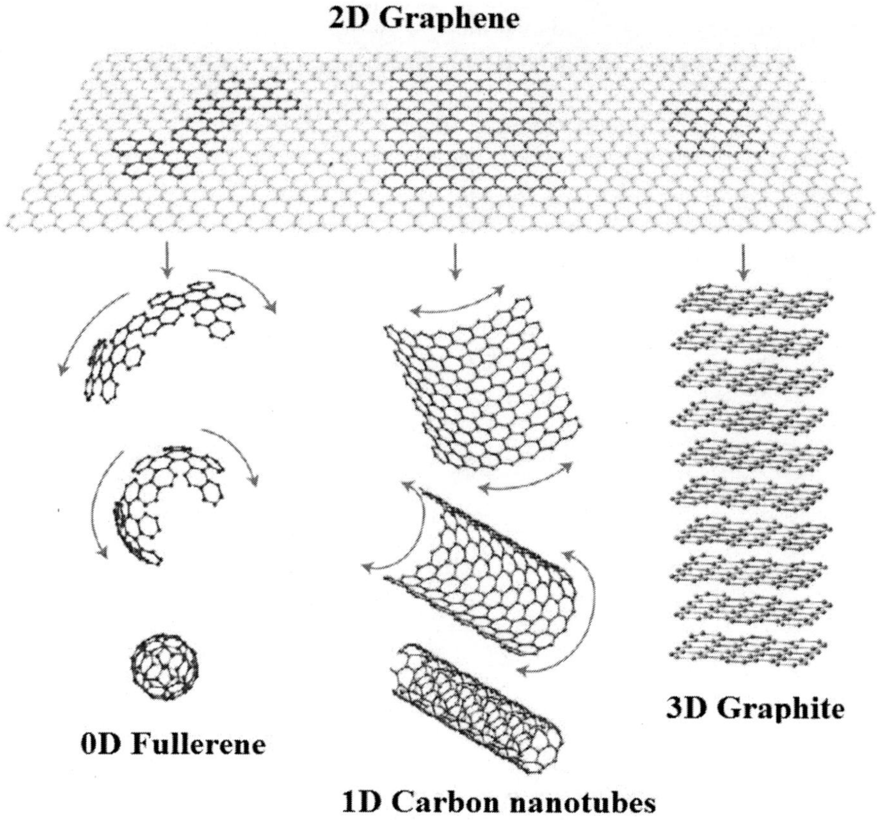

Figure 2. Basis of all graphitic forms. Graphene is a 2D building material for carbon materials of all other dimensionalities. It can be wrapped up into 0D buckyballs, rolled into 1D nanotubes or stacked into 3D graphite [6].

Graphene consists of a single sheet of carbon atoms that bind with each other in a honeycomb lattice. Graphene sheets are one-atom thick and 2D layers of sp^2-bonded carbon [4]. It is the basic building block for graphitic materials of all other dimensionalities. Graphene can be wrapped up into spherical shape to form 0D fullerenes (buckyballs), rolled into a cylindrical shape to form 1D nanotubes or stacked in layers to form 3D graphite as illustrated in Figure 2 [1, 3, 5-8]. The electronic structure of graphene sheets, individual layers of graphite, was first discussed by Wallace in 1947 [2]. Boehm and co-workers separated thin lamellae of carbon by heating and chemical reduction of graphite oxide in 1962 [6]. Until 2004, single layers of graphite were believed to be thermodynamically unstable under ambient conditions [2, 6]. Since the historical observation of single layer graphene in 2004 by two Nobel Laureates in physics, Prof. Andre Geim and Prof. Konstantin Novoselov, this atomically thin carbon film has received ever increasing attention and become the hottest topics and a rapidly "rising star" on the horizon of materials science and condensed-matter physics attracting enormous interests [8-10]. This revolutionary discovery has added a new dimension of research in the fields of physics, chemistry, biotechnology, and materials science [6]. Technologists and materials scientists have rapidly grabbed some of the assets of graphene and are already exploring the ways of incorporating graphene into applied devices and materials [3].

Table 1. Important parameters of carbon materials of different dimensionalities [6].

Dimension	0D	1D	2D	3D
Isomer	Fullerene	Nanotube	Graphene	Diamond
Hybridization	sp^2	sp^2	sp^2	sp^3
Density (g/cm^3)	1.72	1.2-2.0	2.26	3.515
Bond length (Å)	1.40 (C=C) 1.46 (C-C)	1.44 (C=C)	1.42 (C=C)	1.54 (C-C)
Electronic properties	Semiconductor E_g = 1.9 eV	Meta/semiconductor E_g =~ 0.3-1.1 eV	Zero-gap semiconductor	Insulator

2. GRAPHENE: PROMISING MATERIAL FOR DIFFERENT APPLICATIONS

Graphene, a monolayer of sp^2-bonded carbon atoms arranged in a honeycomb lattice, has attracted tremendous attention from both the theoretical and experimental scientific communities in recent years because of its unique nanostructure and extraordinary properties [11, 12]. It has become a novel and very promising material for nanoelectronics, nanocomposites, optoelectronic devices, electrochemical supercapacitor devices, fabricated field-effect transistors, drug delivery systems, solar cells, memory devices, and constructed ultrasensitive chemical sensors such as pH sensors, gas sensors, biosensors, etc. [6, 11, 12]. The rapid adoption of graphene as a material of interest lies in its actual availability by a range of techniques and methods and may also be principally because monolayer and few-layer graphene possess a diverse set of unusual properties. These properties happen to be matching the short-comings of other materials, such as carbon nanotubes, graphite, or indium tin oxide (ITO) that have been studied and used for some time [3]. In the following points, we will give an exhaustive account of the unique physical and chemical properties of graphene.

- Graphene implies large theoretical specific surface area (~ 2630 $m^2 g^{-1}$, for single-layer graphene), exhibiting high electrocatalytic activities and ultrahigh loading capacity for biomolecules and drugs [6, 8, 13].
- Graphene is a unique bendable material with excellent mechanical flexibility and high crystallographic quality. It is strictly 2D and stable under ambient conditions [6, 11, 14].
- Graphene exhibits high optical transparency [6, 11] with an absorption of ~ 2.3% toward visible light [15]. Indeed, in the visible range, thin graphene films have a transparency that decreases linearly with the film thickness. For 2-nm thick films, the transmittance is higher than 95% and remains above 70% for 10-nm thick films. This optical characteristic combined with the excellent conductivity of graphene-based materials holds promises as a replacement to the cost-raising standard ITO [3].
- The breaking strength of graphene is 200 times higher than steel making it the strongest material ever tested. Graphene exhibits high values of Young's modulus (~1.1 Tpa) [6, 11].
- It is one of the thinnest known materials. The carbon atoms are densely packed in a honeycomb crystal lattice with a bond length of 0.141 nm. Different research groups have measured the thickness of graphene from 0.35 nm to 1.00 nm. Novoselov et al. have determined platelet thicknesses of 1.00–1.60 nm [6].
- It is well established that the superior properties of graphene are associated with its single layer [6]. Figure 3 (a) shows the honeycomb lattice of graphene that is formed by a triangular lattice with a basis of 2 atoms per unit cell. Each atom has four valence electrons: one *s* and three *p*orbitals. The *s* and two *p* orbitals hybridize to form strong covalent bonds in the plane. The out of plane *p* orbital contributes to the conductivity [1, 3]. The hexagonal structure of graphene possesses an alternate double bond arrangement that makes it perfectly conjugated in sp^2hybridization. In this case, its *px* and *py* orbitals contain one electron each and the remaining *pz*has only one electron. This *pz* orbital overlaps with the *pz* orbital of a neighbor carbon atom to form a π-bond while the remaining orbitals form σ-bonds with other neighboring carbons. The π-electrons in graphene are delocalized and are largely responsible for its conduction properties while π-orbitals are the most important for determining the solid state properties of graphene [4]. Thus, graphene has extraordinary electronic transport properties and high electrocatalytic activities owing to its subtle electronic characteristics and attractive π–π interaction [12].
- It exhibits excellent room temperature electrical conductivity (~ 7200 $S.m^{-1}$) [6, 11] with amazing intrinsic mobilities of ~ 200,000 $cm^2V^{-1}s^{-1}$. The room

temperature thermal conductivity values have been determined to be ~ $(4.84 \pm 0.44)\times10^3$ to $(5.30 \pm 0.48)\times10^3$ WmK^{-1} for single-layer graphene [5, 6, 16].

- One of the factors that made graphene so attractive in the research field is its low energy dynamics of electrons. The 2D crystal of sp^2 hybridized carbon is a zero band gap semiconductor in which the π and π^* bands touch in a single point at the Fermi energy at the corner of the Brillouin zone and close to this so-called Dirac point, the bands display a linear dispersion (seeFigure 3(b)) [1]. This topology of the bands gives rise to exotic electronic transport properties – the charge carriers behave like relativistic particles – which manifest themselves in unusual phenomena such as room-temperature anomalous quantum Hall effect and electrons behaving like massless (zero effective mass) Dirac fermions [2, 10, 16]. The ballistic charge carrier transport at 300 K and at high charge carrier concentrations makes graphene also interesting for applications in electronic devices [14].
- Graphene exhibits strong adsorptive capability [12] since every atom in a graphene sheet is a surface atom thus molecular interaction and electron transport through graphene can be highly sensitive to adsorbed molecules [17]. The unique electrochemical responses of graphene to target molecules are from the planar geometric structure and special electronic character of graphene [10].
- Easy synthesis, low cost, and non-toxicity of graphene make this material a promising candidate for many technological applications [5, 16].
- Graphene is an ideal material for various applications because of its very large 2D electrical conductivity, large surface area, and low cost. In comparison with carbon nanotubes (CNTs), two advantages of graphene are obvious as follows:
    1. Graphene does not contain metallic impurities as CNTs do. In many cases, such impurities dominate the electrochemistry of CNTs and lead to misleading conclusions.
    2. The production of graphene can be achieved using graphite that is cheap and accessible [18].

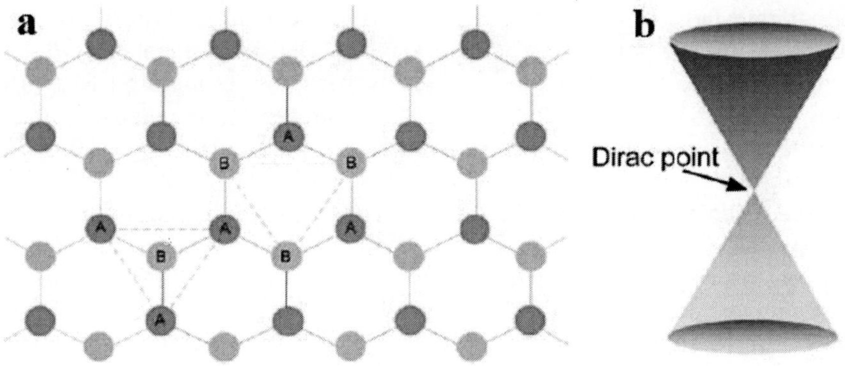

Figure 3. (a) Honeycomb lattice of graphene showing two triangular sublattices A and B. (b) Linear energy bands in the honeycomb lattice close to the Dirac point [1].

3. METHODS OF GRAPHENE PREPARATION

The rapid adoption of graphene as a material of interest lies in its actual availability by a range of techniques and methods [3, 5]. In 2004, Geim and coworkers first reported graphene sheets prepared by mechanical exfoliation (repeated peeling) of highly oriented pyrolytic graphite. This method, which is called the "scotch-tape" method, is still widely used in many laboratories to obtain pristine perfect structured graphene layer(s) for basic scientific research and for making proof-of-concept devices [18,19]. However, this method is not suitable for the large-scale production of graphene necessary to fulfill the requirements in different areas. A large amount of recent effort has been devoted to develop methods to synthesize graphene at large scale for practical electronic applications. These methods include mechanical exfoliation and cleavage of natural graphite, chemical vapor deposition on metals, electric arc discharge, epitaxial growth on electrically insulating surfaces, unzipping of carbon nanotubes, solution-based chemical approaches (chemical or thermal reduction of graphite oxide), and thermal decomposition of SiC wafer under ultrahigh vacuum conditions [2, 5, 6, 19].

3.1. Chemical Vapor Deposition

Chemical vapor deposition (CVD) is a simple, scalable, and cost-efficient method to prepare single and few-layer of graphene on various substrates. It opens a new route to large-area production of high-quality graphene films for practical applications [20-23]. In general, CVD techniques involve the decomposition of fluid (gas and liquid sprays) at high temperature to form either thin films on substrates or powders through filters. There are many forms of CVD like hot wire CVD, thermal CVD [24], plasma-enhanced CVD [25], radio-frequency CVD, and ultrasonic spray pyrolysis. Evaporated Ni film on SiO_2/Si wafers or copper foils are ideal substrates for graphene synthesis. Synthesis of graphene on Ni supported on Si/SiO_2 wafers facilitated the breakthrough approach for large-scale graphene. This is attributed to the excellent geometrical fit of the ordered graphene/graphite phase of carbon to the crystalline metal surface that is provided by Ni films. Another reason is the convenient interactions that favor bond formation between carbon atoms at specific conditions. Carbon atoms dissolve into the Ni crystalline surface and they arrange epitaxially on the Ni (111) surface to form graphene at certain temperatures (Figure 4). CVD synthesis of graphene can be carried out at ambient pressure or vacuum by systematically varying parameters such as gas composition and flow rate, temperature, and deposition time [4, 19]. The carbon precursor in CVD is methane [26-29]. By using diluted methane, single and few-layers of graphene (less than 5 layers) were grown, while using concentrated methane results in multilayer growth [4]. Other carbon precursors can be used in CVD such as ethanol [30], isopropanol [27], ethylene, acetylene, and others [4].

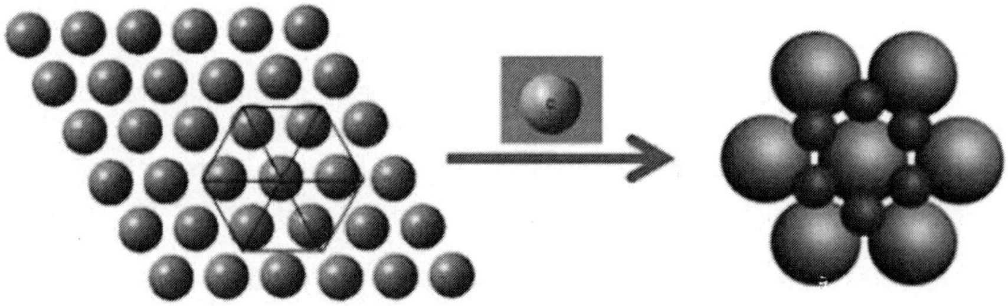

Figure 4. Schematic representation of the atomic arrangement of the hexagonal lattice of the (111) face of nickel (blue spheres) and how carbon atoms (gray spheres) would arrange on the Ni (111) surface to form graphene [4].

3.2. Unzipping of Cnts Into Monolayer Graphene

The method of unzipping single-walled carbon nanotubes to form monolayer graphene has been reported recently accomplishing by harsh acids and the right thermodynamic conditions (Figure 5) [31]. Recently, Dai and coworker have taken a unique approach to fabricate graphene nanoribbons (GNRs) with well-controlled dimensions. Based on the concept that CNTs are considered to be GNRs rolled up into seamless tubes and the fact that the size of CNTs has been well controlled, Dai's group developed a method to produce GNRs through controlled unzipping of CNTs using the Ar plasma etching method. Shimizu and coworkers produced GNRs through the oxidization and longitudinal unzipping of MWCNTs in concentrated sulfuric acid followed by treatment with $KMnO_4$ [32]. Another approach involved the preparation of longitudinal unzipped MWCNTs by controlled oxidation and intercalation through reaction with concentrated H_2SO_4 and HNO_3 at room temperature in order to control the reaction rate. The unzipped MWCNTs are thermally treated up to 1000ºC and 2500ºC in an inert atmosphere to exfoliate and reduce the edge defects. The sonication process is responsible for chemo-mechanical breaking and the separation of multi-layer nanoribbons [33]. Furthermore, graphene nanoribbons were synthesized by the oxidative unzipping of SWCNTs using a mixture of concentrated HNO_3/H_2SO_4 in a 1:3 volume ratio followed by 8 hours of ultrasonic treatment performed at 50 W at 45ºC [34]. In addition, catalytic unzipping of single-, double-, and multi-walled carbon nanotubes (SWCNTs, DWCNTs, and MWCNTs) in the presence of Pd nanoparticles and an oxygen-containing liquid medium yielding few layer graphene sheets was performed under microwave irradiation [35].

Figure 5. Modelling illustration of the SWCNT unzipped into monolayer graphene [31].

3.3. ARC Discharge

Arc discharge is a versatile and low-cost method for producing various kinds of carbon nanomaterials. Compared to other methods, the graphene sheets synthesized by arc discharge method have merits of good dispersibility in organic solvent, few defects, and high thermal stability [36]. In the arc discharge technique, graphite rods (purity: 99.99 %) are used as electrodes for high voltage arcing. At extremely high voltages between the electrodes that are separated by very small distances, very high electric fields can be produced leading to instantaneous sparks like in a welding process. The fall-out during the discharge process is the end product that contains the carbon nanostructures [31]. To prepare pure few-layered graphene, a direct current arc discharge of graphite evaporation was carried out in a water-cooled stainless steel chamber filled with different buffer gases after evacuating the chamber by a mechanical pump [31, 36]. When the arc discharge process ended, the generated soot was collected under ambient conditions. The soot deposited on the inner and top wall of the reaction chamber was only collected in order to exclude relatively heavy products (amorphous carbon nanoparticles or multi-layered graphene sheets) dropped to the bottom of the chamber during the arc discharge process [36]. Different discharge atmospheres may induce different chemical bonds that cause the prepared graphene sheets to have different properties. Direct current hydrogen arc discharge evaporation of pure graphite electrodes in various gas mixtures can be used to mass-produce high-crystalline few-layer graphene sheets [40]. Various gas mixtures were reported such as H_2, He, mixture of H_2 and He, mixture of NH_3 and He, and mixture of CO_2 and He to synthesize few-layered graphene sheets with high electrical conductivity [36-38]. On the other hand, a novel method for the large-scale production of graphene flakes in an arc discharge enhanced with a specially shaped magnetic field and a custom-designed catalyst was reported [39].

3.4. Reduction of Graphene Oxide Into Graphene

Graphene oxide (GO) was synthesized by the strong oxidation of flake graphite using acid via Hummers' method [40, 41] or modified Hummers' method [42-44]. GO, an oxidized form of graphene, is decorated by hydroxyl and epoxy functional groups on the hexagonal network of carbon atoms with carbonyl and carboxyl groups at the edges. GO is highly hydrophilic and forms stable aqueous colloids due to the large number of oxygen-containing functional groups as well as the repulsive electrostatic interactions at the edges of the platelets [41, 45]. It was reported that GO can be reduced via thermal, chemical, electrochemical, and laser-scribing methods.

3.4.1. Thermal reduction

The thermal method is believed to be a green method in which no hazardous reductants are used. This process requires heating up to 1050°C in an oven under argon gas, up to 800°C under hydrogen gas, or up to 700°C in a quartz tube. The thermal reduction of GO was accompanied by the elimination of epoxy and carboxyl groups in the forms of O_2, CO, CO_2, and H_2O as predicted from thermogravimetry data [46-48]. Recently, graphene sheets were prepared by a solvothermal reduction of GO suspension in water, octadecylamine, butanol, or N,N-dimethylacetamide [46]. Wei Huang synthesized high-performance graphene nanosheets by thermal reduction of GO under ethanol atmosphere at 900°C. This method can effectively remove the oxygen-containing functional groups and restore its graphic structure compared to the ones obtained using hydrazine or hydrogen. It can provide a simple, low-cost, and high yield approach for the preparation of high performance graphene nanosheets [49]. Furthermore, graphene sheets with different reduction levels were produced through thermal reduction of GO in the temperature range of 200–900°C under N_2 atmosphere for supercapacitor applications [50, 51]. Thermal reduction can produce few layer graphene with less agglomeration, higher specific surface area, and higher electrical conductivity compared to other reduction routes [51].

3.4.2. Chemical reduction

GO formation was followed by subsequent chemical reduction with a reducing agent such as hydroquinone, sodium borohydride, hydrazine, and hydrazine with NH_3 [42, 47, 52]. Hydrazine hydrate was proved to be the best reducing agent for the chemical reduction of GO into graphene and very thin graphene-like sheets were obtained. Moreover, the hydrazine treatment resulted in the formation of unsaturated and conjugated carbon atoms, which in turn imparts electrical conductivity of the graphene sheets [41]. These reducing agents, particularly hydrazine, are highly toxic, explosive, and their use should be with extreme care and minimized [42, 52]. The chemical reduction may also leave some residual epoxide groups on the reduced GO sheets leading to some loss in electron mobility [52]. On the basis of these observations, a new approach was employed to develop a "green" synthesis method for the production of graphene by using environmentally friendly reducing agents such as vitamin C [44], reducing, sugars and l-glutathione under mild conditions [42].

On the other hand, the chemical reduction methods of exfoliated GO with reducing agents such as hydrazine hydrate provide a promising approach for the efficient large-scale production of chemically converted graphene (CCG) sheets. However, in most cases heating to nearly 100ºC over several hours is required. A novel, facile, convenient, and scalable method for the synthesis of CCG sheets as well as metallic and bimetallic nanoparticles supported on the CCG sheets using a simple household microwave oven was utilized [53-56]. It was reported that graphene is a good microwave-absorbing material and can reach a high temperature in minutes [57]. Microwave irradiation (MWI) has been demonstrated for the synthesis of a variety of nanomaterials with controlled size and shape without the need for high temperature or high pressure. The main advantages of MWI over other conventional heating methods are (i) rapid reaction velocity, (ii) clean and energy efficient, and (iii) uniform heating of the reaction mixture [47, 58-60]. Due to the difference in the solvent and reactant dielectric constants, selective dielectric heating can provide significant enhancement in the transfer of energy directly to the reactants causing an instantaneous internal temperature rise. Dried GO was sonicated in deionized water until homogenous dispersion was obtained. After that it was placed inside a conventional microwave after adding 100 μl of a reducing agent [hydrazine hydrate, ethylenediamine, or ammonium hydroxide] in 30 s cycles for a total reaction time of 60 s. A black color was obtained indicating the complete chemical reduction to graphene. The graphene sheets were separated by using centrifuge operated at 5000 rpm for 15 min and dried overnight under vacuum [53]. Microwave-assisted methods allowed the rapid chemical reduction of GO using a variety of reducing agents in either aqueous or organic media. By using this method, many types of metallic and bimetallic nanoparticles can be dispersed on the graphene sheets via simultaneous reduction of GO and a variety of metal salts to create novel nanocatalysts supported on the large surface area of the thermally stable 2D graphene [61]. Moreover, this method provides a low-cost approach to the facile production of nanoparticles/graphene composites on a large scale for applications [57, 61, 62].

3.4.3. Electrochemical Reduction

As pointed out by Kaner and co-workers, both chemical and thermal reduction techniques have some obvious drawbacks, namely, the highly toxic reductant hydrazine, lack control of the film thickness, and the incompatibility of the thermal reduction process under some conditions. Thus, the electrochemical method was adopted as an effective and controllable alternative technique for the modification of electronic states [63]. It has drawn great attention due to its fast and green nature [17,52]. This is done by adjusting the external power source to change the Fermi energy level of the electrode surface that reduces GO in the presence of direct current bias [63]. Typically, the electrochemical synthesis of graphene was carried out via two steps, namely, GO being first assembled on the electrodes by solution deposition methods then being subjected to electrochemical reduction by scanning the potential (cyclic voltammetry) [45, 47, 52, 64, 65] or by applying constant potential (bulk electrolysis) [66, 67]. Another approach was achieved via a one-step technique in which graphene films can be prepared on electrodes directly from GO dispersions by simultaneous electrodeposition and electroreduction [17]. The reduction of GO can be confirmed from the color of GO electrodes that changes from yellow (before reduction) to black (after reduction) [64], involving the reduction of the high number of functional groups present in the GO sheets such as OH, COOH, and epoxides [52].

The properties of electrochemically converted graphene (ECG) are quite different from other nanostructured carbons (CNTs) and even different from CCG. The morphology of CCG and ECG on glassy carbon sheet was characterized by high field emission-scanning electron microscopy (FE-SEM) (Figure 6 (A, B)), respectively. The micrographs show that CCG sheets are rippled and crumpled with a dimension of several nm to few μm while ECG sheets are more homogeneous and more compact on the surface than CCG

[66]. ECG exhibits greatly enhanced activity for the electrocatalytic reduction of $O_2$and H_2O_2 and much higher electrochemical capacitance for potential application in ultracapacitors [64]. Moreover, the ECG coating is very stable as a result of its poor insolubility in common solvents [17].

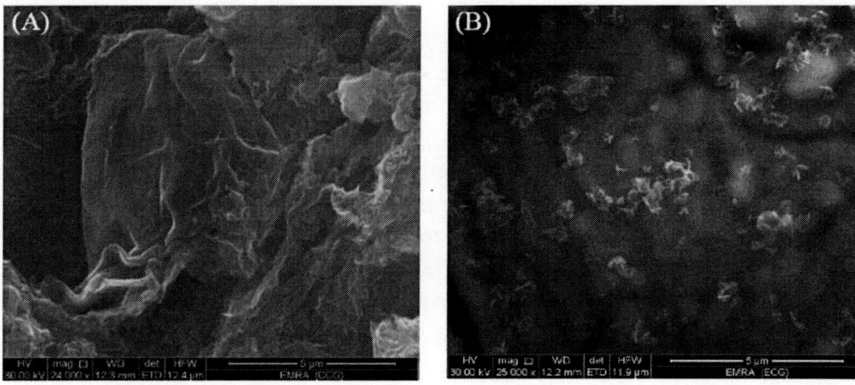

Figure 6. SEM images using FE-SEM for a) CCG, b) ECG [66].

3.4.4. Laser-Scribing Technology

Compared to chemical and thermal methods, pulsed laser irradiation reduction is simple, rapid (shorten the reaction time from several hours to few minutes), energy efficient, and free from poisonous material and high temperature. Also, the pulsed laser reduction of GO in solution makes it particularly attractive for producing graphene composites by pre-mixing with other materials [68]. Moreover, Feng-Shou Xiao fabricated graphene microcircuits by direct reduction and patterning of GO films using femtosecond laser representing well-conductivity for electrical applications [69]. Recently, Kaner and co-workers prepared graphene using standard laser-scribing technology more typically used to etch labels and images onto DVDs. Initially, a thin film of GO dispersed in water was drop-cast onto a flexible substrate (a DVD covered with a layer of plastic). Irradiation of the film with an infrared laser inside an inexpensive commercially available light scribe CD/DVD optical drive reduces the GO to laser-scribed graphene (LSG), as indicated by the change in film color from golden brown to black. Analysis of cross sections of the film with scanning electron microscopy showed that the initially stacked GO sheets were converted into well-exfoliated LSG sheets without any sticking together through laser irradiation [70].

4. FUNCTIONALIZATION OF GRAPHENE

GO contains oxygen functional groups such as epoxides, OH, and COOH that make it hydrophilic and well dispersed in water. However, GO is incompatible with most organic polymers [52]. It is well established that the superior properties of graphene are associated with its single layer. However, the fabrication of single-layer graphene is difficult at ambient temperature. Graphene sheets with a high specific surface area tend to form irreversible agglomerates or even restack to form graphite through π–π stacking and van der Waals interactions if the sheets are not well separated from each other. Aggregation can be reduced by the attachment of other small molecules or polymers to the graphene sheets. The presence of hydrophilic or hydrophobic groups prevents aggregation of graphene sheets by strong polar–polar interactions or by their bulky size. The attachment of functional groups to graphene also aids in dispersion in a hydrophilic or hydrophobic media as well as in the organic polymer [6, 52]. Furthermore, the oxidized rings of functionalized and defective graphene sheets contain abundant C–O–C (epoxide) and C–OH groups while the sheets are terminated with C–OH and –COOH groups. Defects of graphene may change its electronic and chemical properties. The functionalized and defective graphene sheets are more hydrophilic and can be easily dispersed in solvents with long-term stability [12]. Moreover, they are more easily produced in mass quantities as compared with the carbon nanotubes [12, 16, 52]. They may be used to prepare some novel graphene-based nanocomposite films that could facilitate the further manipulation and processing of these materials for developing novel electronic devices such as chemical sensors, biosensors [12], cellular imaging, and drug delivery [52]. On the other hand, various polymer and nanoparticle (metal, metal oxide, semiconductor) composites have been developed based on the unique properties of graphene. The bulk production of GO and graphene has given opportunities to explore this flat structure of carbon with polymer and nanoparticles in composites [32]. Pristine graphene is a hydrophobic material and has no appreciable solubility in most solvents. Nevertheless, the processing of graphene composites concerns itself foremost with the solubilization of graphene. To improve the solubility of graphene, different functional groups have been attached to the carbon backbone by chemical modification, covalent or non-covalent functionalization [15].

- Covalent modification: The structural alteration can take place at the end of the sheets and/or on the surface. Surface functionalization is associated with rehybridization of one or more sp^2carbon atoms of the carbon network into the sp^3 configuration accompanied by simultaneous loss of electronic conjugation. Covalent functional groups can attach to the surface through the reaction between the functional groups (-OH, -COOH, -CO) present on the GO and reduced graphene oxide (rGO) surfaces and edges [15, 32]. The covalent modification of graphene can be achieved in four different ways: nucleophilic substitution, electrophilic addition, condensation, and addition [6].

- Non-covalent modification: It is based on the van der Waals force or the π–π interaction between reduced graphene oxide (rGO) and stabilizers that not only gives less negative impact on the structure of graphene and its derivatives, but also provides the feasibility to tune their solubility and electronic properties [15, 32].

5. APPLICATIONS OF GRAPHENE TO SENSORS AND BIOSENSORS

Graphene has recently attracted tremendous interest because of its unique thermal, mechanical, and electrical properties. One of the promising applications of graphene is electrochemical sensing. Since every atom in a graphene sheet is a surface atom, molecular interaction and thus electron transport through graphene can be highly sensitive to adsorbed molecules [17, 71]. Graphene is a strictly 2D material and as such enables devices based on graphene to have an identical performance on a large scale. In addition, with the 2D structure, the monolayer graphene has its whole volume exposed to the environment that can maximize the sensing effect. The principle of graphene devices is based on changes in device conductance due to chemical or biological species adsorbed on the surface of graphene acting as electron donors or acceptors [19]. On the other hand, graphene-based nanomaterials have recently shown fascinating applications in electrochemical sensors and biosensors. Owing to the extraordinary electronic transport property and high electrocatalytic activity of graphene, the electrochemical reactions of analyte are greatly promoted on graphene film resulting in enhanced voltammetric response. Moreover, the electrochemical properties of graphene can be effectively modified by integration with other functional nanomaterials such as catalyst nanoparticles to produce versatile electrochemical sensing performance. Comparing with CNTs, graphene has shown the advantages of high conductivity, ease of production and function, good biocompatibility, and abundance of inexpensive source material [71, 72]. Thus, graphene was a good candidate of advanced electrode materials and could be combined with other functional materials to fabricate the sensing interface for electroanalysis [8, 10].

5.1. Gas Sensors

Graphene has potential use in gas sensors owing to its 2D structure with extremely high surface area. Gas sensing by graphene generally involves the adsorption and desorption of gaseous molecules (which act as electron donors or acceptors) on the graphene surface leading to change in the conductance of graphene. The high sensitivity of graphene toward different gaseous molecules has led to its use as gas sensor for hydrogen, H_2O, CO, NH_3, NO_2, Cl_2, ethanol, I_2, and O_2 [73]. Graphene/ZnO hybrid [74] and reduced graphene oxide (rGO) [75] were utilized as highly efficient gas sensors. Qingkai Yu et al. synthesized Pd-decorated graphene films to be used as hydrogen sensor. This gas sensor was fabricated on graphene film with a 1 nm Pd film deposited for hydrogen detection in the concentration range of 25–10,000 ppm. H_2 sensor based on Pd-decorated graphene films showed high sensitivity, fast response, and good recovery and can be used with multiple cycles [19]. Besides, Gupta et al. prepared a nanocomposite film based on poly(methyl methacrylate) (PMMA), rGO, and PMMA/rGO to be used as NH_3 gas sensor. The sensor performance in terms of selectivity and sensitivity was much better in the case of PMMA/rGO nanocomposite compared to the individual layers of rGO or PMMA [76]. On the other hand, an acetylene gas sensor was fabricated by synthesizing Ag-loaded ZnO-rGO hybrid via a facile chemical route (Figure 7). A well-structured crystalline nature and mixed phases of Ag, graphene (Gr), and ZnO was obtained. The morphological characterization revealed that particle-like nanostructures of ZnO and Ag were well distributed and closely distributed onto the surface of thin-layer RGO sheets. A low detection limit of 1 ppm, fast response of 25 s, recovery of 80 s, and good repeatability were obtained [77].

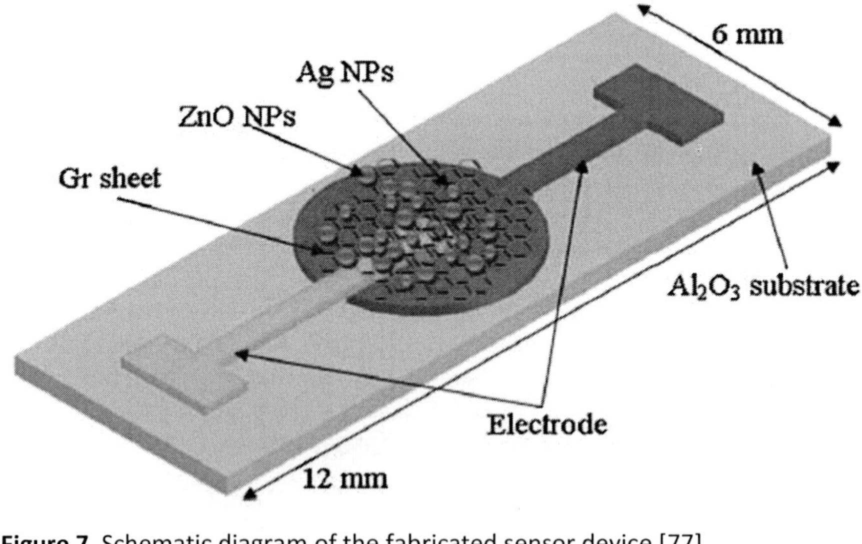

Figure 7. Schematic diagram of the fabricated sensor device [77].

5.2. Graphene-Based Electrochemical Sensors For Heavy Metal Ions

Metal ions such as lead, cadmium, silver, mercury, and arsenic have severe environmental and medical effects so that they require careful monitoring. Therefore, the development of a sensitive and selective detection method would benefit both the environmental and food chemists [5, 73]. A novel stannum film/poly(p-aminobenzene sulfonic acid, p-ABSA)/graphene composite modified glassy carbon electrode (GCE) was used for sensitive determination of trace Cd^{2+} ions by square wave anodic stripping voltammetry. Graphene layer presents enhanced electron transfer and enlarged specific surface area. The resulting matrix offers a good stripping performance for the analysis of Cd^{2+} in the linear range from 1.0 to 70.0 μg L^{-1} with a detection limit of 0.05 μg L^{-1} [78]. Furthermore, Erkang Wang used graphene nanosheets dispersed in Nafion (Nafion-G) solution in combination with in situ plated bismuth film electrode for fabricating the enhanced electrochemical sensing platform for $Pb^{2+}$and Cd^{2+} by differential pulse anodic stripping voltammetry. The prepared Nafion-G composite film exhibited improved sensitivity for the metal ion detections and alleviated the interferences as a result of the synergistic effect of Nafion-G. The stripping current signal is greatly enhanced and well distinguished on graphene electrodes with a wide linear range and 0.02 μg L^{-1} detection limit for both ions (Figure 8) [5, 79]. Moreover, various modified surfaces were constructed for the sensitive and simultaneous determination of Pb^{2+} and Cd^{2+} such as graphene/polyaniline/polystyrene nanoporous fiber [80] and graphene/ionic liquid/bismuth film composite modified screen-printed electrodes [81], functionalized graphene (GNS^{PF6}) with potassium hexafluorophosphate [82], and a hybrid nanocomposite of one-dimensional MWCNTs and graphene sheets [83]. Large surface area, good ionic and electronic conductivity, wide electrochemical window, as well as high stability, sensitivity, low detection limit, and applicability in real samples are the obvious characteristics of the proposed sensors. On the other hand, functionalized graphene sheets were fabricated and used for simultaneous removal of high concentration of inorganic species of arsenic (both trivalent and pentavalent) and sodium from aqueous solution using a supercapacitor-based water filter. Also, they were utilized for the desalination of sea water. High adsorption capacity for both inorganic species of arsenic and sodium along with the desalination ability of a graphene based supercapacitor provides a solution for a commercially feasible water filter. Additionally, the cost effective production of graphene sheets and better performance compared to other adsorbents like CNTs provides a platform for the development of commercially feasible supercapacitor-based water filter [84]. Furthermore, rGO–lead dioxide composite was presented as an excellent material for the detection of a low level of arsenic with a low detection limit of 10 nM. This study opens up the possibility of using the composite for simultaneous detection of arsenic and lead [85]. Furthermore, Zhuangjun Fan synthesized a type of graphene nanosheet/δ-MnO_2 (GNS/MnO_2) composite by a microwave-assisted method to be used as an adsorbent for the removal of nickel ions from waste water [86].

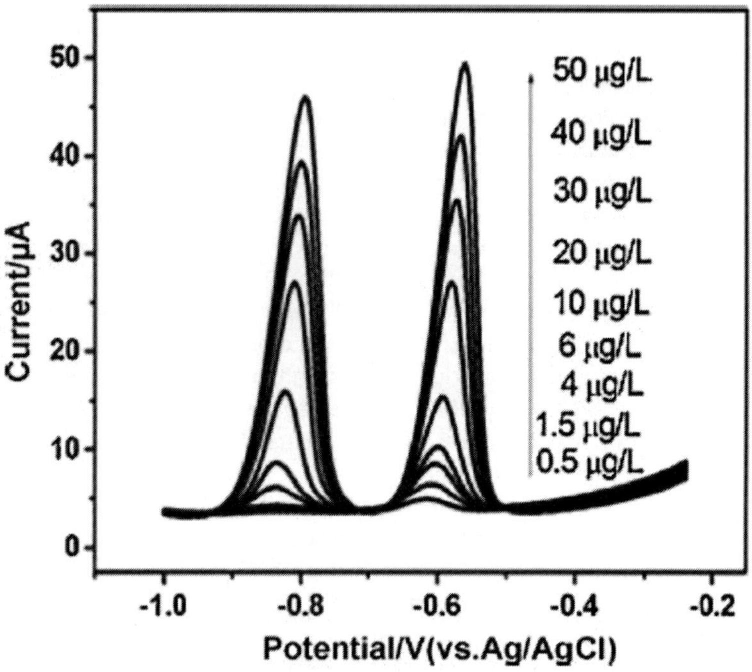

Figure 8. Striping voltammograms for the different concentrations of Cd^{2+} and Pb^{2+} on an in situ plated Nafion-G-BFE (bismuth film electrode) in solution containing 0.4 mg L^{-1} Bi^{3+} [79].

Table 2. Examples of modified electrodes for ACOP detection.

Electrode	Linear range	Detection limit	Reference
Nafion/TiO_2–graphene nanocomposite	1-100 µM	210 nM	[7]
Electrochemically reduced and deposited graphene onto GCE	5.0 nM-800 µM	2.13 nM	[91]
Electrochemically reduced graphene loaded nickel oxides (Ni_2O_3–NiO) nanoparticles coated onto GCE	0.04-100 µM	0.02 µM	[92]
Poly(4-aminobenzoic acid)/ electrochemically reduced graphene oxide composite film modified GCE	0.1–65 µM	0.01 µM	[93]

5.3. Biosensor for Some Biological Compounds

5.3.1. Paracetamol

Paracetamol or acetaminophen (ACOP) is a long-established and one of the most extensively employed "over the counter" drugs in the world. It is non-carcinogenic and an effective substitute for aspirin for patients with sensitivity to aspirin. It is used to reduce fever, cough, and cold and reduce mild to moderate pain. It is also useful in osteoarthritis therapy, protects hardening of arteries, relieves asthma patients, and protects against ovarian cancer. So it is very important to establish a simple, fast, sensitive, and accurate detection method for ACOP [87, 88]. Yuehe Lin et al. presented an electrochemical sensor for ACOP based on the electrocatalytic activity of functionalized graphene. Graphene-modified GCE obviously promotes the sensitivity of the determination of paracetamol with a low detection limit of 32 nM and a satisfied recovery from 96.4% to 103.3% [12]. Moreover, SWCNTs [89] or MWCNTs [90] graphene nanosheet nanocomposite modified GCE (SWCNTs-GNS/GCE or MWCNTs-GNS/GCE) was utilized as a novel, simple, and highly selective electrochemical sensor for tyrosine and paracetamol. The proposed sensor exhibited perfect characteristics like high effective surface area, high porosity, more reactive sites, excellent electrochemical catalytic activity, and applicability in human blood serum and pharmaceutical samples. Very low detection limits of 0.19 μM and 0.10 μM were obtained for tyrosine and paracetamol, respectively [90]. Table 2 contains examples of various modified electrodes for ACOP detection [7, 91-93].

5.3.2. Catecholamine neurotransmitters

Catecholamines are a class of neurotransmitters and their detection in the human body has been of great interest to neuroscientists. They include dopamine (DA), epinephrine (EP), and norepinephrine (NE) playing important roles in various biological, pharmacological, and physical processes. They are widely distributed in the mammalian central nervous system for message transfer [94, 95].

EP, a hormone secreted by the medulla of adrenal glands, is an important catecholamine inhibitory neurotransmitter. The presence of EP in the body affects the regulation of blood pressure and the heart rate, lipolysis, immune system, and glycogen metabolism. Low levels of EP have been found in patients with Parkinson's disease. Therefore, the quantitative determination of EP in different human fluids such as plasma and urine is important for developing nerve physiology, pharmacological research, and life science [96]. Xiaoli Zhang fabricated graphene/Au nanocomposites modified GCE (GR/Au/GCE) for the detection of EP with high sensitivity. The oxidation of EP at this modified electrode occurred at less positive potentials than on bare GCE. In addition, the oxidation peaks of EP and ascorbic acid (AA) were separated from each other by approximately 180 mV. At the same time, this electrode also showed favorable electrocatalytic activity toward some other small biomolecules (such as DA, b-nicotinamide adenine dinucleotide, and uric acid) suggesting the potential applications of GR/Au nanocomposites for constructing biosensors [97]. Moreover, GR modified Pd was prepared to be used as a voltammetric sensor for NE, an important neurotransmitter. GR as well as Pd nanoparticles presented perfect characteristics for the proposed sensor-like excellent electrocatalytic activity for the oxidation of NE in terms of remarkable enhancement in the peak current and lowering of peak potential. Also, low detection and quantification limits were obtained that were 67.44 nM and 224.8 nM, respectively. The application of the proposed sensor in pharmaceutical dosage forms and human urine samples in the presence of high concentration of uric acid was achieved with excellent recovery results [98].

On the other hand, DA, ascorbic acid (AA), and uric acid (UA) are compounds of great biomedical interests playing determining roles in human metabolism. DA is one of the crucial catecholamine neurotransmitters playing an important role in the function of central nervous, renal, hormonal, and cardiovascular systems. Thus, a loss of DA-containing neurons may lead to neurological disorders such as Parkinsonism and schizophrenia [94, 95, 99]. AA (vitamin C), an antioxidant, is a vital vitamin in human diet existing in both animal and plant kingdoms. It has been used for the prevention and treatment of the common cold, mental illness, infertility, and cancer and in some clinical manifestations of human immunodeficiency virus infections. UA is the primary end product of purine metabolism. The extreme abnormalities of UA levels in the body are symptoms of several diseases including gout, hyperuricemia, and Lesch–Nyhan. Therefore, simultaneous detection of DA, AA, and UA is a challenge of critical importance not only in the field of biomedical chemistry and neurochemistry but also for diagnostic and pathological research [94, 95]. Feng Li et al. prepared a graphene doped carbon paste electrode (CPE) by the addition of graphene into the carbon paste mixture for the successful determination of AA. This sensor showed an excellent electrocatalytic oxidation activity toward AA with a lower overvoltage, pronounced current response, and good sensitivity. The proposed sensor exhibited a rapid response to AA within 5 s and a good linear calibration plot from 0.1 to 106 μM with low detection limit of 70 nM [8]. On the other hand, Jinghong Li et al. synthesized graphene chemically by the Hummers and Offeman method and the graphene-modified electrode was applied for the selective determination of DA in the linear range of 5–200 μM in a large excess of AA. Selective detection was realized in completely eliminating AA based on the unique planar structure of graphene. The interesting performance of graphene to DA and AA may be correlated with the following reasons:

1. The high quality of the sp^2 conjugated bond in the carbon lattice, graphene is highly conductive and shows metallic conductance even in the limit of zero carrier density. The unusual density of electronic states of graphene affects its electrochemical reactivity.
2. Molecular structures of DA and AA are distinct from each other. The π–π interaction between the phenyl structure of DA and 2D planar hexagonal carbon structure of graphene makes the electron transfer feasible. AA oxidation is inactive most likely because of its weak π–π interaction with graphene [10].

Several sensors were utilized for the simultaneous determination of DA and AA such as graphene modified electrode [100], graphene sheets/Congo red-molecular imprinted polymers [101], and Pt nanoparticles/polyelectrolyte-functionalized ionic liquid/graphene sheets [102]. On the other hand, TiO_2–graphene nanocomposite modified GCE exhibited remarkable electron transfer kinetics, electrocatalytic activity, and selective determination of DA in the presence of UA and AA. Differential pulse voltammetry (DPV) peaks of DA, AA, and UA are well resolved with peak potentials at 108 mV, 272 mV, and 403 mV for AA, DA, and UA, respectively. The separation of DPV peak potentials for AA–DA and DA–UA are calculated to be 164 mV and 131 mV, respectively, which are large enough to determine DA selectively [103]. Several electrochemical sensors based on graphene-modified surfaces were employed for the simultaneous determination of DA, AA, and UA like polystyrene-grafted-graphene hybrid [104] and nafion covered core–shell structured Fe_3O_4@graphene nanospheres (GNs) [105] modified GCE, graphene-AuAg (Au/Gr-AuAg) composite modified gold electrode [106], Au nanoplates and rGO modified GCE (Au/rGO/GCE) [107], three-dimensional reduced graphene oxide (3D-rGO) material [108], water-soluble sulfonated graphene [109], ferulic acid functionalized electrochemically reduced graphene [110], graphene/nickel hydroxide composite [111], and well-defined flower-like graphene-nanosheet clusters (f-RGO) modified GCE [112]. In addition, Pd–Pt bimetallic nanoparticles anchored on functionalized rGO nanomaterials [113], electrochemically reduced graphene oxide (ERGO) [114], nitrogen-doped graphene (NG) [115], and tryptophan-functionalized graphene nanocomposite (Trp-GR) [116] were fabricated for the simultaneous detection of DA, AA, and UA and multi-walled carbon nanotubes (MWNTs) bridged mesocellular graphene foam (MGF) nanocomposite (MWNTs/MGF) modified GCE for the simultaneous determination of DA, AA, UA, and tryptophan [117]. As well, graphene (Gr)/copper(II) phthalocyanine-tetrasulfonic acid tetrasodium salt/polyaniline nanocomposites [118] and NiO nanoparticles-graphene composite film were employed as electrochemical sensors for the selective determination of AA in the presence of DA and UA [118] and in the presence of folic acid [119], respectively.

On the other hand, serotonin (5-hydroxytryptamine, 5-HT) is an important neurotransmitter and it is important to sensitively determine its concentration. A double-layered membrane of rGO/polyaniline nanocomposites and molecularly imprinted polymers embedded with Au nanoparticles was employed as an electrochemical sensor for 5-HT. The prepared sensor displayed obvious selectivity to 5-HT and a low detection limit of 11.7 nM in the linear range of 0.2–10.0 μM [120]. Moreover, graphene nanosheets [121] and graphene grafted poly(lactic acid) with Pd nanoparticles [122] were applied as 5-HT sensor showing high sensitivity, good selectivity, low detection limit, fast response time, and stability. In addition, electrochemically reduced GO-P (graphene oxide and 5,15-pentafluorophenyl-10,20-p-aminophenyl porphyrin) was employed as an electrochemical sensor for the simultaneous and selective determination of DA and 5-HT in the presence of AA. The detection limits of DA and 5-HT were 35 nM and 4.9 nM, respectively [123].

Levodopa (LD), an important neurotransmitter, was determined at modified CPE modified with graphene (GR), 1-(6,7-dihydroxy-2,4-dimethylbenzofuran-3-yl) ethanone (DE), and ionic liquid (IL) (GR–DE–IL/CPE). GR–DE–IL/CPE showed enhanced electrocatalytic activity toward LD with a lower oxidation potential and good electrochemical performance with higher conductivity and lower electron transfer resistance. Very low detection limit of 5.0 nM in the linear range of 0.015 to 1000 μM was obtained [124]. An assembly of rGO, Au nanoparticles, and 2-(3,4-dihydroxy phenyl) benzothiazole [125] and graphene nanosheets modified GCE [126] and reduced graphene nanoribbons modified carbon screen-printed electrode [127] were employed for the simultaneous determination of LD, UA and folic acid, LD and carbidopa and AA, LD and UA, respectively.

5.3.3. Hydroquinone and Catechol

Graphene has the great potential for distinguishing a diverse range of aromatic isomers. Chengbin Liu et al. had utilized graphene modified electrode (EG/GCE) for the simultaneous detection of hydroquinone HQ (1,4-dihydroxybenzene) and catechol CC (1,2-dihydroxybenzene). It showed enhanced electron transfer properties, good potential separation of oxidation peaks between HQ and CC (about 110 mV), wide linear concentration ranges, low detection limits, excellent reproducibility and stability, and high resolution capacity to the HQ and CC isomers [17]. On the other hand, a highly stable (pyridine-NG) was used as an electrochemical sensor for simultaneous determination of HQ and CC. Excellent electrocatalysis of pyridine-NG for simultaneous determination of HQ and CC was achieved due to the π–π interactions between the benzene ring of CC and graphene layer and the hydrogen bonds formed between hydroxyl in HQ molecule and pyridinic nitrogen atoms within graphene layers, especially the less density distribution of π electron cloud in pyridinic-NG in acidic condition [128]. Several novel sensors like poly(3,4-ethylenedioxythiophene)/nitrogen-doped graphene composite [129],

MWCNTs-poly(diallyldimethylammonium chloride)-graphene [130], and laser reduced graphene [131] modified GCE were fabricated for the simultaneous detection of CC and HQ with low detection limit, high sensitivity, excellent potential peak separation, and anti-interference ability. Table 3 contains examples of modified electrodes for HQ and CC detection [132-135].

Table 3. Examples of modified electrodes for HQ and CC detection.

Electrode	Linear range of HQ	Detection limit of HQ	Linear range of CC	Detection limit of CC	Reference
Graphene and TiO_2 nanocomposite	0.5–100 μM	0.082 μM	0.5–100 μM	0.087 μM	[132]
Electrochemically reduced graphene oxide- MWCNTs and terthiophene	---	35 nM	---	4.9 nM	[133]
1D MWCNTs and 2D graphene modified GCE using 1-butyl-3-methyl-imidazolium hexafluoro-phosphate ionic liquid	---	0.1 μM	---	0.06 μM	[134]
Au nanoparticle and graphene composite film modified carbon ionic liquid electrode	0.06-800 μM	0.018 μM	---	---	[135]

5.4. Drug Biosensor

Morphine, noscapine, and heroin are three major alkaloids in heroin samples. Morphine and noscapine that are directly extracted from poppy are the most abundant constituents in opium. The widespread use of illicit drugs has led to an increased effort toward developing and improving methods for their determination in biological samples, which is still a very challenging task. Graphene nanosheets (GNSs) modified GCE was employed for the simultaneous determination of morphine, noscapine, and heroin exhibiting excellent electrocatalytic activity at reduced overpotentials in wide pH range. Fast response time, high sensitivity and stability, low cost, low detection limit, and ease of preparation method without using any specific electron-transfer mediator or specific reagent are the advantages of the proposed sensor [136]. In addition, electrochemically reduced MWNTs-doped GO (ER-MWNTs-doped-GO) composite film modified GCE was employed as morphine sensor combining the excellent conductivity of MWNTs and ER-GO with the film forming properties of GO [137]. Also, rGO–Pd (rGO–Pd) hybrid had higher current response toward morphine oxidation compared to the unmodified rGO with a low detection limit of 12.95 nM [138].

On the other hand, graphene and Nafion film modified GCE was fabricated for the detection of codeine displaying an excellent analytical performance and enhanced applicability for codeine detection in urine samples and cough syrup. The high electrocatalytic activity of the proposed sensor toward codeine was attributed to the outstanding electric conductivity of graphene and the high codeine loading capacity on the electrode surface [139]. Furthermore, a nanocomposite of Gr and $CoFe_2O_4$nanoparticles modified CPE was proved to be an ultrasensitive electrochemical sensor for codeine and ACOP with low detection limits of 0.011 and 0.025 µM, respectively. The proposed method was free from interference effects of glucose, ascorbic acid, caffeine, naproxen, alanine, phenylalanine, glycine, and others [140]. On the other hand, rapid and simultaneous determination of tramadol and ACOP was achieved based on $NiFe_2O_4$/graphene nanoparticles modified CPE with a low detection limit of 0.0036 and 0.0030 µM, respectively [141]. Table 4 contained a summary of some modified electrodes for drug sensing.

Table 4. Examples of modified electrodes for drug detection.

Electrode	Drug	Linear range	Detection limit	Reference
Graphene nanosheets modified GCE	morphine, noscapine and heroin	up to 65, 40 and 100 µM, respectively	0.4, 0.2 and 0.5 µM, respectively	[136]
electrochemically reduced MWNTs-doped GO modified GCE	morphine	---	---	[137]
rGO-Pd modified GCE	morphine	0.34-12 µM	12.95 nM	[138]
graphene and Nafion film modified GCE	codeine	5×10^{-8} to 3×10^{-5} M	15 nM	[139]
a nanocomposite of Gr and $CoFe_2O_4$ nanoparticles modified CPE	codeine and ACOP	0.03 to 12.0 µM for both	0.011 µM and 0.025 µM, respecively	[140]
$NiFe_2O_4$/graphene nanoparticles modified CPE	tramadol and ACOP	0.01–9 µM	0.0036 and 0.0030 µM, respectively	[141]

5.5. Nadh Biosensor

B-nicotinamide adenine dinucleotide (NAD) and its reduced form (NADH) are a cofactor of many dehydrogenases, which have received considerable interest in developing amperometric biosensors, biofuel cells, and bioelectronic devices [2, 5, 18, 73]. The oxidation of NADH serves as the anodic signal and regenerates the NAD cofactor, which is of great significance in biosensing important substrates such as lactate, alcohol, or glucose. Problems inherent to such anodic detection are the large overvoltage for NADH oxidation and surface fouling associated with the accumulation of reaction products. Graphene shows promise in addressing these problems [73, 142]. Jinghong Li studied the oxidation of NADH at GCE and reduced graphene sheet films (rGSF)/GC. The oxidation of NADH occurs at bare GC electrode at 0.75 V vs. Ag/AgCl, which decreased to 0.42 V at rGSF/GC [2]. Table 5contained examples of several sensors for the sensitive and selective determination of NADH with low detection limit.

Table 5. Examples of modified electrodes for NADH detection.

Electrode	Linear range	Detection limit	Reference
methylene green functionalized graphene modified GCE	---	---	[5]
ionic liquid functionalized graphene modified GCE	---	---	[73]
chemically reduced graphene oxides modified GCE	---	---	[18]
graphene assembled electrode modified GCE	---	---	[142]
electroreduced graphene oxide and polythionine modified GCE	0.01–3.9 mM	0.1 μM	[144]
nitrogen-doped graphene modified GCE	---	---	[145]
graphene–Au nanorods hybrid nanosheets modified GCE	5 to 377 μM	1.5 μM	[146]
Au nanoparticles /reduced graphene oxide nanocomposites modified GCE	50 nM to 500 μM	1.13 nM	[147]
graphene paste electrode modified GCE	---	---	[148]
GCE modified with deoxyribonucleic acid, graphene methylene blue	10 μM to 1.50 mM	1.0 μM	[143]

5.6. DNA Biosensors

Electrochemical DNA sensors offer high sensitivity, good selectivity, and low cost for the detection of selected DNA sequences or mutated genes associated with human disease and promise to provide a simple, accurate, and inexpensive platform of patient diagnosis. Electrochemical DNA sensors also allow device miniaturization for samples with a very small volume. Among all kinds of electrochemical DNA sensors, the one based on the direct oxidation of DNA is the simplest [5, 13]. Zhou et al. reported an electrochemical DNA sensor based on CR-GO. The current signals of the four free bases of DNA (i.e., guanine (G), adenine (A), thymine (T) and cytosine (C)) on the CR-GO/GC electrode are all separated efficiently indicating that CR-GO/GC can simultaneously detect four free bases but neither graphite nor GCE can. This is attributed to the antifouling properties and the high electron transfer kinetics for bases oxidation on CR-GO/GC electrode resulting from a high density of edge-plane-like defective sites and oxygen containing functional groups on CR-GO. These functional groups are beneficial for accelerating electron transfer between the electrode and species in solution providing many active sites. CR-GO/GC is also able to efficiently separate all four DNA bases in both single-stranded DNA (ssDNA) and double-stranded DNA (ds-DNA), which are more difficult to oxidize than free bases at physiological pH without the need of a prehydrolysis step. This is attributed to the unique physicochemical properties of CR-GO (the single sheet nature, high conductivity, large surface area, antifouling properties, high electron transfer kinetics, etc.) [5, 73]. Furthermore, a zinc sulfur-coated poly (3,4-ethylenedioxythiophene)-reduced graphene oxide hybrid film was developed for the simultaneous determination of three deoxyribonucleic acid (DNA) bases: G, A, and T. This electrochemical sensor displayed a good photoelectronic effect and the peak currents of G, A, and T increased obviously. The proposed sensor was successfully applied to the analysis of G, A, and T contents in real-life samples such as herring sperm DNA samples, milk powder, and urine sample from human beings with satisfactory results [149]. As well, an electrochemical DNA biosensor for the detection of ssDNA sequence related to transgenic maize MON810 was fabricated based on electrochemical reduced graphene (ERG) modified carbon ionic liquid electrode (CILE) and methylene blue (MB). The presence of ERG increased the adsorption amounts of probe ssDNA sequence on the electrode resulting in the enhancement of the reduction peak current of MB that used as the hybridization indicator. Under optimal conditions, the linear range of ssDNA sequences was 1.0×10^{-11} to 1.0×10^{-6} M with the detection limit of 4.52×10^{-12} mol/L [150]. In addition, the simultaneous determination of adenine and guanine was achieved at the ionic liquid-functionalized graphene-modified carbon paste electrode (IL-graphene/CPE) with very low detection limits of 65 nM for guanine and 32 nM for adenine [13]. Moreover, TiO_2-graphene nanocomposite modified GCE exhibited high electrocatalytic activity, wide linear range, and low detection limit for the simultaneous determination of adenine and guanine. The excellent performance of this electrochemical sensor was attributed to the high adsorptivity and conductivity of TiO_2-graphene nanocomposite providing an efficient microenvironment for the electrochemical reaction of these purine

Table 6. Examples of modified electrodes for DNA detection.

Electrode	DNA sequences and mutated genes	Linear range	Detection limit	Reference
zinc sulfur-coated poly (3,4-ethylenedioxythiophene)-reduced graphene oxide hybrids film	guanine, adenine and thymine	---	---	[149]
Chemically reduced graphene oxide modified GCE	guanine, adenine, thymine and cytosine	---	---	[5, 73]
electrochemical reduced graphene modified carbon ionic liquid electrode and methylene blue	ssDNA sequence related to transgenic maize MON810	1.0×10^{-11} to 1.0×10^{-6} M	4.52×10^{-12} mol/L	[150]
ionic liquid-functionalized graphene modified carbon paste electrode	adenine and guanine	---	65 nM and 32 nM for guanine and adenine, respectively	[13]
TiO_2-graphene nanocomposite modified GCE	adenine and guanine	0.5–200 μM	0.10 and 0.15 μM for adenine and guanine, respevtively	[151]
graphene-based electrochemical DNA sensor	breast cancer-related BRCA1 gene	---	---	[152]
graphene sheets/polyaniline/gold nanoparticles based DNA sensor	BCR/ABL fusion gene in chronic myelogenous leukemia	---	2.11 pM	[153]
homoadenine single-stranded DNA/molybdenum disulfide–graphene nanocomposite modified gold electrode	riboflavin	0.025 to 2.25 μM	20 nM	[154]

bases [151]. On the other hand, a graphene-based electrochemical DNA sensor was developed for the detection of low concentrations of breast cancer-related BRCA1 gene. The sensor was stable, reproducible, and sensitive and it could detect up to 1 femtomolar BRCA1 gene (5.896 femtogram/ml) [152]. Moreover, a graphene sheets/polyaniline/gold nanoparticle based DNA sensor was employed for the electrochemical determination of BCR/ABL fusion gene in chronic myelogenous leukemia. Under optimum conditions, the DNA sensor exhibited a detection limit as low as 2.11 pM (S/N=3) with an excellent differentiation ability. It has been applied for assay of BCR/ABL fusion gene from real samples with satisfactory results [153]. Also, a riboflavin electrochemical sensor based on homoadenine single-stranded DNA/molybdenum disulfide–graphene nanocomposite modified gold electrode was fabricated. This sensor possessed a wide linear range from 0.025 to 2.25

μM with a low detection limit of 20 nM [154]. A brief summary of DNA biosensors is given in Table 6.

5.7. Graphene-Based Immunosensor

Increasing attention has been focused on the development of immunoassay because it has become a major analytical tool in clinical diagnosis. In immunoassay, the determination of cancer markers associated with certain tumors in patients plays an important role in diagnosing cancer diseases. Carcinoembryonic antigen (CEA) is one of the most extensively used tumor markers. An elevated CEA level in serum may be an early indication of lung cancer, ovarian carcinoma, colon cancer, breast cancer, and cystadenocarcinoma. Hence, Sheng Yu used gold nanoparticles (AuNPs), reduced graphene oxide (R-GO) and poly(L-Arginine) composite material modified CILE as the platform for the construction of a new electrochemical CEA immunosensor. The poly(L-Arginine)/R-GO composite film was used to modify CILE to fabricate Arg/R-GO/CILE through electropolymerization of L-Arginine on R-GO/CILE. AuNPs were adsorbed on the modified electrode to immobilize the CEA antibody and to construct the immunosensor. By combining the specific properties such as the biocompatibility and big surface area of AuNPs, the excellent electron transfer ability of R-GO and the high conductivity of CILE, the synergistic effects of composite increased the amounts of CEA antibody adsorbed on the electrode surface resulting in the greatly increase of the electrochemical responses. The proposed immunosensor showed good reproducibility, selectivity, low detection limit, and acceptable stability [155]. As well, ultrathin Au-Pt nanowire-decorated thionine/reduced graphene oxide (AuPtNWs/THI/rGO) [156] and Au nanoparticles–graphene–chitosan nanocomposite cryogel electrode [157] were developed for CEA detection showing high sensitivity, reproducibility, stability, and applicability in real samples (Table 7). Moreover, Jiadong Huang developed an ultrasensitive electrochemical immunosensor based on nanogold particles (nano-Au), Prussian Blue (PB), polyaniline/poly(acrylic acid) (PANI (PAA)), and Au-hybrid graphene nanocomposite (AuGN) for the detection of salbutamol (SAL). SAL is the most widely used β_2-adrenergic receptor agonist that induces bronchodilation making it highly useful for curing bronchial asthma, chronic obstructive pulmonary disease, and other allergic diseases associated with respiratory pathway. Nano-Au, PB, and PANI (PAA)-incorporated film was used to enhance the electroactivity, stability, and catalytic activity for hydrogen reduction of the electrode. AuGN was used to immobilize chitosan, nano-Au, and horseradish peroxidase–anti-SAL antibody (HRP–AAb). The resulting nanostructure (AuGN–HRP–AAb) was used as the label for the immunosensor. This is attributed to the high surface-to-volume ratio of graphene that allows the immobilization of a high level of chitosan, nano-Au, and HRP–AAb and its good electrical conductivity that can improve the electron transfer among HRP,

Table 7. Examples of modified electrodes used as immunosensors.

Electrode	Immunosensor	Linear range	Detection limit	Reference
gold nanoparticles, reduced graphene oxide and poly(L-Arginine) composite material modified CILE	Carcinoembryonic antigen (CEA)	0.5 to 200 ng mL^{-1}	0.03 ng mL^{-1}	[155]
ultrathin Au-Pt nanowire-decorated thionine/reduced graphene oxide	CEA	50 fg/mL to 100 ng/mL	6 fg/mL	[156]
Au nanoparticles–graphene–chitosan nanocomposite cryogel electrode	CEA	1.0×10^{-6} to 1.0 ng/mL	2.0×10^{-7} ng/mL	[157]
nanogold particles, Prussian Blue, polyaniline/poly(acrylic acid) and Au-hybrid graphene nanocomposite	salbutamol	0.08 ng/mL and 1000 ng/mL	0.04 ng/mL	[158]
graphene sheet and cobalt hexacyanoferrate nanoparticle	prostate specific antigen (PSA)	---	0.01 ng/mL	[159]
graphene-gold composite modified electrode	PSA	0–10 ng/mL	0.59 ng/mL	[160]
amino-functionalized graphene sheet–ferrocenecarboxaldehyde composite materials and silver hybridized mesoporous silica nanoparticles	PSA	0.01–10.0 ng/mL	2 pg/mL	[161]
Au nanoparticles functionalized porous graphene and Au@Pd core shell bimetallic functionalized graphene nanocomposites	carbohydrate antigen 19-9	0.015 to 150 U/mL	0.006 U/mL	[162]
ionic liquid functionalized graphene and Cd^{2+}-functionalized nanoporous TiO_2	carbohydrate antigen 15-3	0.02–60 U/mL	0.008 U/mL	[163]

H_2O_2, and electrode. The proposed immunosensor showed low detection limit, excellent selectivity, good reproducibility, long-term stability, and high accuracy [158]. Furthermore, the synergistic effect between graphene sheet (GS) and cobalt hexacyanoferrate nanoparticle (CoNP) was investigated showing that the electroactivity of CoNP was greatly enhanced in the presence of GS due to the great electron-transfer ability of GS. A label-free electrochemical immunosensor for the sensitive detection of prostate specific antigen (PSA), it was widely used in prostate cancer diagnosis and screening, was fabricated. Molecule 1-pyrenebutanoic acid, succinimidyl ester (PBSE) was adsorbed onto GS and the colloidal solution containing GS–CoNP–PBSE was added onto the GCE surface to form a stable thin film with high electroactivity. After anti-PSA antibody was conjugated onto PBSE, the modified electrode could be used as an amperometric immunosensor for the detection of PSA. This immunosensor has the advantages of high sensitivity with a low detection limit of 0.01 ng/mL, good selectivity, and stability and could become a promising technique for cancer marker detection [159]. In addition, highly conductive graphene (GR)-gold (Au) composite modified electrode [160] and amino-functionalized graphene sheet–ferrocenecarboxaldehyde composite materials (NH_2-GS@FCA) and silver hybridized mesoporous silica nanoparticles (Ag@NH_2-MCM48) [161] were utilized as PSA immunosensor displaying an enhanced electron transfer and high sensitivity toward PSA (Table 7). On the other hand, ultrasensitive electrochemical immunosensor for carbohydrate antigen 19-9 was developed using Au nanoparticles functionalized porous graphene (Au-PGO) as sensing platform and Au@Pd core shell bimetallic functionalized graphene nanocomposites (Au@Pd-Gra) as signal enhancers [162]. Moreover, an electrochemical immunosensor for the sensitive detection of carbohydrate antigen 15-3 (CA15-3) was fabricated based on ionic liquid functionalized graphene and Cd^{2+}-functionalized nanoporous TiO_2. The good performance of the immunosensor was attributed to (i) high surface-to-volume ratio of graphene, (ii) excellent biocompatibility and electron transfer rate originating from ionic liquid functionalized graphene, and (iii) the highly specific surface area of nanoporous TiO_2facilitating the adsorption of a high amount of Cd^{2+} for signal amplification [163].

5.8. Hemoglobin and Myoglobin Biosensors

Hemoglobin (Hb) is the most important part of blood responsible for transporting O_2 throughout the circulatory system. Change of Hb concentration in blood can cause several diseases and even death. Therefore, accurate determination of Hb content in blood is medically very essential [73]. Chunming Wang and Jun-Jie Zhu synthesized functionalized graphene nanosheets (PDDA-G) with poly (diallyldimethylammonium chloride) (PDDA) and used to combine with room temperature ionic liquid (RTIL). The resulting RTIL/PDDA-G composite displayed an enhanced capability for the immobilization of Hb to realize its direct electrochemistry. The proposed biosensor showed a fast direct electron transfer of Hb. Moreover, the RTIL/PDDA-G based biosensor exhibited excellent electrocatalytic activity for the detection of nitrite with wide linear range and low detection limit. The present strategy definitely paves a way for the functionalization of graphene sheets with good biocompatibility and solubility, thus provides a novel and promising platform for the study of the biological application of graphene [164]. Moreover, Ke-Jing Huang fabricated a promising biosensor, hemoglobin (Hb)/chitosan (Chit)–ionic liquid (IL)–ferrocene (Fc)/graphene (Gr)/GCE, that exploited the synergistic beneficial characteristics of Fc, Gr, and IL for Hb detection. The proposed biosensor showed a strong electrocatalytic activity toward the reduction of H_2O_2, which could be attributed to the favored orientation of Hb in the well-confined surface as well as the high electrical conductivity of the resulting Chit–IL–Fc/Gr inorganic hybrid composite. The developed biosensor exhibited a fast amperometric response, a good linear response toward H_2O_2 over a wide range of concentration, and a low detection limit of 3.8 μM. High sensitivity and stability, technically simple, and the possibility of preparation at short period of time are of great advantage of the developed biosensors [165]. Moreover, GR and Mg_2Al layered double hydroxide composite was prepared and used for the immobilization of Hb on a CILE to obtain an electrochemical biosensor exhibiting electrocatalytic reduction to trichloroacetic acid [166]. As well, a biocompatible biosensor based on immobilized Hb on carboxymethyl cellulose functionalized reduced graphene oxide (rGO) nanocomposite was developed. The proposed Hb biosensor exhibited excellent bioelectrocatalytic activity toward the reduction of NO and H_2O_2. This biosensor opened up a promising platform for protein immobilization and biosensor preparation [167]. Also, a mediator-free biosensor was fabricated by the immobilization of Hb on TiO_2NPs–rGO with a double-layered structure. TiO_2NPs–rGO nanocomposite was an excellent matrix for immobilization and biocompatibility for redox protein presenting good protein bioactivity and stability. The direct electron transfer of Hb is facilitated due to the special double-layered structure of TiO_2NPs–rGO. The prepared biosensor showed good performance for H_2O_2 detection with a low detection limit of 10 nM in the linear range of 0.1–140 μM [168]. On the other hand, a novel biocompatible sensing strategy based on a graphene (GR), ionic liquid (IL) 1-ethyl-3-methylimidazolium tetrafluoroborate ($EMIMBF_4$) and chitosan (CTS) composite film for the immobilization of myoglobin (Mb) was adopted. The CTS–Mb–GR–IL modified electrode exhibited an excellent enzyme-like catalytic activity for the reduction of trichloroacetic acid [169]. Another nanocomposite based on Pt nanoparticle decorated GR nanosheet was developed for the direct electrochemistry and electrocatalysis of Mb. GR–Pt nanocomposite was integrated with Mb on the surface of CILE. Nafion/Mb–GR–Pt/CILE electrochemical biosensor had potential applications for the electrocatalytic reduction of trichloroacetic acid in the linear range of 0.9 to 9.0 mM with the detection limit of 0.32 mM [170]. Moreover, 1-hexylpyridinium hexafluorophosphate based CILE was used as a substrate for the in situ electrodeposition of GR and nickel oxide (NiO) to get NiO/GR/CILE. Nafion/Mb/NiO/GR/CILE electrochemical sensor was constructed by the immobilization of Mb on the surface of NiO/GR/CILE with a Nafion film. The presence of the NiO/GR nanocomposite on the electrode promoted the direct electron transfer of Mb. A third-generation electrochemical Mb biosensor based on NiO/GR/CILE was constructed for the electrocatalytic reduction of trichloroacetic acid and H_2O_2 [171]. Table 8 contained a brief summary of some hemoglobin and myoglobin biosensors.

Table 8. Examples of modified electrodes used as Hemoglobin and Myoglobin Biosensors.

Electrode	Hb and Mb biosensor	Linear range	Detection limit	Reference
poly(diallyldimethylammonium chloride) functionalized graphene nanosheets combined with room temperature ionic liquid and modified with Hb	nitrite	0.2 to 32.6 μM	0.04 μM	[164]
Hb/chitosan–ionic liquid–ferrocene/graphene/GCE	H_2O_2	50 μM to 1200 μM	3.8 μM	[165]
Hb/ GR and Mg_2Al layered double hydroxide composite	trichloroacetic acid	1.6 to 25.0 mM	0.534 mM	[166]
immobilized Hb on carboxymethyl cellulose functionalized reduced graphene oxide (rGO) nanocomposite	NO and H_2O_2	0.864 μM-19.8 μM and 0.083–13.94 μM for NO and H_2O_2, respectively	0.37 μM and 0.08 μM for NO and H_2O_2, respectively	[167]
Hb/ TiO_2NPs–rGO	H_2O_2	0.1–140 μM	10 nM	[168]
myoglobin (Mb)/graphene , 1-ethyl-3-methyl-imidazolium tetrafluoroborate and chitosan composite film	trichloroacetic acid	---	---	[169]
Mb/ Pt nanoparticle decorated GR nanosheet	trichloroacetic acid	0.9 to 9.0 mM	0.32 mM	[170]
Nafion/Mb/NiO/GR/1-hexylpyridinium hexafluorophosphate CILE	trichloroacetic acid and H_2O_2	---	---	[171]

5.9. Glucose and H_2O_2 Biosensors

Glucose biosensors are becoming increasingly important due to their applications in biological and chemical analyses, clinical detection, and environmental monitoring. The immobilization of enzymes on electrodes is generally the first step in fabrication attracting significant efforts because enzymes are highly selective and quickly responsive to specific substrates. Glucose oxidase (GOD) is employed as the enzyme in most of the glucose biosensors. It catalyzes the electron transfer from glucose to oxygen accompanied by the production of gluconolactone and hydrogen peroxide. The quantification of glucose can be achieved via electrochemical detection of the enzymatically liberated H_2O_2 [172]. Hua Zhang proposed a direct electrochemical method to reduce single-layer GO adsorbed on the 3-aminopropyltriethoxysilane (APTES) modified GCE. The reduced GO was modified with GOD and the obtained GCE-APTES-rGO-GOD was successfully used to detect glucose opening up a potential application in biosensing [52]. In addition, a bionanocomposite film consisting of GOD/Pt/functional graphene sheets/chitosan (GOD/Pt/FGS/chitosan) was employed for glucose sensing. With the electrocatalytic synergy of FGS and Pt nanoparticles to H_2O_2, a sensitive biosensor with a detection limit of 0.6 μM for glucose was achieved displaying good reproducibility, long-term stability, and anti-interfering ability from AA and UA. The large surface area and good electrical conductivity of graphene suggests that graphene is a potential candidate as a sensor material [173]. Also, AuNPs–graphene nanocomposite was prepared and its electrocatalytic activity toward glucose oxidation was investigated. The results indicated that AuNPs–graphene nanocomposite had a higher catalytic activity than AuNPs or graphene alone indicating the synergistic effect of graphene and AuNPs [174]. Furthermore, Li Niu developed a novel glucose biosensor based on graphene/AuNPs/chitosan at gold electrode exhibiting good electrocatalytic activity toward H_2O_2 and O_2. With GOD as a model, the graphene/AuNPs/GOD/chitosan composite-modified electrode was constructed through a simple casting method. The resulting biosensor exhibited good amperometric response to glucose with linear range, good reproducibility, and low detection limit of 180 μM [175]. Also, amperometric glucose biosensors were fabricated by using Pt–Au and Au nanoparticle spacers decorated functionalized graphene (f-G) nanosheets. The immobilization of GOD over nafion-solubilized metal nanoparticles dispersed graphene f-G-(Pt–Au) and f-G-(Au) electrodes has been achieved by physical adsorption. The resultant bioelectrode retains its biocatalytic activity offering fast and sensitive glucose quantification. The fabricated f-G-(Au) based glucose biosensor exhibited best sensing performance with a linear response up to 30 mM with an excellent detection limit of 1 μM [172]. Moreover, a dual biosensor platform based on hemin-functionalized graphene nanosheets (H-GNs) was presented. The nanomaterial combines the features of both graphene (high conductivity and surface area) and hemin (excellent catalysis and intrinsic peroxidase-like activity). Based on these features, the simple, economical, and highly sensitive amperometric and colorimetric biosensors for H_2O_2 and glucose have been developed. The detection limits for H_2O_2 and glucose reached 0.2 μM and 0.3 μM by the amperometric method, respectively. By the colorimetric methods, the detection limits for H_2O_2 and glucose were as low as 20 nM and 30 nM, respectively. The rapid, simple, and sensitive sensing platform showed great promising applications in the pharmaceutical, clinical, and industrial detection of H_2O_2 and glucose [176]. Also, GCE modified with electrochemically reduced graphene oxide/sodium dodecyl sulfate (GCE/ERGO/SDS) was further modified with GOD to be used as an enzymatic glucose biosensor with a low detection limit of 40.8 μM [177]. On the other hand, graphene functionalized with poly(diallyldimethylammonium chloride) (PDDA) can be used as non-enzymatic H_2O_2 biosensor showing enhanced electrocatalytic activity toward H_2O_2 reduction. As well, this modified surface can be used as an enzymatic glucose sensor by the immobilization of the negatively charged GOD onto the positively charged PDDA-G matrix through electrostatic interaction.

GOD/PDDA-G bionanocomposite can detect glucose with a low detection limit of 8 μM in the linear range of 0.02 to 1.8 mM [178]. In addition, a graphene and cobalt oxide nanoparticle composite was employed as non-enzymatic H_2O_2 and enzymatic glucose biosensor [179]. Several modified surfaces were utilized as non-enzymatic H_2O_2 biosensors, which were summarized in Table 9. On the other hand, a stable and sensitive non-enzymatic glucose biosensor based on CuO and rGO nanocomposite was developed. The proposed biosensor presented a high sensitivity of 2221 μA/mM.cm^2 and a wide linear range from 0.4 μM to 12 mM toward glucose with good selectivity and stability. The TEM of rGO sheets showed a typical wrinkled and crumpled shape with thin layers. Upon CuO modification, there is a uniform distribution of rod-like nanoparticles with an average length of 20–30 nm on the surface of rGO indicating the successful synthesis of CuO-rGO nanocomposite [186]. Also, $Ni(OH)_2$/electroreduced graphene oxide-MWCNTs film was employed as non-enzymatic H_2O_2 and glucose biosensors [187].

Table 9. Examples of modified electrodes used as Glucose and H_2O_2 Biosensors.

Electrode	H_2O_2 and glucose biosensor	Linear range	Detection limit	Reference
GOD modified reduced GO adsorbed on the 3-aminopropyltriethoxysilane (APTES) modified GCE	enzymatic glucose	---	---	[52]
GOD/Pt/functional graphene sheets/chitosan	enzymatic glucose	---	0.6 μM	[173]
AuNPs–graphene nanocomposite	enzymatic glucose	---	---	[174]
graphene/AuNPs/GOD/chitosan composite-modified electrode	enzymatic glucose	2 to 10 mM	180 μM	[175]
GOD modified Pt–Au and Au nanoparticle spacers decorated functionalized graphene (f-G) nanosheets	enzymatic glucose	up to 30 mM	1 μM	[172]
hemin-functionalized graphene nanosheets (H-GNs)	H_2O_2 and glucose	---	0.2 μM for H_2O_2 and 0.3 μM for glucose	[176]
GCE modified with electrochemically reduced graphene oxide/sodium dodecyl sulfate and GOD	enzymatic glucose	---	40.8 μM	[177]
graphene functionalized with poly(diallyldimethylammonium chloride)	Non-enzymatic H_2O_2 and enzymatic glucose	0.02 to 1.8 mM	8 μM	[178]

graphene and cobalt oxide nanoparticle composite	non-enzymatic H_2O_2 and enzymatic glucose	0.5 mM to 16.5 mM for glucose. 0.2 to 211.5 μM for H_2O_2	0.06 μM for H_2O_2	[179]
prussian blue nanocubes on nitrobenzene-functionalized reduced graphene oxide	non-enzymatic H_2O_2	1.2 μM to 15.25 mM	0.4 μM	[180]
layer-by-layer deposited low dimensional building blocks of graphene-prussian blue onto graphite screen-printed electrodes	non-enzymatic H_2O_2	---	---	[181]
Ag nanoparticles/graphene nanosheet	non-enzymatic H_2O_2	0.1 to 100 mM	0.5 μM	[61]
prussian blue-graphene modified GCE	non-enzymatic H_2O_2	10–1440 μM	3 μM	[182]
graphene/Nafion/azure I/Au nanoparticles composites modified GCE	non-enzymatic H_2O_2	30 μM to 5 mM	10 μM	[183]
rGO and Fe_3O_4 nanocomposites	non-enzymatic H_2O_2	0.1 mM to 6 mM	3.2 μM	[184]
cationic polyelectrolyte-functionalized ionic liquid decorated graphene sheets nanocomposite	non-enzymatic H_2O_2	---	---	[185]
CuO and rGO nanocomposite	non-enzymatic glucose	0.4 μM to 12 mM	---	[186]
$Ni(OH)_2$/electroreduced graphene oxide-MWCNTs film	non-enzymatic H_2O_2 and non-enzymatic glucose	10-1500 μM and 10 μM-9050 μM for glucose and H_2O_2, respectively	2.7 μM and 4.0 μM for glucose and H_2O_2, respectively	[187]

6. CONCLUSIONS

Graphene exhibits a unique chemical structure and outstanding properties making it very attractive for various engineering applications. Graphene and graphene-based materials have a profound impact on electroanalysis, electrocatalysis, sensors, and biosensors. In sensing applications, graphene-based materials featured with good conductivity and large specific surface area have demonstrated accurate, rapid, selective, sensitive, and even single-molecular sensing abilities. Graphene-based materials are ideally used as gas sensors, electrochemical sensors for heavy metal ions, immunosensors and NADH, DNA, catecholamine neurotransmitters, paracetamol, glucose, H_2O_2, hemoglobin, and myoglobin biosensors. Future investigations on graphene-based sensing platforms, combined with versatile sensing strategies, are expected to continuingly lower the detection threshold.

7. PERMISSIONS

Figure number (in chapter)	Figure Source (reference number)	Permission number
2	[6]	3556440341708
3	[1]	3556440556169
4	[4]	Open access
5	[31]	Open access
6	[66]	Open access
7	[77]	3556441025155
8	[79]	3556441165370
Table 1	[6]	3556460926032

REFERENCES

1. Deshpande A, LeRoy B. Scanning probe microscopy of graphene. Physica E 2012;44:743–759.
2. Tang L, Wang Y, Li Y, Feng H, Lu J, Li J. Preparation, structure, and electrochemical properties of reduced graphene sheet films. Advanced Functional Materials 2009;19:2782–2789.
3. Soldano C, Mahmood A, Dujardin E. Production, properties and potential of graphene. Carbon 2010;48:2127–2150.
4. Arco L, Zhang Y, Zhou C. Large Scale Graphene by Chemical Vapor Deposition: Synthesis, Characterization and Applications. In: Gong J, editor. Graphene–Synthesis, Characterization, Properties and Applications. Croatia: InTech; 2011. pp. 161–184.
5. Shao Y, Wang J, Wu H, Liu J, Aksay I, Lin Y. Graphene based electrochemical sensors and biosensors: A review. Electroanalysis 2010;22:1027–1036.
6. Kuila T, Bose S, Mishra A, Khanra P, Kim N, Lee J. Chemical functionalization of graphene and its applications. Progress in Materials Science 2012;57:1061–1105.
7. Fan Y, Liu J, Lu H, Zhang Q. Electrochemical behavior and voltammetric determination of paracetamol on Nafion/TiO2–graphene modified glassy carbon electrode. Colloids and Surfaces B: Biointerfaces 2011;85:289–292.
8. Li F, Li J, Feng Y, Yang L, Du Z. Electrochemical behavior of graphene doped carbon paste electrode and its application for sensitive

determination of ascorbic acid. Sensors and Actuators B 2011;157:110–114.

9. Odahara G, Ishikawa T, Fukase K, Otani S, Oshima C, Suzuki M, Yasue T, Koshikawa T. Self-Standing Graphene Sheets Prepared with Chemical Vapor Deposition and Chemical Etching. In: Gong J, editor. Graphene–Synthesis, Characterization, Properties and Applications. Croatia: InTech; 2011. pp. 3–20.

10. Wang Y, Li Y, Tang L, Lu J, Li J. Application of graphene-modified electrode for selective detection of dopamine. Electrochemistry Communications 2009;11:889–892.

11. Craciun M, Russo S, Yamamoto M, Tarucha S. Tuneable electronic properties in graphene. Nano Today 2011;6:42–60.

12. Kang X, Wang J, Wu H, Liu J, Aksay I, Lin Y. A graphene-based electrochemical sensor for sensitive detection of paracetamol. Talanta 2010;81:754–759.

13. Du M, Yang T, Ma S, Zhao C, Kui J. Ionic liquid-functionalized graphene as modifier for electrochemical and electrocatalytic improvement: Comparison of different carbon electrodes. Analytica Chimica Acta 2011;690:169–174.

14. Wintterlin J, Bocquet M. Graphene on metal surfaces. Surface Science 2009;603:1841–1852.

15. Huang X, Yin Z, Wu S, Qi X, He Q, Zhang Q, Yan Q, Boey F, Zhang H. Graphene-based materials: Synthesis, characterization, properties, and applications. Small 2011;7:1876–1902.

16. Wang X, Chen S. Graphene-Based Nanocomposites. In: Mikhailov S, editor. Physics and Applications of Graphene – Experiments. Croatia: InTech; 2011. pp. 135–168.

17. Chen L, Tang Y, Wang Ke, Liu C, Luo S. Direct electrodeposition of reduced graphene oxide on glassy carbon electrode and its electrochemical application. Electrochemistry Communications 2011;13:133–137.

18. Pumera M, Ambrosi A, Bonanni A, Chng E, Poh H. Graphene for electrochemical sensing and biosensing. Trends in Analytical Chemistry 2010;29:954–965.

19. Wu W, Liu Z, Jauregui L, Yu Q, Pillai R, Cao H, Bao J, Chen Y, Pei S. Wafer-scale synthesis of graphene by chemical vapor deposition and its application in hydrogen sensing. Sensors and Actuators B 2010;150:296–300.

20. Kalita G, Qi L, Namba Y, Wakita K, Umeno M. Femtosecond laser induced micropatterning of graphene film. Materials Letters 2011;65:1569–1572.

21. Feng T, Xie D, Tian H, Peng P, Zhang D, Fu D, Ren T, Li X, Zhu H, Jing Y. Multi-layer graphene treated by O2 plasma for transparent conductive electrode applications. Materials Letters 2012;73:187–189.

22. Choi Y, Kang S, Kim H, Choi W, Na S. Multilayer graphene films as transparent electrodes for organic photovoltaic devices. Solar Energy Materials & Solar Cells 2012;96:281–285.

23. Somani P, Somani S, Umeno M. Planer nano-graphenes from camphor by CVD. Chemical Physics Letters 2006;430:56–59.

24. An H, Lee W, Jung J. Graphene synthesis on Fe foil using thermal CVD. Current Applied Physics 2011;11:S81–S85.

25. Terasawa T, Saiki K. Growth of graphene on Cu by plasma enhanced chemical vapor deposition. Carbon 2012;50:869–874.

26. Juang Z, Wu C, Lu A, Su C, Leou K, Chen F, Tsai C, Graphene synthesis by chemical vapor deposition and transfer by a roll-to-roll process. Carbon 2010;48:3169–3174.

27. Guermoune A, Chari T, Popescu F, Sabri S, Guillemette J, Skulason H, Szkopek T, Siaj M. Chemical vapor deposition synthesis of graphene on copper with methanol, ethanol, and propanol precursors. Carbon 2011;49:4204–4210.

28. Gautam M, Jayatissa A. Gas sensing properties of graphene synthesized by chemical vapor deposition. Materials Science and Engineering C 2011;31:1405–1411.

29. Liu W, Chung C, Miao C, Wang Y, Li B, Ruan L, Patel K, Park Y, Woo J, Xie Y. Chemical vapor deposition of large area few layer graphene on Si catalyzed with nickel films. Thin Solid Films 2010;518:S128–S132.

30. Negishi R, Hirano H, Ohno Y, Maehashi K, Matsumoto K, Kobayashi Y. Layer-by-layer growth of graphene layers on graphene substrates by chemical vapor deposition. Thin Solid Films 2011;519:6447–6452.

31. Mwakikunga B, Hillie K. Graphene Synthesis, Catalysis with Transition Metals and Their Interactions by Laser Photolysis. In: Gong J, editor. Graphene–Synthesis, Characterization, Properties and Applications. Croatia: InTech; 2011. pp. 59–78.

32. Singh V, Joung D, Zhai L, Das S, Khondaker S, Seal S. Graphene based materials: Past, present and future. Progress in Materials Science 2011;56:1178–1271.

33. Dhakate S, Chauhan N, Sharma S, Mathur R. The production of multi-layer graphene nanoribbons from thermally reduced unzipped multi-walled carbon nanotubes. Carbon 2011;49:4170–4178.

34. Cataldo F, Compagnini G, Patané G, Ursini O, Angelini G, Ribic P, Margaritondo G, Cricenti A, Palleschi G, Valentini F. Graphene nanoribbons produced by the oxidative unzipping of single-wall carbon nanotubes. Carbon 2010;48:2596–2602.

35. Janowska I, Ersen O, Jacob T, Vennégues P, Bégin D, Ledoux M, Pham-Huu C. Catalytic unzipping of carbon nanotubes to few-layer graphene sheets under microwaves irradiation. Applied Catalysis A: General 2009;371:22–30.

36. Shen B, Ding J, Yan X, Feng W, Li J, Xue Q. Influence of different buffer gases on synthesis of few-layered graphene by arc discharge method. Applied Surface Science 2012;258:4523–4531.

37. Chen Y, Zhao H, Sheng L, Yu L, An K, Xu J, Ando Y, Zhao X. Mass-production of highly-crystalline few-layer graphene sheets by arc discharge in various H2-inert gas mixtures. Chemical Physics Letters 2012;538:72–76.

38. Li N, Wang Z, Zhao K, Shi Z, Gu Z, Xu S. Large scale synthesis of N-doped multi-layered grapheme sheets by simple arc-discharge method. Carbon 2010;48:255–259.

39. Levchenko I, Volotskova O, Shashurin A, Raitses Y, Ostrikov K, Keidar M. The large-scale production of graphene flakes using magnetically-enhanced arc discharge between carbon electrodes. Carbon 2010;48:4556–4577.

40. Hummers W, Offeman R. Preparation of graphitic oxide. Journal of the American Chemical Society 1958;80(6):1339–1339.

41. Stankovich S, Dikin D, Piner R, Kohlhaas K, Kleinhammes A, Jia Y, Wu Y, Nguyen S, Ruoff R. Synthesis of graphene-based nanosheets via chemical reduction of exfoliated graphite oxide. Carbon 2007;45:1558–1565.

42. Phama T, Kimb J, Kima J, Jeong Y. One-step reduction of graphene oxide with l-glutathione. Colloids and Surfaces A: Physicochemical and Engineering Aspects 2011;384:543–548.

43. Marcano D, Kosynkin D, Berlin J, Sinitskii A, Sun Z, Slesarev A, Alemany L, Lu W, Tour J. Improved synthesis of graphene oxide. AcsNano 2010;4:4806–4814.

44. Kai-xuan S, Yu-xi X, Chun L, Gao-quan S. High-performance self-assembled graphene hydrogels prepared by chemical reduction of graphene oxide. New Carbon Materials 2011;26(1):9–15.

45. Ramesha G, Sampath S. Electrochemical reduction of oriented graphene oxide films: An in situ Raman spectroelectrochemical study. The Journal of Physical Chemistry C 2009;113:7985–7989.

46. Chen W, Yan L. Preparation of graphene by a low-temperature thermal reduction at atmosphere pressure. Nanoscale 2010;2:559–563.

47. Pei S, Cheng H. The reduction of graphene oxide. Carbon 2012;50:3210–3228.

48. Cuong T, Pham V, Tran Q, Chung J, Shin E, Kim J, Kim E. Optoelectronic properties of graphene thin films prepared by thermal reduction of graphene oxide. Materials Letters 2010;64:765–767.

49. Wei A, Wang J, Long Q, Liu X, Li X, Dong X, Huang W. Synthesis of high-performance graphene nanosheets by thermal reduction of graphene oxide. Materials Research Bulletin 2011;46:2131–2134.

50. Le L, Ervin M, Qiu H, Fuchs B, Lee W. Graphene supercapacitor electrodes fabricated by inkjet printing and thermal reduction of graphene oxide. Electrochemistry Communications 2011;13:355–358.

51. Zhao B, Liu P, Jiang Y, Pan D, Tao H, Song J, Fang Tao, Xu W. Supercapacitor performances of thermally reduced graphene oxide. Journal of Power Sources 2012;198:423–427.

52. Wang Z, Zhou X, Zhang J, Boey F, Zhang H. Direct electrochemical reduction of single-layer graphene oxide and subsequent functionalization with glucose oxidase. The Journal of Physical Chemistry C 2009;113:14071–14075.

53. Hassan H, Abdelsayed V, Khder A, AbouZeid K, Terner J, El-Shall M, Al-Resayes S, El-Azhary A. Microwave synthesis of graphene sheets supporting metal nanocrystals in aqueous and organic media. Journal of Materials Chemistry 2009;19:3832–3837.

54. Wang S, Jiang S, Wang X. Microwave-assisted one-pot synthesis of metal/metal oxide nanoparticles on graphene and their electrochemical applications. Electrochimica Acta 2011;56:3338–3344.

55. Hu H, Zhao Z, Zhou Q, Gogotsi Y, Qiu J. The role of microwave absorption on formation of graphene from graphite oxide. Carbon 2012;50:3267–3273.

56. Liao C, Liao C, Tso C, Shy H. Microwave-polyol synthesis and electrocatalytic performance of Pt/graphene nanocomposites. Materials Chemistry and Physics 2011;130:270–274.

57. Xin Y, Liu J, JiX, Liu W, Liu F, Yin Y, Gu J, Zou Z. Preparation and electrochemical characterization of nitrogen doped graphene by microwave as supporting materials for fuel cell catalysts. Electrochimica Acta 2012;60:354–358.

58. Galal A, Atta N, Ali S. Optimization of the synthesis conditions for LaNiO3 catalyst by microwave assisted citrate method for hydrogen production. Applied Catalysis A: General 2011;409–410:202–208.

59. Galal A, Atta N, Ali S. Investigation of the catalytic activity of LaBO3 (B = Ni, Co, Fe or Mn) prepared by the microwave-assisted method for hydrogen evolution in acidic medium. Electrochimica Acta 2011;56:5722–5730.

60. Galal A, Darwish S, Atta N, Ali S, Abd El Fatah A. Synthesis, structure and catalytic activity of nano-structured Sr–Ru–O type perovskite for hydrogen production. Applied Catalysis A: General 2010;378:151–159.

61. Liu S, Tian J, Wang L, Sun X. Microwave-assisted rapid synthesis of Ag nanoparticles/graphene nanosheet composites and their application for hydrogen peroxide detection. Jornal of Nanoparticle Research 2011;13:4539–4548.

62. Siamaki A, Khder A, Abdelsayed V, El-Shall M, Gupton B. Microwave-assisted synthesis of palladium nanoparticles supported on graphene: A highly active and recyclable catalyst for carbon–carbon cross-coupling reactions. Journal of Catalysis 2011;279:1–11.

63. Liu S, Wang J, Zeng J, Ou J, Li Z, Liu X, Yang S. "Green" electrochemical synthesis of Pt/graphene sheet nanocomposite film and its electrocatalytic property. Journal of Power Sources 2010;195:4628–4633.

64. Shao Y, Wang J, Engelhard M, Wang C, Lin Y. Facile and controllable electrochemical reduction of graphene oxide and its applications. Journal of Materials Chemistry 2010;20:743–748.

65. Harima Y, Setodoi S, Imae I, Komaguchi K, Ooyama Y, Ohshita J, Mizota H, Yano J. Electrochemical reduction of graphene oxide in organic solvents. Electrochimica Acta 2011;56:5363–5368.

66. Galal A, Atta N, Hassan H. Graphene supported-Pt-M (M = Ru or Pd) for electrocatalytic methanol oxidation. International Journal of Electrochemical Science 2012;7:768–784

67. Peng X, Liu X, Diamond D, Lau K. Synthesis of electrochemically-reduced graphene oxide film with controllable size and thickness and its use in supercapacitor. Carbon 2011;49:3488–3496.

68. Huang L, Liu Y, Ji L, Xie Y, Wang T, Shi W. Pulsed laser assisted reduction of graphene oxide. Carbon 2011;49:2431–2436. 7

69. Zhang Y, Guo L, Wei S, He Y, Xia H, Chen Q, Sun H, Xiao F. Direct imprinting of microcircuits on grapheme oxides film by femtosecond laser reduction. Nano Today 2010;5:15–20.

70. El-Kady M, Strong V, Dubin S, Kaner R, Laser scribing of high-performance and flexible graphene-based electrochemical capacitors. Science 2012;335:1326–1330.

71. Artiles M, Rout C, Fisher T. Graphene-based hybrid materials and devices for biosensing. Advanced Drug Delivery Reviews 2011;63:1352–1360.

72. Fan Y, Liu J, Yang C, Yu M, Liu P. Graphene–polyaniline composite film modified electrode for voltammetric determination of 4-aminophenol. Sensors and Actuators B 2011;157:669–674.

73. Kuila T, Bose S, Khanra P, Mishra A, Kim N, Lee J. Recent advances in graphene-based biosensors. Biosensors and Bioelectronics 2011;26:4637–4648.

74. Khadem S, Abdi Y, Darbari S, Ostovari F. Investigating the effect of gas absorption on the electromechanical and electrochemical behavior of graphene/ZnO structure, suitable for highly selective and sensitive gas sensors. Current Applied Physics 2014;14(11):1498–1503.

75. Zhou Y, Jiang Y, Xie T, Tai H, Xie G. A novel sensing mechanism for resistive gas sensors based on layered reduced graphene oxide thin films at room temperature. Sensors and Actuators B 2014;203:135–142.

76. Mishra S, Tripathi S, Choudhary V, Gupta B. SPR based fibre optic ammonia gas sensor utilizing nanocomposite film of PMMA/reduced graphene oxide prepared by in situ polymerization. Sensors and Actuators B 2014;199:190–200.

77. Uddin A, Phan D, Chung G. Low temperature acetylene gas sensor based on Ag nanoparticles-loaded ZnO-reduced graphene oxide hybrid. Sensors and Actuators B 2015;207:362–369.

78. Wang Z, Wang H, Zhang Z, Yang X, Liu G. Sensitive electrochemical determination of trace cadmium on astannum film/poly(p-aminobenzene sulfonic acid)/electrochemically reduced graphene composite modified electrode. Electrochimica Acta 2014;120:140–146.

79. Li J, Guo S, Zhai Y, Wang E. High-sensitivity determination of lead and cadmium based on the nafion-graphene composite film. Analytica Chimica Acta 2009;649:196–201.

80. Promphet N, Rattanarat P, Chailapakul O, Rangkupan R, Rodthongkum N. An electrochemical sensor based on graphene/polyaniline/polystyrene nanoporous fibers modified electrode for simultaneous determination of lead and cadmium. Sensors and Actuators B 2015;207:526–534.

81. Wang Z, Wang H, Zhang Z, Liu G. Electrochemical determination of lead and cadmium in rice by a disposable bismuth/electrochemically reduced graphene/ionic liquid composite modified screen-printed electrode. Sensors and Actuators B 2014;199:7–14.

82. Deng X, Lu L, Li H, Luo F. The adsorption properties of Pb(II) and Cd(II) on functionalized grapheme prepared by electrolysis method. Journal of Hazardous Materials 2010;183:923–930.

83. Huang H, Chen T, Liu X, Ma H. Ultrasensitive and simultaneous detection of heavy metal ions based on three-dimensional graphene-carbon nanotubes hybrid electrode materials. Analytica Chimica Acta 2014;852:45–54.

84. Mishra A, Ramaprabhu S. Functionalized graphene sheets for arsenic removal and desalination of sea water. Desalination 2011;282:39–45.

85. Ramesha G, Sampath S. In-situ formation of graphene-lead oxide composite and its use in trace arsenic detection. Sensors and Actuators B 2011;160:306–311.

86. Ren Y, Yan N, Wen Q, Fan Z, Wei T, Zhang M, Ma J. Graphene/δ-MnO2 composite as adsorbent for the removal of nickel ions from waste water. Chemical Engineering Journal 2010;175:1–7.

87. Atta N, Galal A, Abu-Attia F, Azab S. Simultaneous determination of paracetamol and neurotransmitters in biological fluids using a carbon paste sensor modified with gold nanoparticles. Journal of Materials Chemistry 2011;21:13015–13024.

88. Atta N, El-Kady M, Galal A. Palladium nanoclusters-coated polyfuran as a novel sensor for catecholamine neurotransmitters and paracetamol. Sensors and Actuators B 2009;141:566–574.

89. Chen X, Zhu J, Xi Q, Yang W. A high performance electrochemical sensor for acetaminophen based on single-walled carbon nanotube–graphene nanosheet hybrid films. Sensors and Actuators B: Chemical 2012;161(1):648–654.

90. Arvand M, Gholizadeh T. Simultaneous voltammetric determination of tyrosine and paracetamol using a carbon nanotube-graphene nanosheet nanocomposite modified electrode in human blood serum and pharmaceuticals. Colloids and Surfaces B: Biointerfaces 2013;103:84–93.

91. Adhikari B, Govindhan M, Chen A. Sensitive detection of acetaminophen with graphene-based electrochemical sensor. Electrochimica Acta.

92. Liu G, Chen H, Lin G, Ye P, Wang X, Jiao Y, Guo X, Wen Y, Yang H. One-step electrodeposition of graphene loaded nickel oxides nanoparticles for acetaminophen detection. Biosensors and Bioelectronics 2014;56:26–32.

93. Zhu W, Huang H, Gao X, Ma H. Electrochemical behavior and voltammetric determination of acetaminophen based on glassy carbon electrodes modified with poly(4-aminobenzoic acid)/electrochemically reduced graphene oxide composite films. Materials Science and Engineering: C 2014;45:21–28.

94. Atta N, El-Kady M, Galal A. Simultaneous determination of catecholamines, uric acid and ascorbic acid at physiological levels using poly(N-methylpyrrole)/Pd-nanoclusters sensor. Analytical Biochemistry 2010;400:78–88.

95. Atta N, El-Kady M. Poly(3-methylthiophene)/palladium sub-micro-modified sensor electrode. Part II: Voltammetric and EIS studies, and analysis of catecholamine neurotransmitters, ascorbic acid and acetaminophen. Talanta 2009;79:639–647.

96. Atta N, Galal A, El-Ads E. A novel sensor of cysteine self-assembled monolayers over gold nanoparticles for the selective determination of epinephrine in presence of sodium dodecyl sulfate. Analyst 2012;137:2658–2668.

97. Cui F, Zhang X. Electrochemical sensor for epinephrine based on a glassy carbon electrode modified with graphene/gold nanocomposites. Journal of Electroanalytical Chemistry 2012;669:35–41.

98. Rosy, Yadav S, Agrawal B, Oyama M, Goyal R. Graphene modified palladium sensor for electrochemical analysis of norepinephrine in pharmaceuticals and biological fluids. Electrochimica Acta 2014;125:622–629.

99. Atta N, El-Kady M. Novel poly(3-methylthiophene)/Pd, Pt nanoparticle sensor: Synthesis, characterization and its application to the

simultaneous analysis of dopamine and ascorbic acid in biological fluids. Sensors and Actuators B 2010;145:299–310.

100. Kim Y, Bong S, Kang Y, Yang Y, Mahajan R, Kim J, Kim H. Electrochemical detection of dopamine in the presence of ascorbic acid using graphene modified electrodes. Biosensors and Bioelectronics 2010;25:2366–2369.

101. Mao Y, Bao Y, Gan S, Li F, Niu L. Electrochemical sensor for dopamine based on a novel graphene-molecular imprinted polymers composite recognition element. Biosensors and Bioelectronics 2011;28:291–297.

102. Li F, Jia Chai, Yang H, Han D, Niu Li. Synthesis of Pt/ionic liquid/graphene nanocomposite and its simultaneous determination of ascorbic acid and dopamine. Talanta 2010;81:1063–1068.

103. Fan Y, Lu H, Liu J, Yang C, Jing Q, Zhang Y, Yang X, Huang K. Hydrothermal preparation and electrochemical sensing properties of TiO2–graphene nanocomposite. Colloids and Surfaces B: Biointerfaces 2011;83:78–82.

104. Liu W, Xiao J, Wang C, Yin H, Xie H, Cheng R. Synthesis of polystyrene-grafted-graphene hybrid and its application in electrochemical sensor of dopamine. Materials Letters 2013;100:70–73.

105. Zhang W, Zheng J, Shi J, Lin Z, Huang Q, Zhang H, Wei C, Chen J, Hu S, Hao A. Nafion covered core–shell structured Fe3O4@graphene nanospheres modified electrode for highly selective detection of dopamine. Analytica Chimica Acta 2015;853:285–280.

106. Pruneanu S, Biris A, Pogacean F, Socaci C, Coros M, Rosu M, Watanabe F, Biris A. The influence of uric and ascorbic acid on the electrochemical detection of dopamine using graphene-modified electrodes. Electrochimica Acta 2015;154:197–204.

107. Wang C, Du J, Wang H, Zou C, Jiang F, Yang P, Du Y. A facile electrochemical sensor based on reduced graphene oxide and Au nanoplates modified glassy carbon electrode for simultaneous detection of ascorbic acid, dopamine and uric acid. Sensors and Actuators B: Chemical 2014;204:302–309.

108. Yu B, Kuang D, Liu S, Liu C, Zhang T. Template-assisted self-assembly method to prepare three-dimensional reduced graphene oxide for dopamine sensing. Sensors and Actuators B: Chemical 2014;205:120–126.

109. Li S, He J, Zhang M, Zhang R, Lv X, Li S, Pang H. Electrochemical detection of dopamine using water-soluble sulfonated graphene. Electrochimica Acta 2013;102:58–65.

110. Han H, Seol H, Kang D, Ahmed M, You J, Jeon S. Electrochemical oxidation and determination of dopamine in the presence of AA using ferulic acid functionalized electrochemically reduced graphene. Sensors and Actuators B: Chemical 2014;204:289–296.

111. Nancy T, Kumary V. Synergistic electrocatalytic effect of graphene/nickel hydroxide composite for the simultaneous electrochemical determination of ascorbic acid, dopamine and uric acid. Electrochimica Acta 2014;133:233–240.

112. Wang H, Ren F, Yue R, Wang C, Zhai C, Du Y. Macroporous flower-like graphene-nanosheet clusters used for electrochemical determination of dopamine. Colloids and Surfaces A: Physicochemical and Engineering Aspects 2014;448:181–185.

113. Yan J, Liu S, Zhang Z, He G, Zhou P, Liang H, Tian L, Zhou X, Jiang H. Simultaneous electrochemical detection of ascorbic acid, dopamine and uric acid based on graphene anchored with Pd–Pt nanoparticles. Colloids and Surfaces B: Biointerfaces 2013;111:392–397.

114. Yang L, Liu D, Huang J, You T. Simultaneous determination of dopamine, ascorbic acid and uric acid at electrochemically reduced graphene oxide modified electrode. Sensors and Actuators B: Chemical 2014;193:166–172.

115. Li S, Yang S, Wang Y, Lien C, Tien H, Hsiao S, Liao W, Tsai H, Chang C, Ma C, Hu C. Controllable synthesis of nitrogen-doped graphene and its effect on the simultaneous electrochemical determination of ascorbic acid, dopamine, and uric acid. Carbon 2013;59:418–429.

116. Lian Q, He Z, He Q, Luo A, Yan K, Zhang D, Lu X, Zhou X. Simultaneous determination of ascorbic acid, dopamine and uric acid based on tryptophan functionalized graphene. Analytica Chimica Acta 2014;823:32–39.

117. Li H, Wang Y, Ye D, Luo J, Su B, Zhang S, Kong J. An electrochemical sensor for simultaneous determination of ascorbic acid, dopamine, uric acid and tryptophan based on MWNTs bridged mesocellular graphene foam nanocomposite. Talanta 2014;127:255–261.

118. Pakapongpan S, Mensing J, Phokharatkul D, Lomas T, Tuantranont A. Highly selective electrochemical sensor for ascorbic acid based on anovel hybrid graphene-copper phthalocyanine-polyanilinenanocomposites. Electrochimica Acta 2014;133:294–301.

119. Liu B, Luo L, Ding Y, Si X, Wei Y, Ouyang X, Xu D. Differential pulse voltammetric determination of ascorbic acid in the presence of folic acid at electro-deposited NiO/graphene composite film modified electrode. Electrochimica Acta 2014;142:336–342.

120. Xue C, Wang X, Zhu W, Han Q, Zhu C, Hong J, Zhou X, Jiang H. Electrochemical serotonin sensing interface based on double-layered membrane of reduced graphene oxide/polyaniline nanocomposites and molecularly imprinted polymers embedded with gold nanoparticles. Sensors and Actuators B: Chemical 2014;196:57–63.

121. Kim S, Kim D, Jeon S. Electrochemical determination of serotonin on glassy carbon electrode modified with various graphene nanomaterials. Sensors and Actuators B: Chemical 2012;174:285–291.

122. Han H, You J, Jeong H, Jeon S. Synthesis of graphene oxide grafted poly(lactic acid) with palladium nanoparticles and its application to serotonin sensing. Applied Surface Science 2013;284:438–445.

123. Han H, Lee H, You J, Jeong H, Jeon S. Electrochemical biosensor for simultaneous determination of dopamine and serotonin based on electrochemically reduced GO-porphyrin. Sensors and Actuators B: Chemical 2014;190:886–895.

124. Mazloum-Ardakani M, Khoshroo A, Hosseinzadeh L. Application of graphene to modified ionic liquid graphite composite and its enhanced electrochemical catalysis properties for levodopa oxidation. Sensors and Actuators B: Chemical 2014;204:282–288.

125. Benvidi A, Dehghani-Firouzabadi A, Mazloum-Ardakani M, Mirjalili B, Zare R. Electrochemical deposition of gold nanoparticles on reduced graphene oxide modified glassy carbon electrode for simultaneous determination of levodopa, uric acid and folic acid. Journal of Electroanalytical Chemistry 2015;736:22–29.

126. Wang Q, Das M, Li M, Boukherroub R, Szunerits S. Voltammetric detection of l-dopa and carbidopa on graphene modified glassy carbon interfaces. Bioelectrochemistry 2013;93:15–22.

127. Martín A, Hernández-Ferrer J, Martínez M, Escarpa A. Graphene nanoribbon-based electrochemical sensors on screen-printed platforms. Electrochimica Acta 2014; in press.

128. Guo H, Peng S, Xu J, Zhao Y, Kang X. Highly stable pyridinic nitrogen doped graphene modified electrode in simultaneous determination of hydroquinone and catechol. Sensors and Actuators B: Chemical 2014;193:623–629.

129. Si W, Lei W, Han Z, Hao Q, Zhang Y, Xia M. Selective sensing of catechol and hydroquinone based on poly(3,4-ethylenedioxythiophene)/nitrogen-doped graphene composites. Sensors and Actuators B: Chemical 2014;199:154–160.

130. Song D, Xia J, Zhang F, Bi S, Xiang W, Wang Z, Xia L, Xia Y, Li Y, Xia L. Multiwall carbon nanotubes-poly(diallyldimethylammonium chloride)-

graphene hybrid composite film for simultaneous determination of catechol and hydroquinone. Sensors and Actuators B: Chemical 2015;206:111–118.

131. Lai T, Cai W, Dai W, Ye J. Easy processing laser reduced graphene: A green and fast sensing platform for hydroquinone and catechol simultaneous determination. Electrochimica Acta 2014;138:48–55.

132. Zhang Y, Xiao S, Xie J, Yang Z, Pang P, Gao Y. Simultaneous electrochemical determination of catechol andhydroquinone based on graphene–TiO2 nanocomposite modified glassy carbon electrode. Sensors and Actuators B 2014;204:102–108.

133. Han H, You J, Seol H, Jeong H, Jeon S. Electrochemical sensor for hydroquinone and catechol based on electrochemically reduced GO–terthiophene–CNT. Sensors and Actuators B: Chemical 2014;194:460–469.

134. Wang X, Wu M, Li H, Wang Q, He P, Fang Y. Simultaneous electrochemical determination of hydroquinone and catechol based on three-dimensional graphene/MWCNTs/BMIMPF6 nanocomposite modified electrode. Sensors and Actuators B: Chemical;192:452–458.

135. Hu S, Wang Y, Wang X, Xu L, Xiang J, Sun W. Electrochemical detection of hydroquinone with a gold nanoparticle and graphene modified carbon ionic liquid electrode. Sensors and Actuators B 2012;168:27–33.

136. Navaee A, Salimi A, Teymourian H. Graphene nanosheets modified glassy carbon electrode for simultaneous detection of heroine, morphine and noscapine. Biosensors and Bioelectronics 2012;31:205–211.

137. Li Y, Zou L, Li Y, Li K, Ye B. A new voltammetric sensor for morphine detection based on electrochemically reduced MWNTs-doped graphene oxide composite film. Sensors and Actuators B: Chemical 2014;201:511–519.

138. Atta N, Hassan H, Galal A. Rapid and simple electrochemical detection of morphine on graphene–palladium-hybrid-modified glassy carbon electrode. Analytical and Bioanalytical Chemistry 2014;406:6933–6942.

139. Li Y, Li K, Song G, Liu J, Zhang K, Ye B. Electrochemical behavior of codeine and its sensitive determination on graphene-based modified electrode. Sensors and Actuators B: Chemical 2013;182:401–407.

140. Afkhami A, Khoshsafar H, Bagheri H, Madrakian T. Facile simultaneous electrochemical determination of codeine and acetaminophen in pharmaceutical samples and biological fluids by graphene–CoFe2O4

nancomposite modified carbon paste electrode. Sensors and Actuators B: Chemical 2014;203:909–918.

141. Afkhami A, Khoshsafar H, Bagheri H, Madrakian T. Preparation of NiFe2O4/graphene nanocomposite and its application as a modifier for the fabrication of an electrochemical sensor for the simultaneous determination of tramadol and acetaminophen. Analytica Chimica Acta 2014;831:50–59.

142. Guo K, Qian K, Zhang S, Kong J, Yu C, Liu B. Bio-electrocatalysis of NADH and ethanol based on graphene sheets modified Electrodes. Talanta 2011;85:1174–1179.

143. Ferreira G, Oliveira F, Leite F, Maroneze C, Kubota L, Damos F, Luz R. DNA and graphene as a new efficient platform for entrapment of methylene blue (MB): Studies of the electrocatalytic oxidation of β-nicotinamide adenine dinucleotide. Electrochimica Acta 2013;111:543–551.

144. Li Z, Huang Y, Chen L, Qin X, Huang Z, Zhou Y, Meng Y, Li J, Huang S, Liu Y, Wang W, Xie Q, Yao S. Amperometric biosensor for NADH and ethanol based on electroreduced graphene oxide–polythionine nanocomposite film. Sensors and Actuators B: Chemical 2013;181:280–287.

145. Gai P, Zhao C, Wang Y, Abdel-Halim E, Zhanga J, Zhua J. NADH dehydrogenase-like behavior of nitrogen-doped graphene and its application in NAD+-dependent dehydrogenase biosensing. Biosensors and Bioelectronics 2014;62:170–176.

146. Li L, Lu H, Deng L. A sensitive NADH and ethanol biosensor based on graphene–Au nanorods nanocomposites. Talanta 2013;113:1–6.

147. Govindhan M, Amiri M, Chen A. Au nanoparticle/graphene nanocomposite as a platform for the sensitive detection of NADH in human urine. Biosensors and Bioelectronics 2015;66:474–480.

148. Gasnier A, Pedano M, Rubianes M, Rivas G. Graphene paste electrode: Electrochemical behavior and analytical applications for the quantification of NADH. Sensors and Actuators B: Chemical 2013;176:921–926.

149. Ye X, Du Y, Duan K, Lu D, Wang C, Shi X. Fabrication of nano-ZnS coated PEDOT-reduced graphene oxidehybrids modified glassy carbon-rotating disk electrode and its application for simultaneous determination of adenine, guanine, and thymine. Sensors and Actuators B 2014;203:271–281.

150. Sun W, Lu Y, Wu Y, Zhang Y, Wang P, Chen Y, Li G. Electrochemical sensor for transgenic maize MON810 sequence with electrostatic

adsorption DNA on electrochemical reduced graphene modified electrode. Sensors and Actuators B 2014;202:160–166.

151. Fan Y, Huang K, Niu D, Yang C, Jing Q. TiO2-graphene nanocomposite for electrochemical sensing of adenine and Guanine. Electrochimica Acta 2011;56:4685–4690.

152. Abdul Rasheed P, Sandhyarani N. Graphene-DNA electrochemical sensor for the sensitive detection of BRCA1 gene. Sensors and Actuators B: Chemical 2014;204:777–782.

153. Wang L, Hua E, Liang M, Ma C, Liu Z, Sheng S, Liu M, Xie G, Feng W. Graphene sheets, polyaniline and AuNPs based DNA sensor for electrochemical determination of BCR/ABL fusion gene with functional hairpin probe. Biosensors and Bioelectronics 2014;51:201–207.

154. Wang Y, Zhuang Q, Ni Y. Fabrication of riboflavin electrochemical sensor based on homoadenine single-stranded DNA/molybdenum disulfide–graphene nanocomposite modified gold electrode. Journal of Electroanalytical Chemistry 2015;736:47–54.

155. Yu S, Cao X, Yu M. Electrochemical immunoassay based on gold nanoparticles and reduced graphene oxide functionalized carbon ionic liquid electrode. Microchemical Journal 2012;103:125–130.

156. Lu W, Ge J, Tao L, Cao X, Dong J, Qian W. Large-scale synthesis of ultrathin Au-Pt nanowires assembled on thionine/graphene with high conductivity and sensitivity for electrochemical immunosensor. Electrochimica Acta 2014;130:335–343.

157. Samanman S, Numnuam A, Limbut W, Kanatharana P, Thavarungkul P. Highly-sensitive label-free electrochemical carcinoembryonic antigen immunosensor based on a novel Au nanoparticles–graphene–chitosan nanocomposite cryogel electrode. Analytica Chimica Acta 2015;853:521–532.

158. Huang J, Lin Q, Zhang X, He X, Xing X, Lian W, Zuo M, Zhang Q. Electrochemical immunosensor based on polyaniline/poly (acrylic acid) and Au-hybrid graphene nanocomposite for sensitivity enhanced detection of salbutamol. Food Research International 2011;44:92–97.

159. Li T, Yang M, Li H. Label-free electrochemical detection of cancer marker based on graphene–cobalt hexacyanoferrate nanocomposite. Journal of Electroanalytical Chemistry 2011;655:50–55.

160. Jang H, Kim S, Chang H, Choi J. 3D label-free prostate specific antigen (PSA) immunosensor based on graphene–gold composites. Biosensors and Bioelectronics 2015;63:546–551.

161. Li Y, Han J, Chen R, Ren X, Wei Q. Label electrochemical immunosensor for prostate-specific antigen based on graphene and silver hybridized mesoporous silica. Analytical Biochemistry 2015;469:76–82.

162. Yang F, Yang Z, Zhuo Y, Chai Y, Yuan R. Ultrasensitive electrochemical immunosensor for carbohydrate antigen 19-9 using Au/porous graphene nanocomposites as platform and Au@Pd core/shell bimetallic functionalized graphene nanocomposites as signal enhancers. Biosensors and Bioelectronics 2015;66:356–362.

163. Zhao L, Wei Q, Wu H, Dou J, Li H. Ionic liquid functionalized graphene based immunosensor for sensitive detection of carbohydrate antigen 15-3 integrated with Cd2+-functionalized nanoporous TiO2 as labels. Biosensors and Bioelectronics 2014;59:75–80.

164. Liu K, Zhang J, Yang G, Wang C, Zhu J. Direct electrochemistry and electrocatalysis of hemoglobin based on poly(diallyldimethylammonium chloride) functionalized graphene sheets/room temperature ionic liquid composite film. Electrochemistry Communications 2010;12:402–405.

165. Huang K, Miao Y, Wang L, Gan T, Yu M, Wang L. Direct electrochemistry of hemoglobin based on chitosan–ionic liquid–ferrocene/graphene composite film. Process Biochemistry 2012;47:1171–1177.

166. Sun W, Guo Y, Lu Y, Hu A, Shi F, Li T, Sun Zhenfan. Electrochemical biosensor based on graphene, Mg2Al layered double hydroxide and hemoglobin composite. Electrochimica Acta 2013;91:130–136.

167. Cheng Y, Feng B, Yang X, Yang P, Ding Y, Chen Y, Fei J. Electrochemical biosensing platform based on carboxymethyl cellulose functionalized reduced graphene oxide and hemoglobin hybrid nanocomposite film. Sensors and Actuators B: Chemical 2013;182:288–293

168. Liu H, Duan C, Su X, Dong X, Shen W, Zhu Z. Titania nanoparticles modified reduced graphene oxide nanocomposite with a double-layered structure encapsulating hemoglobin for a mediator-free biosensor. Ceramics International 2014;40(7):9867–9874.

169. Ruan C, Li T, Niu Q, Lu M, Lou J, Gao W, Sun W. Electrochemical myoglobin biosensor based on graphene–ionic liquid–chitosan bionanocomposites: Direct electrochemistry and electrocatalysis. Electrochimica Acta 2012;64:183–189.

170. Suna W, Li L, Lei B, Li T, Ju X, Wang X, Li G, Sun Z. Fabrication of graphene–platinum nanocomposite for the direct electrochemistry and electrocatalysis of myoglobin. Materials Science and Engineering: C 2013;33(4):1907–1913.

171. Suna W, Gong S, Deng Y, Li T, Cheng Y, Wang W, Wang L. Electrodeposited nickel oxide and graphene modified carbon ionic liquid electrode for electrochemical myglobin biosensor. Thin Solid Films 2014;562:653–658.

172. Baby T, Aravind S, Arockiadoss T, Rakhi R, Ramaprabhu S. Metal decorated graphene nanosheets as immobilization matrix for amperometric glucose biosensor. Sensors and Actuators B 2010;145:71–77.

173. Wu H, Wang J, Kang X, Wang C, Wang D, Liu J, Aksay I, Lin Y. Glucose biosensor based on immobilization of glucose oxidase in platinum nanoparticles/graphene/chitosan nanocomposite film. Talanta 2009;80:403–406.

174. Hu Y, Jin J, Wu P, Zhang H, Cai C. Graphene–gold nanostructure composites fabricated by electrodeposition and their electrocatalytic activity toward the oxygen reduction and glucose oxidation. Electrochimica Acta 2010;56:491–500.

175. Shan C, Yang H, Han D, Zhang Q, Ivaska A, Niu L. Graphene/AuNPs/chitosan nanocomposites film for glucose biosensing. Biosensors and Bioelectronics 2010;25:1070–1074.

176. Guo Y, Li J, Dong S. Hemin functionalized graphene nanosheets-based dual biosensor platforms for hydrogen peroxide and glucose. Sensors and Actuators B 2011;160:295–300.

177. Shamsipur M, Tabrizi M. Achieving direct electrochemistry of glucose oxidase by one step electrochemical reduction of graphene oxide and its use in glucose sensing. Materials Science and Engineering C 2014;45:103–108.

178. Jia L, Liu J, Wang H. Preparation of poly(diallyldimethylammoniumchloride)-functionalized graphene and its applications for H2O2 and glucose sensing. Electrochimica Acta 2013;111:411–418.

179. Karuppiah C, Palanisamy S, Chen S, Veeramani V, Periakaruppan P. A novel enzymatic glucose biosensor and sensitive non-enzymatichydrogen peroxide sensor based on graphene and cobalt oxidenanoparticles composite modified glassy carbon electrode. Sensors and Actuators B 2014;196:450–456.

180. Wang L, Ye Y, Lu X, Wu Y, Sun L, Tan H, Xu F, Song Y. Prussian blue nanocubes on nitrobenzene-functionalized reducedgraphene oxide and its application for H2O2 biosensing. Electrochimica Acta 2013;114:223–232.

181. Michopoulos A, Kouloumpis A, Gournis D, Prodromidis M. Performance of layer-by-layer deposited low dimensional building blocks of graphene-prussian blue onto graphite screen-printed electrodes as sensors for hydrogen peroxide. Electrochimica Acta 2014.

182. Jiang Y, Zhang X, Shan C, Hua S, Zhang Q, Bai X, Dan L, Niu L. Functionalization of graphene with electrodeposited Prussian blue towards amperometric sensing application. Talanta 2011;85:76–81.

183. Zhang Y, Liu Y, He J, Pang P, Gao Y, Hu Q. Electrochemical behavior of graphene/Nafion/Azure I/Au nanoparticles composites modified glass carbon electrode and its application as nonenzymatic hydrogen peroxide sensor. Electrochimica Acta 2013;90:550–555.

184. Ye Y, Kong T, Yu X, Wu Y, Zhang K, Wang X. Enhanced nonenzymatic hydrogen peroxide sensing with reduced graphene oxide/ferroferric oxide nanocomposites. Talanta 2012;89:417–421.

185. Mao Y, Bao Y, Wang W, Li Z, Li F, Niu L. Layer-by-layer assembled multilayer of graphene/Prussian blue toward simultaneous electrochemical and SPR detection of H2O2. Talanta 2010;85:2106–2112.

186. Wang X, Liu E, Zhang X. Non-enzymatic glucose biosensor based on copper oxide-reducedgraphene oxide nanocomposites synthesized from water-isopropanolsolution. Electrochimica Acta 2014;130:253–260.

187. Gao W, Tjiu W, Wei J, Liu T. Highly sensitive nonenzymatic glucose and H2O2 sensor based on Ni(OH)2/electroreduced graphene oxide Multiwalled-carbon nanotube film modified glassy carbon electrode. Talanta 2014;120:484–490.

Chapter 6

Nanomaterials for Advancing the Health Immunosensor

Blanca A.G. Rodriguez[1], Erika K.G. Trindade[1], Diego G.A. Cabral[1], Erika C.L. Soares[1], Cayo E.L. Menezes[1], Danielle C.M. Ferreira[2], Renata K. Mendes[3] and Rosa F. Dutra[1]

[1] Laboratory of Biomedical Engineering, Federal University of Pernambuco, Recife, Brazil

[2] National Center for Research in Energy and Materials (CNPEM), Campinas, Brazil

[3] Pontifical Catholic University of Campinas, São Paulo, Brazil

ABSTRACT

Nanotechnology has exerted a significant impact in the development of biosensors allowing more sensible analytical methods. In health applications, the main challenge of the immunoassay is to reach the suitable limit of detection, recognizing different analytes in complex samples like whole blood, serum, urine, and other biological fluids. Different nanomaterials, including metallic, silica and magnetic nanoparticles, quantum dots, carbon nanotubes, and graphene, have been applied, mainly to improve charge electron transfer, catalytic activity, amount of immobilized biomolecules, low-background current, signal-to-noise ratio that consequently increase the sensitivity of immunosensors. Given the great impact of nanotechnology, this chapter intends to discuss new aspects of nanomaterials relating to immunosensor advancement.

Keywords: Immunosensors, immunoassay, nanosensor, nanomaterial

1. INTRODUCTION

A major challenge faced by health programs is the maintenance and availability of diagnostic tests that are required not only in inpatient or outpatient hospital but also for an improved epidemiological survey. In many cases, the absence of laboratory testing or delay of diagnosis generates negative economic impacts, resulting in unnecessary hospitalization, intercurrence, and in some cases implications on the global life quality of patients and underreporting of surveillance. In this context, the development of practical, fast, and reliable analytical methods is imperative.

Biosensors have been considered one of the more attractive analytical methods. They are biodevices capable of transforming an interaction with specific analytes into an electrical signal by a transducer, including a biorecognition element. [1] Pharmaceutical industries and users of rapid tests from the United States and Europe are unanimous in stating that biosensors, mainly those based on point-of-care testing (POCT), or bedside testings are a practical technology, regarded as a short-, medium-, and long-term trend. Among several advantages, POCT can provide immediate responses (results in few minutes or in real time), samples do not need to be transported to the analytical phase (in situ monitoring) and require generally small volumes of samples, and the users can be skilled or unskilled and present better cost-effective analyses compared with conventional technologies used in clinical diagnostic (user-friendly technology). One of the most widely useful POCT is the glucometer, which measures glucose levels with accuracy by requiring a single drop of blood. The rapid glucose measurement is very important in trials to avoid serious adverse effects stemming from diabetes, including seizures, coma, or even death. Worldwide, some diabetic outpatients have been benefited by POCTs.

Although there is a great promising market dedicated to health for the detection of diseases and therapeutic monitoring, biosensors are not yet entirely broadcast, especially those devoted to nonenzymatic reactions, i.e., biosensors based on the affinity between antigen–antibodies, DNA–RNA, DNA–DNA, etc. So far, there are focused studies to develop affinity biosensors for a wide number of applications. Some of these include environmental, agriculture, veterinary, safety food analysis, and health diagnostic in attempting to detect pesticides in water, in monitoring of environmental pollutants in soil, and in determining contaminants and pathogens in food and many others.[2]

Regarding the health diagnostic, affinity biosensors devoted to immunoserological diagnosis have demonstrated to be more accurate, feasible, practical, and advantageous for POCTs than nucleic acid biosensors. First, the levels of antibodies or antigens circulating in whole blood, serum, or other biological fluids are in higher amounts compared to RNA or DNA sequences.

Second, blood samples of immunosensors do not need cell lysis before measurements to release the analytes. Third, antigen or antibody samples do not need pretreatment before measurements as amplification by polymerase chain reaction (PCR) or transcriptase reverse polymerase chain reaction (TR-PCR). Fourth, antibodies are more chemically stable than RNA or DNA sequences that are easily contaminated by attacking the RNases or DNases enzymes present in digital samples.

Due especially to nanotechnology, biosensors dedicated to immunoserological diagnosis have emerged, in the last decade, with the possibility of very promising point-of-care diagnosis. The contribution of nanomaterials has made possible the development of new immobilization matrices with improved features, increased sensor surface area, greater amount of biomolecules per area/volume, and major electrical conductivity, making it possible to achieve a lower limit of detection compared to existing bulk biosensors. Currently, several studies have highlighted the following nanomaterials: metallic, silica and magnetic nanoparticles, quantum dots (semiconductor nanoparticle), carbon nanotubes, graphene, and nanostructured surface.

Selective monoclonal antibodies, recombinant antigens, fragments, and aptamers associated with the nanomaterial advancements to mediate the antigen–antibody responses have also allowed several nanostructured devices with optical, piezoelectric, and electrochemical improved transductions, besides the integration of microfluidics and portable approaches.

Given the great impact of nanotechnology, this chapter intends to discuss new aspects of nanomaterials concerning to the development of immunosensors that resulted in more accurate, reliable, and practical analytical methods for health.

2. IMPORTANT ASPECTS FOR IMMUNOSENSOR DEVELOPMENT

Immunosensor technology has shown an exponential growth in the number of publications over the last decade (Figure 1(a)). Although, there were significant advances in all the areas mainly in the food analyses, immunosensors devoted to health still have huge challenges to overcome in order to yield commercial uses (Figure 1(b)). Some difficulties can be attributed to the biomolecules specificity, immobilization matrix stability, transduction mode employed and pretreatment of complex samples like whole blood, serum, or other fluidic biologics for a reliable detection.

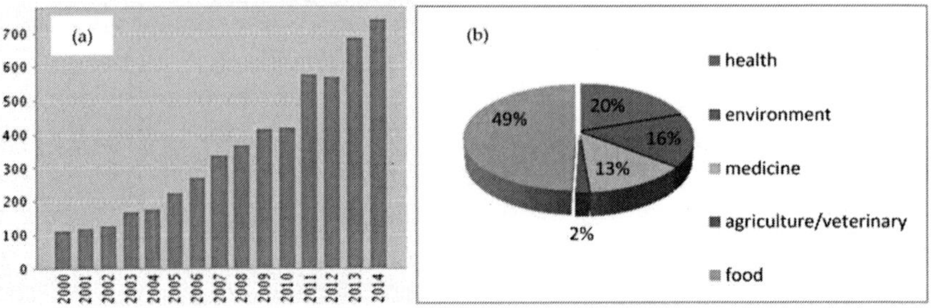

Figure 1. Immunosensors research published over the last decade (a) and main application areas (b). (Extracted from ISI of Knowledge base)

Three aspects are considered crucial in the development of an ideal immunosensor: (a) the bioreceptor, i.e., biomolecule used to recognize the antigen or antibodies in the sample; (b) the matrix assembled for immune compound immobilization; and (c) the transducer type employed.

The choosing of bioreceptors for analyte recognition is a fundamental aspect to ensure an optimal selectivity of immunosensor. Different immunomolecules have been used to detect antigens or antibodies in different samples, besides monoclonal or polyclonal antibodies, and antigens, recently recombinant antibodies, [3] aptamers, [4] and antibody fragments [5] have been also assayed. Immunoglobulin classes IgG and IgM are the most commonly employed in immunosensors. IgG is a Y-shaped structure with two binding sites for antigens recognizing (two paratopes), with approximately 150 kDa. Meanwhile, IgM immunoglobulins are pentamers comprising of ten antigen sites, called natural antibodies. However, due to IgG being more prevalent and most abundant in the circulation (73%), this immunglobulin is more frequently used in all immunoassays. IgM immunoglobulins are detected in specific assays when is important to identify diseases in their acute Phase, Kidwai et al. [6] developed a rapid immunochromatographic (ICT) assay detection for IgM and IgG detection in serum.

Immunosensor performance is directly dependent on the immobilization matrix used and orientation and density of affinity biomolecules (antibodies and antigens) on the electrode surfaces. There are different strategies used to immobilize the recognition element, either directly on the electrode surface or on other solid supports. [7] Conventionally, there are noncovalent and covalent techniques employed to immobilize antibodies, which are based on adsorption, encapsulation, and entrapment in polymers, covalent binding, and cross-linking of antibodies aggregates (Figure 2). Developments in these techniques have great interest and potential application in other areas of biotechnology, including purification of proteins, [8] medicine and drug delivery, [9] regenerative medicine, tissue engineering, and many other applications. [10]

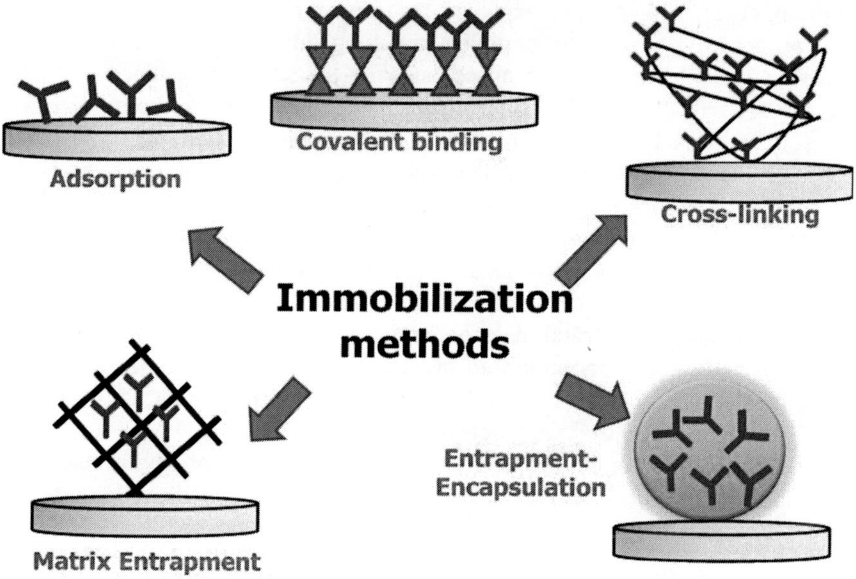

Figure 2. Illustration of different methods of antibodies immobilization.

Although sensor surfaces prepared with antibodies immobilized in a random manner yield satisfactory results, the site-directed immobilization of the sensing molecules significantly improves the immunosensor sensitivity. [11] In this sense, antibodies should be immobilized with optimal capability to recognize the antigens, while fully maintaining their preserved structures. The Fab region needs to be sufficiently free in order to be exposed to the medium, i.e., epitopes of antigens. The best approach is to immobilize antibodies by their Fc regions. [12] This configuration has been achieved using different strategies, including by the use of protein A [13] or protein G, [14] via covalent immobilization through the oxidized sugar chains of the antibody, [15] and others. Besides orientation, it's important to consider antibody density on the electrode surface. A higher density increases the sensor response, however is likely to increase the steric hindrance on planar substrates causing a low immobilization efficiency and low assay sensitivity. To solve these problems, researchers are focused on modifying the substrates for forming the 3D network, which ensures high percentage availability of antibody binding sites. Nanomaterials contribute to increase the amount of protein immobilization because of their capability to form 3D nanostructured surfaces with innumerous cavities and valleys.

Choosing the transducer is another important and fundamental aspect to achieve the sensitivity and response time desired. Bioaffinity sensors (immunosensor) have been explored by using different transduction modes: optical, acoustic and electrochemical by using different approaches. Surface plasmon resonance, [16] localized surface plasmon resonance [17] and fluorescence resonance energy transfer (FRET) [18] are examples of optical transducers. Quartz crystal microbalance (QCM), also entitled mass-sensitive, is the most explored as acoustic transducer. Electrochemical transducers are comprised of different ways to generate an electrical signal, for instance, by amperometric, impedimetric, potentiometric and capacitive changes. [19] Regarding the response time, two classes of immunosensor operation mode are distinguished by, a) Label-free or nonlabeled immunosensors that readily convert the species interaction response with the complementary species into an electrical signal, denominated as direct transduction, and b) Labeled immunosensors that need a second antibody or antigen conjugated to chemical species to generate the analytical response, such as enzymes, fluorescent labels, etc. [20] Although labeled immunosensors are more time consuming than label-free immunosensors, they provide more specificity due to the second antibody which minimize the nonspecific binding negative effects.

The design of label-free affinity biosensors has been extensively studied in academy and industry. One source of stimulation is the demand of POCTs for health, which requires rapid response, lower cost-effective analyses and simplicity for potential analysis. The main technologies of label-free immunosensors currently in use or under development are: surface plasmon resonance (SPR) devices, mass-sensitive, field effect transistor (FET), and electrochemical sensors, including impedimetric and capacitive. Recently, due to advances of nanomaterial based-immunosensors, new categories of label-free amperometric sensors using screen printed electrodes have been successfully developed. [21, 22,23, 24].

3. IMMUNOSENSOR BASED ON NANOMATERIALS

Nanomaterial is composed of unique functional materials that display incomparable characteristics related to their shape, structure and size (in the order of 1 to 100 nm). Nanostructured materials are interesting because they can bridge the gap between the bulk and molecular levels and lead to entirely new avenues for applications, especially in electronics, optoelectronics and biology. The contribution of nanomaterials has allowed powerful immunosensor assemblies, creating platforms with increasing detection limit. [25]

In recent years, various nanomaterials with different physical and chemical properties have been applied to achieve the immobilization of immunocompounds. They can modify the sensing surface, improving the immobilization of procedures and transduction properties of immunosensors. A great number of electrochemical advantages have been mentioned, such as possessing low-background current, high signal-to-noise ratio, and fast electron transfer, including an increased amount of immobilized biomolecules, with consequent increase on the sensitivity of sensors. Nanomaterials with zero-dimensional space (metallic, silica, and magnetic nanoparticles and quantum dots or semiconductor nanoparticles), one-dimensional space (carbon nanotubes), and two-dimensional space (graphene) have been show as potential for different transducers in many immunosensor applications.

3.1. Metallic, Silica, and Magnetic Nanoparticles

Nanoparticles (NPs) obtained from commercial sources or properly produced in laboratories have attracted much attention in biological studies due to their low toxicity, biocompatibility, and unique optical properties. Nanoparticles and nanospheres can be divided into magnetic, metallic, semiconducting, or insulating nanoparticles based on their conductivity.

NPs have high surface areas and unique physical–chemical properties that can be easily tuned, making them ideal candidates for developing immunosensors devices. The basic function of nanoparticles in an immunosensor can be summarized as follows: immobilizing the biomolecules on the electrode surface, catalyzing electrochemical reaction, enhancing electron transfer charge, and acting as a reactant or labeling biomolecules for further experiments, among others. [26]

Biological tests measuring the presence or activity of selected analytes become quicker, more sensitive, and flexible when nanoscale particles are combined, with numerous advantages over more traditional procedures. In recent years, various nanomaterials with distinct physical and chemical properties have been applied to improve the immobilization of immunocompounds. [27] These have many electrochemical advantages, such as possessing low-background current, high signal-to-noise ratio, and fast electron transfer, besides increased amount of immobilized biomolecules, with consequent increase on the sensitivity of sensors. [28]

Surface modification using nanoparticles composites have shown an increase of sensitivity and help adsorb a large amount of antibodies on electrode surface. Lu et al. [29] constructed an immunosensor based on a nanocomposite formed with CeO_2 and gold nanoparticles on the Au electrode via cysteine to detect a cardiac marker, the myeloperoxidase.[29] Thereat, the nanoparticles enhanced the active surface area available for antibody binding. The high stability of this sensor was attributed to the good biocompatibility of the composite. Another study shows an increase in immunosensor response. Fe_2O_3nanoparticles were used in the construction of an electrochemical device to detect cancer biomarker prostate antigen (PSA) via horseradish peroxidase (HRP) signal. The high amount of nitrodopamine (film coated on nanomaterial to immobilize the anti-PSA) anchored onto Fe_3O_4 increased the loading of biomolecules onto the surface, which increased the electrochemical immunosensor sensitivity. [30]

The carcinoembryonic antigen (CEA) is a protein used as a tumor marker and has been frequently investigated in immunoreactions. An elevated CEA level in serum may be an early indication of lung cancer, ovarian carcinoma, colon cancer, breast cancer, and cystadenocarcinoma. Recently, an interesting work was reported involving this protein investigation using an immunosensor constructed by Pt hollow nanospheres modified with anti-CEA as label for a 3D Au-TiO_2 hybrid platform. [31] The immunoassay exhibited a high sensitivity and a low detection limit compared with conventional label methods. Another way to detect CEA antigen was developed by Gao et al. [32] using a label-free voltammetric sensor with chitosan and gold nanoparticles (AuNPs) to immobilize anti-CEA on carbon surface. The detection is based on the variation of current responses before and after the immunoreaction. When the immobilized antibodies have bounded with antigens, the antigen–antibody complex formed on the surface inhibited the electron transfer. Then a decrease of the electrochemical signal was verified as the concentration of antigen on surface increased. Another method also using a composite of chitosan and AuNPs for CEA determination, but with multiwalled carbon nanotubes, was described by Huang et al. [33] The nanocomposite film exhibited high current response intensity, good biocompatibility, and high stability. Similar CEA detection was also performed using a gold nanoparticle–thionine-reduced graphene oxide composite that possesses as advantage fast electron transfer kinetics and large specific surface area. [34] Another work for CEA analysis described the use of silver nanoparticles on SiO_2 surfaces. [35] The high stability of the immunosensor was assigned to the stable nanocomposite produced.

The sensitivity of electrochemical immunosensors can also be improved by using the association of AuNPs and dendrimers that are three-dimensional macromolecules, with hundreds of functional groups at the periphery, for surface modification. This architecture was employed by An et al. [36] to detect α-synuclein, a very important neuron protein. The dendrimer (PAMAM)-encapsulated AuNPs were covalently bound on the poly-o-aminobenzoic acid (ABA) electropolymerized on a glassy carbon electrode surface to achieve abundant carboxyl groups, which allowed a highly dense antigen immobilization and facilitated the improvement of electrochemical responses as well. Subsequently, the enhanced gold nanoparticle labels were fabricated by immobilizing a horseradish peroxidase secondary antibody (HRP-Ab_2) on the AuNPs surface. After an immunoassay process, the labels were introduced onto the electrode surface to produce an electrocatalytic response with hydrogen peroxide. The presence of dendrimer Au not only increased the covalent coupling of more protein but also accelerated electron transfer when compared to immunosensor without dual signal amplification strategy.

The picogram detection limit of estradiol was achieved using an immunosensor constructed with AuNPs and protein G scaffold to modify a gold electrode. [37] Coupled with the amperometric determination of the hormone in a flow system, the device exhibited superior linear range, sensitivity, and stability in blood serum samples spiked with estradiol.

Other applications for metallic nanoparticle have included optical transduction. Krishnan constructed an optical immunosensor in a quartz glass surface for the detection of *Escherichia coli*, using core-shell nanoparticles (silver-silica) anchoring labeled antibodies. The results show that changes in photoluminescent standards are consistent with the immobilization of various species. Thus, the optical immunosensor demonstrated improved sensitivity and specificity in comparison to the usual methods, detecting as low as 5 CFU/mL.

Using a great number of luminescence molecules as stabilizers coated on the surface of the AuNPs, Shen et al. [40] developed an electrochemiluminescence immunosensor to detect human cardiac troponin, an important acute myocardial infarction biomarker. First, the sensor was constructed by using streptavidin-coated gold nanoparticles as the immobilization matrix for biotinylated antibody. Meanwhile, the three-dimensional nanostructures increased the surface-to-volume ratio, allowing more biomolecules to be immobilized. The sandwich-type immunosensor was fabricated by reacting with antigen and AuNPs modified with luminescence molecules labeled with the secondary antibody, forming a nanoprobe. The enhanced sensitivity of the proposed apparatus mainly derives from the novel nanoprobe, which achieves a large amount of luminescence molecules loading toward each sandwich immunological reaction event.

Another strategy in immunocomplex detection involves the use of magnetic nanoparticles as solid support for biomolecule immobilization. The magnetic particles offer the convenience of magnetic separation. These particles respond to a magnetic field but demagnetize completely when the field is removed. Thus, the nanoparticles can easily be separated from the liquid phase with a small magnet but can be dispersed again immediately after the magnet is removed. The use of magnetic nanoparticles as solid phase for the immunosensor development improves the bioreaction performance due to surface area increase and has better immunoassay kinetics because the particles are in suspension and the target species does not need to diffuse very far. [41]

An interesting work was described by Shen et al., [42] who developed a device to detect *E. coli*, an intestinal pathogenic bacterium, using a quartz crystal microbalance (QCM) immunosensor based on beacon magnetic nanoparticles. A polyclonal antibody was immobilized on iron nanoparticles with subsequent addition of *E. coli*. AuNPs were inserted in the system to amplify the signal. Weakly bound biomolecules were removed with a magnetic plate. Finally, the crystal was modified with protein A and monoclonal antibody. The frequency shift of the QCM immunosensor is amplified using *E. coli*immobilized on to magnetic particles and enlarged gold particles for the bacterium detection. The signal was amplified three times, and the crystal was regenerated without difficulty and could be used at least 10 times. In a recent work, the use of magnetic nanoparticles as an amplification means for QCM signal for avian influenza H5N1 virus detection has been reported. [43] Polyclonal antibodies against the virus were immobilized on the gold surface of the crystal through self-assembled monolayer (SAM). Target H5N1 viruses were then captured by the immobilized antibodies, resulting in a change in the frequency. Magnetic nanoparticles coated with anti-H5 antibodies were used for further amplification of the binding reaction between antibody and antigen (virus).

AuNPs have a remarkably high extinction coefficient and strong distance-dependent optical properties. Different aggregation states of AuNPs correspond to distinctive color, which can be appreciably discerned with the naked eye and be used in immunoassay. Based on this, Yuan et al. [44] developed a label-free colorimetric immunoanalysis for the simple detection of neurogenin3, a marker for pancreatic endocrine precursor cells, using glutathione functionalized gold nanoparticles. The antibody-conjugated AuNPs were formed through electrostatic interaction upon the addition of the antibody to the modified AuNPs solution. The antigen positively charged to the negatively charged AuNP antibody will minimize the electrostatic repulsion between nanoparticles by neutralizing the surface charge and then agglomeration is induced by an increasing NaCl salt concentration, noticeably revealed by the color change of the solution from red to purple or blue. The concentration of neurogenin3 can be conveniently accessed by the optical absorption spectra. Another important property of the AuNPs is that they could catalyze silver reduction and act as the nuclei for silver precipitation. [45] In this interesting work, the core mechanism of the method to quantify cardiac troponin is that the catalytic capability of the AuNPs was inhibited by immunocompounds covering their surface. This covering is influenced by the amount of reduced silver of the reaction, resulting in a color difference.

In state-of-the-art improved sensor devices for health applications, the possibility of assembling nanoparticles and biomolecules in different ways by using different sizes, formats, and compound types allow more sensitive, simple, robust, and especially faster analysis.

3.2. Quantum Dots

Quantum dots (QDs) represent one class of nanostructured materials. They are spherical nanocrystals of semiconductor, 1–10 nm in diameters, made of elements of the IIB–VIA or IIIA–VA groups. The use of QD properties requires sufficient control during their synthesis because their intrinsic properties are determined by different factors, such as size, shape, defect, impurities, and crystallinity. [46, 47] Two of the most widely used commercial QDs come with a core of CdSe or CdTe and a shell of ZnS and emissions from 405 to 805 nm. [48, 49] The shell stabilizes the structure, helping to overcome quenching compared to a QD made only from a core and provides a large surface area available for further modification.

Analogous dimensions of QD and biological materials, such as enzymes, antigens/antibodies, protein receptors, or nucleic acids, show great promise as photonic labels for bioanalytical applications and suggest that electronic communication between the QD and the specific recognition site or biocatalytic processes of the biomaterials can exist. These electronic interactions may lead to the optical or photoelectrochemical transduction of the biological events. [47, 50]

Generally, monodispersed QDs are developed by introducing organic molecules that adsorb on the surface and act as capping agents. The efficacy of QD in a biological application is critically dependent on coating properties. The liabilities of these initial methods require the continued development of QD coatings. Important criteria for an ideal QD coating include high-affinity for the QD surface, long-term colloidal stability across a broad range of pH and ionic strengths capacity for bioconjugation, minimization of hydrodynamic size, and biocompatibility with nonspecific binding. [51] However, the selection of organic ligands that bond with surface atoms of the QD is a very delicate issue. In general, phosphenes or mercaptans (-SH) are the most widely used ligands. [52]

In order to make QD suitable for biological imaging and use in a biological environment, they also have to be rendered water soluble. This is done by capping the shell with a polymer layer that contains a hydrophobic segment facing inward, the shell, and a hydrophilic segment facing outward. The hydrophilic layer can be modified to include functional groups such as –COOH and $-NH_2$ groups for further conjugation to proteins and antibodies or oligonulceotides. [49, 53] The single-step synthesis of thiolated cyclodextrin-modified CdSe/CdS core-shell QD resulted in a water-soluble QD, keeping the luminescence properties of the QD in aqueous systems. This is an important aspect since biorecognition events require aqueous environments for reaction. [54]

Thiol ligands and amphiphilic polymers are the most common types of QD coating available. They allow two essential design elements: a moiety that anchors to the QD surface and a hydrophilic functionality for aqueous dispersion. The selection of these groups determines the degree to which a QD coating can approach the ligand/amphiphilic polymer structures. [51] For example, small molecules with thiol groups can bind to the quantum dot surface, and distal carboxylated group provides aqueous colloidal stability. [55] Another strategy for QD coating that provides aqueous dispersion, improves the biocompatibility, and minimizes nonspecific binding was developed by Mattoussi et al. (2000). They combined dihydrolipoic acid, a dithiol ligand that binds the QD more closely, which is attached to a poly(ethylene glycol) oligomer.

Biomolecule conjugation on to the QD is achieved by different ways like electrostatic binding, noncovalent biotin–streptavidin bonding, or covalent bonding. The most widely used conjugation technique of all is the covalent bond formation between the QD surface and the biomolecules. Surface modifications on QD allow easier covalent bond formation. In one of the most widely used methods, amine-terminated QDs are used for conjugating antibodies. The amine-terminated QDs are activated with maleimide containing a cross-linker molecule, which can then be conjugated to a fragment or whole antibody molecule. Some of the most commonly employed QD conjugation methods are based on cross-linking reactions between amine and sulfhydryl groups, carboxylic acid, and amine and aldehyde and hydrazide groups. The carboxylic-amine bond has one advantage over all other methods, seeing as this method does not require any antibody modification before QD conjugation. In the case of amine and sulfhydryl bond formation, the antibody should be reduced to expose their interchain -SH bonds. In relation to aldehyde and hydrazide bonds, carbohydrate groups on the antibody Fc portion are oxidized. These modifications on antibodies may affect their performance to a certain extent. [56]

Functionalized semiconductor quantum dots have been used as fluorescence labels in numerous biorecognition events. For example, Liu et al. (2004) developed an immunosensor with simultaneous measurements of four proteins based on antibodies linked to the inorganic nanocrystal. Stripping voltammetric immunoassay was used to observe the response of a mixture containing microglobulin, IgG, bovine serum albumin, and C-reactive protein connected with ZnS-, CdS-, PbS-, and CuS-labeled antibodies, respectively. The system was obtained by using carbamate linkage for conjugating the hydroxyl-terminated nanocrystal with the secondary antibody. [57]

Li et al. (2011) [58] used a novel strategy to modify the surface of graphene quantum dots composites. A layer-by-layer assembling process was employed via electrostatic interactions between negatively charged thioglycolic acid modified CdSe QD and positively charged graphene, which was noncovalently functionalized with poly (diallyldimethylammonium chloride) (PDDA) via an exfoliation in situ reduction of graphene oxide in the presence of PDDA. This process allowed excellent conductivity, extraordinary electron transport properties, and large specific surface area, which resulted in high electroluminescence (ECL) intensity and excellent film-forming ability and made it a promising candidate for the development of ECL immunosensors.

Luminescent quantum dots are viable optical markers and have been used in a direct assay for IgG. Protein A was labeled with CdSe/Zn QD (λ_{max} of 655 nm) and then was immobilized at the tip of an optical fiber. Once the immunoreaction with IgG occurs, a decrease in fluorescence intensity is observed as a result of the fluorescence resonance energy transfer from the QD to the bound protein. [59] Lingerfelt et al. [60] reported the preparation of QD-biotin conjugates and their use in immunochromatographic assays. The detection of immunoglobulin G was carried out on a glass chip through a sandwich assay approach using a secondary antibody conjugated to the QD. [61] A sandwich immunoassay for the detection of staphylococcal enterotoxin B was run using polyclonal sheep anti-staphylococcal enterotoxin B antibody conjugated with QD and microtiter plates coated with monoclonal staphylococcal enterotoxin B antibody. [62]

Kerman et al. (2007) applied conjugated QD streptavidin in a model immunoassay system for the detection of a total prostate-specific antigen cancer marker from the spiked and undiluted serum samples. Immunorecognition was carried out on a carbon substrate using a sandwich assay approach. After the recognition event, the substrate was exposed to the biotinylated secondary antibodies and, subsequently, fluorescence imaging of the substrate surface illuminated the QD. [63]

QDs based on narrow photoemission spectra, with high resistance to photobleaching and broad excitation spectra, are widely used as tags in immunoassay. A carcinoembryonic antigen immunosensor was fabricated using biofunctionalized QD probes. This immunosensor array was designed to detect a wide range of analytes using the inherent characteristics of QD and the flexibility of engineered elastin-like polypeptides. [64]

There are some studies based on thioalkyl-functionalized QD, which are pH sensitive, [65] suggesting many different biological applications. In this context, mercaptoacetic acid-CdSe/ZnSe/ZnS QDs have been used as an intracellular pH sensor by observing a quenching of fluorescent QDs in acidic pH. [66]

Another approach was based on the direct conjugation of CdSe/ZnS QD–IgG complexes using a genetically engineered tripartite fusion protein. This fusion protein was made up of a histidine tag for QD conjugation, an elastin-like peptide for stimuli-responsive purification and the protein L (cell-wall component of *Peptostreptococcus magnus*) that has high affinity to IgG. The functionality of this sensitive immunofluorescent probe was demonstrated in the detection of a representative tumor antigen. [64]

Despite recent progress, more work still needs to be done to achieve reproducible and robust surface functionalization and develop flexible bioconjugation techniques. The potential of QD in biology has just begun.

3.3. Carbon Nanotubes

Carbon nanotubes (CNTs) can be highlighted as the most important nanomaterials for biosensors. Their excellent optical and mechanical conductivity, high surface-to-volume ratio, good chemical stability, biocompatibility, and easy functionality have revolutionized the biosensors area for the last decade.

Since their discovery by Ijima in 1991, CNTs are being used in large volumes for different purposes in many industrial areas, i.e., in nanocomposites for sporting materials, as a battery in supercapacitors, transparent films, and liquid crystal displays. Other limited-volume carbon applications include their use as components in wind turbine blades, scanning probe tips, membrane filters and sorbents, flat panel displays, memory devices, transistors, drug delivery systems, and other medical and analytical chemistry applications. [67]

Carbon nanotubes can be described as hollow cylindrical tubes of graphene sheets with high aspect ratios (length/diameter). [68] The structure of graphene is a planar atomic sheet consisting of covalently bonded carbon atoms. The atoms in graphene are sp^2 carbon units, forming a two-dimensional (2D) network with a hexagonal lattice. A graphene layer wrapped as a cylinder forms a single-walled carbon nanotube (SWNT). A multiwalled carbon nanotube (MWNT) is nothing more than multiple SWNTs packed in a tight concentric frame. All the carbon nanotubes have several nanometers in diameter and many microns in length. SWNTs have the smallest diameter (0.8–5 nm), whereas MWNTs have a larger diameter (~3 to >100 nm), both variable in length (from millimeters to tens of nanometers). The proper architecture is reflected in the highly anisotropic properties. Most of the extraordinary electrical, thermal, and mechanical characteristics are localized specifically along the axial direction. The strong sp^2 bonding between the carbon atoms in CNTs yields remarkable mechanical strength, making them one of the most resilient materials. Moreover, it is known that an SWNT presents metallic and semiconducting properties where such electronic features depend on its chirality. [69] They have three different structures: armchair, zigzag, and chiral. [70]

The applications of CNTs in biosensors have been hindered for a long time due to the drawback of insolubility. CNTs present a high molecular weight, an ability to entangle (tendency to individually interact with each other through van der Waals forces), aggregating into bundles and ropes. However, these bundles can be quite large that they become insoluble in any solvent; thereby, it can be difficult to disperse them in either aqueous or nonaqueous medium. [71]

Ultrasonication is an effective method to disperse CNTs in liquids that have low viscosity, such as water, acetone, and ethanol. However, most polymers are either in a solid or viscous liquid state, which require the polymer to be dissolved or diluted using a solvent to reduce the viscosity before dispersion of

CNTs. [72] The simplest stable dispersions have been achieved by using a solvent able to efficiently interact with CNTs, such as phenylethyl alcohol, *N*-methylpyrrolidone (NMP), *N,N*-dimethylformamide (DMF), and *N,N*-diethylacetamide (DEA). An additional strategy to favor dispersion in organic solvents is to coat CNTs with a molecule characterized by a high affinity toward nanotube sidewalls and at the same time soluble in the selected solvent. Both small molecules and polymers formed by repetitive units of alkyl chains and aromatic compounds have been used as dispersants. Thus, the adsorption of different polycyclic aromatic hydrocarbons such as pyrene, anthracene, tetracene, and phenanthrene on SWCNTs has been extensively investigated. In order to favor the dispersion of CNTs in water, the widely and most used approach is the adsorption of surfactants. These small molecules typically have a hydrophobic tail and a hydrophilic head group—the former is intended to favor adsorption onto the hydrophobic carbon nanotube and the latter to promote affinity with the aqueous solvent. Over the years, stable aqueous CNT dispersions were obtained with differently charged and nonionic surfactants such as sodium dodecylbenzene sulfonate (SDBS), cetyltrimethylammonium p-toluenesulfonate (CTAT), cetyltrimethylammonium bromide (CTAB), and sodium cholate (SC) enhanced by sonication. Additionally, polymers have been employed for CNT dispersion in water. The majority of polymers and block copolymers have been used to wrap CNT by exposing their polar domains toward the aqueous environments while favoring the contact of their hydrophobic domains with the nanotube surface. [73]

Strategies for the immobilization of biomolecules on CNTs have been widely explored aiming to improve sensitivity on an immunosensor. The high aspect-ratio of CNT allows a great amount of anchored biomolecules by noncovalent and covalent functionalization for different types of transduction (Figure 3).

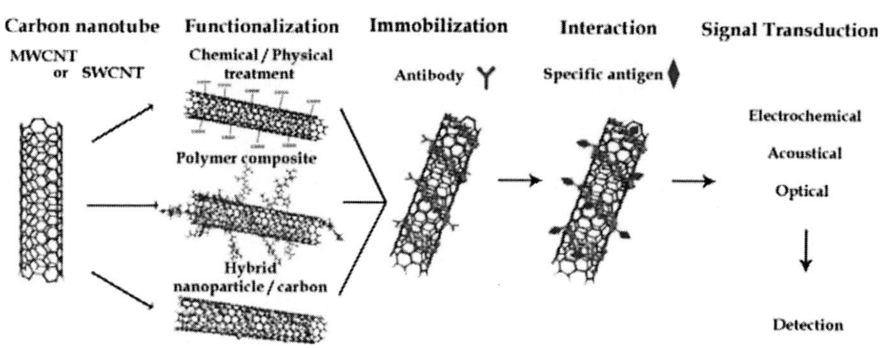

Figure 3. The schematic diagram steps and transducer types of immunosensors based on carbon nanotubes.

Noncovalent functionalization enables reversible adsorption of biomolecules on the CNT surface. For this purpose, CNTs are added to a dispersant solution, and the mixture is agitated in an ultrasound bath. The CNTs are mechanically debundled and then stabilized by dispersant molecules through noncovalent interactions. This does not cause changes in the chemical structures, electronic, and mechanical properties of the carbon nanotubes, and therefore it is a very attractive method. Surfactants, biomolecules, and polymers are widely used as dispersants and noncovalent modifiers. Among them, the polymers are quite efficient dispersants because of their long chain structure that can wrap themselves around CNTs by disrupting the van der Waals interactions between the walls of nanotubes. In biosensors, polymers are particularly interesting and have been widely employed to prepare CNT composites for electrochemical detection, especially for conductive polymers due to their native electron trans-mediation, high conductivity, good environmental stability, and specific organic groups. Furthermore, they can be overoxidized to create an electrically insulating layer. Many reports have demonstrated that CNTs coated with polymers, including polypyrrole, poly(methylene blue), poly(neutral red), poly(acrylic acid), and poly(3-methylthiophene) have become a popular strategy [74]

The covalent chemical functionalization arises mainly from organic molecules reacting with carboxyl groups of CNTs treated by oxidation, which depends on the hydrophilicity/hydrophobicity of the species attached, which can make carbon nanotubes soluble in water or organic solvents. The modification of carbon nanotube surfaces by covalent attachments of soluble groups usually alters intrinsic properties such as conductivity, mechanical strength, and optical properties. [73, 75] Nevertheless, the functionalization involving the introduction of carboxyl, amine, thiol, and other reactive groups are attractive strategies because antibodies or antigens can be covalently immobilized, improving the stability and, in some cases, the sensitivity and selectivity of the immunosensors. Figure 4 exhibits different covalent and noncovalent methods of functionalization of the carbon nanotubes.

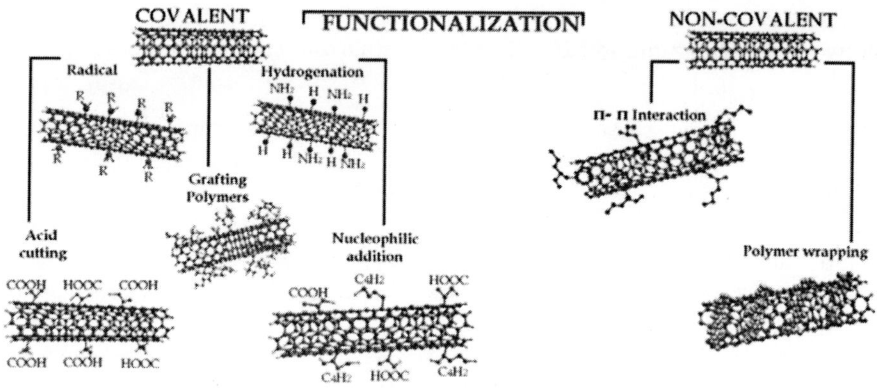

Figure 4. Illustration of different carbon nanotube functionalization methods.

In some practical applications, Sánchez and coworkers [76] have constructed immunosensors where the biomolecules are immobilized on an MWCNT–polysulfone composite film. The layer was applied onto screen printed working electrodes to provide a suitable immunosensor for the rapid determination of human chorionic gonadotropin hormone. The detection limit was 14.6 mIU/mL with a linear range up to 600 mIU/mL. Viswanathan et al. [77] developed another disposable electrochemical immunosensor based on CNTs for the detection of carcinoembryonic antigen with a detection limit of 1 pg/mL in saliva and serum.

For better detecting performance toward interleukin-6, in cases of oral cancer, Malhotra et al. [78] made an ultrasensitive immunosensor sandwich assay on an electrically conductive and high surface area platform, featuring densely packed and upright SWCNTs with capture antibodies attached to their ends. This biosensor had the highest sensitivity at 19.3 nA/mL (pg IL-6)$^{-1}$ cm^{-2} and the best detection limit (DL) of 0.5 pg/mL (25 fM) for IL-6 in 10 μL of calf serum. Similarly, Munge et al. [79] have presented a novel electrochemical sensor using a sandwich immunoassay for the detection of metalloproteinase-3, a cancer biomarker, based on vertically aligned SWCNT arrays. The multilabeled polymeric bead amplification method demonstrated a detection limit of 0.4 ng/mL in 10 μL of calf serum. This showed great potential for these elements in future cancer diagnostics.

The self-assembly of oxidative SWCNTs on gold was attempted for the detection of bovine serum albumin, BSA, by cyclic voltammetry. This sensor has shown excellent sensitivity and dynamic linear response at the range of 0.1 to 1.2 μM. [80] A conductive multilayer composed of Nafion-coated MWCNTs, thionine (Thi), and AuNPs was prepared using an innovative self-assembly strategy to form an immunosensor for α-1-fetoprotein. This reagentless amperometric sensor presented broader linear response in two ranges between 0.5–20 ng/mL and 20–200 ng/mL with a detection limit of 0.26 ng/mL. [81]

There are many studies demonstrating that CNTs can provide high electrocatalytic activity to the electrochemical devices and minimize surface fouling effect. Their unique properties enable them to promote a fast electron transfer, play the role of a biomolecular immobilization platform, and be compatible with different materials for construction of different electrodes. The sensitivity of electrochemical sensors has been greatly enhanced due to these materials, which promotes high active surface area and conductivity. CNTs play an important role in recent trends for immunosensor fabrication. They can function as transducers, act as carriers and labels of immunoassay due to the transfer of large amounts of electroactive species for amplifying electrochemical signals, and also offer an easy way to protect and stabilize these bioactive species. [82] In this section, different strategies were described like the easy adsorption of CNT on the electrode surface, biomolecule immobilization by simple adsorptions and covalent binding, and preparation of screen printed electrode.

Based on a simple amino-functionalization method for MWCNT, Dutra's group developed an electrochemical immunosensor for the detection of human cardiac troponin T (cTnT), an important marker for acute myocardial infarction. It showed a broad linear range (0.02 to 0.32 ng/mL) and a low limit of detection, 0.016 ng/mL. [83] Another sandwich-type immunosensor for the detection of cTnT based on carbon nanotubes supported by a conductive polyethyleneimine film has achieved a low limit of the detection of 0.033 ng/mL and a linear range between 0.1 and 10 ng/mL. [24] Amperometric response is generated by peroxidase reaction with substrate in chronoamperometry detection. The high electronic transfer and catalytic response helped by the CNT was essentially important to dispense the mediator in order to generate the analytical responses. Due to the high conductivity achieved by incorporation of CNTs in screen printed electrodes, a label-free amperometric immunosensor was fabricated, presenting new strategies based on differential pulse amperometry. The immunosensing device for cTnT, with amine-functionalized carbon nanotubes incorporated in screen-printed electrode ink, reached a lower detection limit of 0.0035 ng/Ml, better than any previously described immunologic sensors. [21]

3.4. Graphene

Graphene is a two-dimensional material, formed by carbon atoms that are densely packed in a regular sp^2-bonded atomic scale as a hexagonal pattern, [84] which was produced in laboratory for the first time in 2004. [85] This is the base construction block for other carbon allotropes such as fullerene, carbon nanotubes, graphite, nanoribbons, and others. [86] It is a transparent (optical transmittance of ~97.7%), very thin sheet with large theoretical surface area (2630 $m^2 g^{-1}$), one atom thick, stronger than steel (mechanical stiffness of 1TPa). In addition, it is a good heat conductor (thermal conductivity of 500 W $m^{-1} K^{-1}$), chemically inert, and a semimetal with high electron transfer (charge-carrier mobility of 250 000 $cm^2 V^{-1} s^{-1}$ at room temperature). [87, 88] These properties make them attractive for many applications. [84] There are a variety of synthesis methods for obtaining graphene such as chemical vapor deposition, chemical vapor deposition by plasma, the graphite intercalation of metal sheets, mechanical or thermal exfoliation of graphite oxide, intercalation, and exfoliation of graphite, among other variants of these. Despite all these syntheses methods, the mass production is still difficult, making it hard to develop some applications. [88]

Graphene oxide (GO) synthesis has been an alternative to graphene mass production. It is produced from the oxidation of graphite and has polar oxygen functional groups. GO is rich in carboxylic acids at its edges, and epoxy and hydroxyl groups at basal planes, which grants many functionalization routes and good dispersion in water. [89] Furthermore, the functional groups are responsible for the exfoliation of graphite, seeing as they increase the interplanar distance due to the formation of hydrogen bonds between the graphite sheets. The hydrogen bonds are weak and can be easily broken by ultrasound bath, resulting in monolayer or a few sheets of carbon, known as GO. This is an excellent material for biological applications attributed to the functional groups that readily interact with nucleic acids, proteins, cells, and other organic molecules. However, GO is not a good electrical conductor because of the disruption of its sp^2 bonding as functional groups increase, which can narrow its nanobiotechnology applications. To overcome these difficulties, the reduced form of GO has been chosen as an alternative.

Reduced graphene oxide (RGO) has more commonly been used to form nanocomposites with nanoparticles or polymers to develop biomedical applications such as biosensors, controlled drug delivery, therapeutic modalities for cancer treatment, substrates for antibacterial effects, scaffolds for mammalian cell culture, and gene delivery among others. [90]

In the RGO synthesis, functional groups are removed, and the conductivity is increased again. This removal can be done in different ways such as electrochemical, optical, hydrothermal, microwave, or heating procedures. These methods for removing the functional groups form different shapes and therefore the conductivity recovery is variable. Also, they form different functional groups, becoming favorable in a wide number of applications. [91]

The most common method for obtaining RGO is chemical reduction, which is used in colloidal dispersing of GO. Hydrazine monohydrate is the most used reduction agent, seeing as it does not react with water, which makes it attractive for aqueous dispersions. The reduction process mediated by hydrazine normally occurs through the addition of H_2 groups and removal of

N_2, and it is gentle enough not to affect the cyano and nitro groups. The second most used reducing agent is sodium borohydride ($NaBH_4$), which is more effective than hydrazine and easily hydrolyzed in water. The hydrolysis process should be slow enough so it does not affect the reduction process. The $NaBH_4$ reduces C=O species and has a low effect in epoxy and carboxylic acid groups. Other reduction agents such as hydroquinone, alkaline solutions, and gaseous hydrogen are also being described as mediators. [92]

Another low-cost mean of producing RGO is by thermally reducing GO, heating it in a furnace at 1050 °C, which creates thermodynamically stable oxide carbon species. Electrochemistry can also be used in the reduction process of GO, removing oxygen functionalities. Thermal and electrochemical reduction techniques have the advantage of avoiding dangerous reducers and the problem with their disposal, but they are still less used than chemical reduction. The reduction processes frequently provide RGO with functional groups, but in some cases, its functionalization is still necessary prior to use. Covalent and noncovalent methods for functionalization of RGO have been studied, whereas noncovalent bonds are the most common used, for instance, the physical adsorption of both polymers and small molecules via van der Waals interactions onto the basal planes of RGO sheets. [92]

An initial and successful approach using RGO to create biosensors was its combination with nanoparticles. An example is the work of Shan et al., [93] who used Au nanoparticles associated with RGO and chitosan as a nanocomposite film onto a gold electrode for developing an electrochemical glucose sensor obtaining a linear response range from 2 to 10 mM. Copper nanoparticles were also used to modify RGO sheets to create an electrochemical sensor for glucose obtaining a detection limit of 0.5 μM. [94]

Afterward, RGO was applied to the production of immunosensors, with and without nanoparticles. An example is the work of Mao et al., [95] who reported the use of RGO sheets coated with Au nanoparticles, which were initially functionalized with human immunoglobulin G (IgG) to create conjugates. These conjugates were immobilized onto a field effect transistors (FETs) biosensor platform for the detection of human proteins.

A developing area for immunosensors is the detection of cancer markers. It is a recent and very attractive field, with growing publication numbers, including the use of RGO for these. An example is the work of Zhong et al., [96] who used a gold nanoparticle enwrapped graphene nanocomposite on a glassy carbon electrode in a sandwich-type immunoassay format. The detection limit obtained for this assay was 10.0 pg mL^{-1}. Another CEA immunosensor was developed by Huang et al. [97] using Ag/Au nanoparticles coated with RGO in a clinical immunoassay for the detection of carcinoembryonic antigen (CEA). The nanoparticles were used as means for amplification of the signal and the method showed a detection limit of 8.0pg mL^{-1} in human serum.

Different cancer markers were the focus of other works, such as the one developed by Tang et al., [98] which aimed to create an electrochemical immunosensor for the simultaneous detection of alpha fetoprotein (AFP) and carcinoembryonic antigen (CEA), using biofunctionalized magnetic RGO nanosheets (MGO) coated with iron oxide nanoparticles as immunosensing probes, obtaining detection limits of 1 pg mL^{-1} for CEA and 1 pg mL^{-1} for AFP. Also, Teixeira et al. [99] created a chemically modified epitaxial graphene diagnostic sensor for the detection of human chorionic gonadotropin, which is a main marker for pregnancy and can also indicate some types of tumors. They obtained a detection limit of 0.62 ng/mL.

For optical transductors in cancer marker detection, RGO was used by Xu et al. [100] in a modified glassy carbon electrode using luminol to create a electrogenerated chemiluminescence (ECL) immunosensor for prostate specific antigen, using two antibodies in a sandwich immunoassay, which achieved a detection limit 8.0pg mL^{-1}. RGO was also used in the development of an ECL immunosensor using CdTe quantum dots (semiconductor nanocrystals) along with Au nanoparticles for signal amplification in the detection of human IgG with detection limit of 0.005 pg/mL. [101]

4. CONCLUSIONS

The different concepts of nanomaterials applied to immunosensors have been discussed. Nanomaterials can be utilized for a wide variety of immobilization matrices intending to improve the immunosensor sensitivity, allowing lower limit of detection. The potential of nanomaterials on immunosensors has resulted in a positive impact on the clinical outcome of various diseases, including cancer, cardiac injuries, parasitic infections, and viruses, among others. It is well known that carbon nanotubes and grapheme nanostructures are more favorable to amperometric transducers due to their electrochemical proprieties, which increase the electronic transfer charge and electrocatalytic activity. Metallic and magnetic nanoparticles have successfully been applied to different transducers, especially electrochemical, by enlarging the electroactive surface area. Quantum dots, a semiconductor nanoparticle, present a promising potential for many transducers mainly due to their photostability and luminescence characteristics. Nevertheless, more challenging studies involving nanomaterial sciences, biochemistry, electronic, and molecular engineering should be done in attempting to achieve faster, more practical, and more reliable biosensors. More specifically, biomolecules and a deeper knowledge in nanomaterial science associated to new electronic designs represent a promising field in the development of portable and integrated point of care devices for health applications and other areas of diagnostic.

5. ACKNOWLEDGEMENTS

The authors thank the financial support of the National Council for Scientific and Technological Development (CNPq), CAPES (Coordination for the Improvement of Higher Level Personnel), and the Foundation for Research of the State of Pernambuco (FACEPE).

REFERENCE

1. Morgan, C.; Newman, D.; Price, C., Immunosensors: technology and opportunities in laboratory medicine. *Clinical Chemistry* 1996, *42* (2), 193–209.

2. Holford, T. R. J.; Davis, F.; Higson, S. P. J., Recent trends in antibody based sensors. *Biosensors and Bioelectronics* 2012, *34* (1), 12–24.

3. Zeng, X. Q.; Shen, Z. H.; Mernaugh, R., Recombinant antibodies and their use in biosensors. *Analytical and Bioanalytical Chemistry* 2012, *402* (10), 3027–3038.

4. Lim, Y. C.; Kouzani, A. Z.; Duan, W., Aptasensors: a review. *Journal of Biomedical Nanotechnology* 2010, *6* (2), 93–105.

5. Rousseaux, J.; Rousseaux-Prevost, R.; Bazin, H., Optimal conditions for the preparation of Fab and F(ab)2 fragments from monoclonal IgG of different rat IgG subclasses. *Journal of Immunological Methods* 1983, 64, 141–146.

6. Kidwai, A. A.; Jamal, Q.; Mehrunnisa, S.; Farooqi, F. U. R.; Saleem, U., Serodiagnosis of dengue infection using rapid immunochromatography test in patients with probable dengue infection. *Journal of the Pakistan Medical Association* 2010, *60* (11), 936–939.

7. Pei, X.; Zhang, B.; Tang, J.; Liu, B.; Lai, W.; Tang, D., Sandwich-type immunosensors and immunoassays exploiting nanostructure labels: a review.*Analytica Chimica Acta* 2013, *758*, 1–18.

8. Pfaunmiller, E.; Moser, A. C.; Hage, D. S., Biointeraction analysis of immobilized antibodies and related agents by high-performance immunoaffinity chromatography. *Methods* 2012, *56* (2), 130–135.

9. Sanchez-Moreno, P.; Ortega-Vinuesa, J. L.; Boulaiz, H.; Marchal, J. A.; Peula-Garcia, J. M., Synthesis and characterization of lipid immuno-nanocapsules for directed drug delivery: selective antitumor activity against HER2 positive breast-cancer cells. *Biomacromolecules* 2013, *14* (12), 4248–4259.

10. Shalumon, K. T.; Chen, J. P., Scaffold-based drug delivery for cartilage tissue regeneration. *Current Pharmaceutical Design* 2015.

11. Makaraviciute, A.; Ramanaviciene, A., Site-directed antibody immobilization techniques for immunosensors. *Biosensors and Bioelectronics* 2013, *50*, 460–471.

12. Bergstrom, G.; Mandenius, C.-F., Orientation and capturing of antibody affinity ligands: applications to surface plasmon resonance biochips. *Sensors and Actuators B: Chemical* 2011, *158* (1), 265–270.

13. Cavalcanti, I. T.; Silva, B. V. M.; Peres, N. G.; Moura, P.; Sotomayor, M. D. P. T.; Guedes, M. I. F.; Dutra, R. F., A disposable chitosan-modified carbon fiber electrode for dengue virus envelope protein detection. *Talanta* 2012, *91*, 41–46.

14. Lee, J.; Park, H.; Jung, Y.; Kim, J.; Jung, S.; Chung, B., Direct immobilization of protein G variants with various numbers of cysteine residues on a gold surface. *Analytical Chemistry* 2007, *79* (7), 2680–2687.

15. Batalla, P.; Fuentes, M.; Grazu, V.; Mateo, C.; Fernandez-Lafuente, R.; Guisan, J. M., Oriented covalent immobilization of antibodies on physically inert and hydrophilic support surfaces through their glycosidic chains. *Biomacromolecules* 2008, *9* (2), 719–723.

16. Dutra, R. F.; Kubota, L. T., An SPR immunosensor for human cardiac troponin T using specific binding avidin to biotin at carboxymethyldextran-modified gold chip. *Clin Chim Acta* 2007, *376* (1–2), 114–120.

17. Lee, J. H.; Kim, B. C.; Oh, B. K.; Choi, J. W., Highly sensitive localized surface plasmon resonance immunosensor for label-free detection of HIV-1.*Nanomedicine* 2013, *9* (7), 1018–1026.

18. Grant, S. A.; Pierce, M. E.; Lichlyter, D. J.; Grant, D. A., Effects of immobilization on a FRET immunosensor for the detection of myocardial infarction. *Analytical and Bioanalytical Chemistry* 2005, *381* (5), 1012–1028.

19. Mehrvar, M.; Abdi, M., Recent developments, characteristics, and potential applications of electrochemical biosensors. *Analytical Sciences* 2004, *20*(8), 1113–1126.

20. Mistry, K. K.; Layek, K.; Mahapatra, A.; RoyChaudhuri, C.; Saha, H., A review on amperometric-type immunosensors based on screen-printed electrodes. *Analyst* 2014, *139* (10), 2289–2311.

21. Silva, B.; Cavalcanti, I.; Silva, M.; Dutra, R., A carbon nanotube screen-printed electrode for label-free detection of the human cardiac troponin T.*Talanta* 2013, *117*, 431–437.

22. Qi, T.; Liao, J.; Li, Y.; Peng, J.; Li, W.; Chu, B.; Li, H.; Wei, Y.; Qian, Z., Label-free alpha fetoprotein immunosensor established by the facile synthesis of a palladium-graphene nanocomposite. *Biosens Bioelectron* 2014, *61*, 245–250.

23. Pandiaraj, M.; Sethy, N. K.; Bhargava, K.; Kameswararao, V.; Karunakaran, C., Designing label-free electrochemical immunosensors for cytochrome *c*using nanocomposites functionalized screen printed electrodes. *Biosensors and Bioelectronics* 2014, *54*, 115–121.

24. Gomes-Filho, S. L. R.; Dias, A. C. M. S.; Silva, M. M. S.; Silva, B. V. M.; Dutra, R. F., A carbon nanotube-based electrochemical immunosensor for cardiac troponin T. *Microchemical Journal* 2013, *109*, 10–15.

25. Dolatabadi, J.; de la Guardia, M., Nanomaterial-based electrochemical immunosensors as advanced diagnostic tools. *Analytical Methods* 2014, *6* (12), 3891–3900.

26. Luo, X.; Morrin, A.; Killard, A. J.; Smyth, M. R., Application of Nanoparticles in Electrochemical Sensors and Biosensors. 2006; *319*–326.

27. Vidotti, M.; Carvalhal, R. F.; Mendes, R. K.; Ferreira, D. C. M.; Kubota, L. T., Biosensors based on gold nanostructures. *Journal of the Brazilian Chemical Society* 2011, *22* (1), 3–20.

28. Campbell, F.; Compton, R., The use of nanoparticles in electroanalysis: an updated review. *Analytical and Bioanalytical Chemistry* 2010, *396* (1), 241–259.

29. Lu, L.; Liu, B.; Li, S.; Zhang, W.; Xie, G., Improved electrochemical immunosensor for myeloperoxidase in human serum based on nanogold/cerium dioxide-BMIMPF6/L-Cysteine composite film. *Colloids and Surfaces B: Biointerfaces* 2011, *86* (2), 339–344.

30. Li, H.; Wei, Q.; Wang, G.; Yang, M.; Qu, F.; Qian, Z., Sensitive electrochemical immunosensor for cancer biomarker with signal enhancement based on nitrodopamine-functionalized iron oxide nanoparticles. *Biosensors and Bioelectronics* 2015, *26* (6), 3044–3049.

31. Yang, H.; Yuan, R.; Chai, Y.; Mao, L.; Su, H.; Jiang, W.; Liang, M., Electrochemical immunosensor for detecting carcinoembryonic antigen using hollow Pt nanospheres-labeled multiple enzyme-linked antibodies as labels for signal amplification. *Biochemical Engineering Journal* 2011, *56* (3), 116–124.

32. Gao, X.; Zhang, Y.; Wu, Q.; Chen, H.; Chen, Z.; Lin, X., One step electrochemically deposited nanocomposite film of chitosan-carbon nanotubes-gold nanoparticles for carcinoembryonic antigen immunosensor application. *Talanta* 2011, *85* (4), 1980–1985.

33. Huang, K.; Niu, D.; Xie, W.; Wang, W., A disposable electrochemical immunosensor for carcinoembryonic antigen based on nano-Au/multi-walled carbon nanotubes-chitosans nanocomposite film modified glassy carbon electrode. *Analytica Chimica Acta* 2010, *659* (1–2), 102–108.

34. Kong, F.; Xu, M.; Xu, J.; Chen, H., A novel label-free electrochemical immunosensor for carcinoembryonic antigen based on gold nanoparticles-thionine-reduced graphene oxide nanocomposite film modified glassy carbon electrode. *Talanta* 2011, *85* (5), 2620–2625.

35. Yuan, Y.; Yuan, R.; Chai, Y.; Zhuo, Y.; Mao, L.; Yuan, S., A novel label-free electrochemical immunosensor for carcinoembryonic antigen detection based on the [Ag-Ag2O]/SiO2 nanocomposite material as a redox probe. *Journal of Electroanalytical Chemistry* 2010, *643* (1–2), 15–19.

36. An, Y.; Jiang, X.; Bi, W.; Chen, H.; Jin, L.; Zhang, S.; Wang, C.; Zhang, W., Sensitive electrochemical immunosensor for alpha-synuclein based on dual signal amplification using PAMAM dendrimer-encapsulated Au and enhanced gold nanoparticle labels. *Biosensors and Bioelectronics* 2012, *32* (1), 224–230.

37. Liu, X.; Wong, D., Picogram-detection of estradiol at an electrochemical immunosensor with a gold nanoparticle vertical bar protein G-(LC-SPDP)-scaffold. *Talanta* 2009, *77* (4), 1437–1443.

38. Krishnan, S.; Chinnasamy, T.; Veerappan, S.; Senthilkumar, K.; Kannaiyan, D., Dual labeled Ag@SiO2 core-shell nanoparticle based optical immunosensor for sensitive detection of *E. coli*. *Materials Science and Engineering C: Materials for Biological Applications* 2014, *45*, 337–342.

39. Zhang, S.; Ma, H.; Yan, L.; Cao, W.; Yan, T.; Wei, Q.; Du, B., Copper-doped titanium dioxide nanoparticles as dual-functional labels for fabrication of electrochemical immunosensors. *Biosensors and Bioelectronics* 2014, *59*, 335–341.

40. Shen, W.; Tian, D.; Cui, H.; Yang, D.; Bian, Z., Nanoparticle-based electrochemiluminescence immunosensor with enhanced sensitivity for cardiac troponin I using *N*-(aminobutyl)-*N*-(ethylisoluminol)-functionalized gold nanoparticles as labels. *Biosensors and Bioelectronics* 2011, *27* (1), 18–24.

41. Mendes, R.; Laschi, S.; Stach-Machado, D.; Kubota, L.; Marrazza, G., A disposable voltammetric immunosensor based on magnetic beads for early diagnosis of soybean rust. *Sensors and Actuators B-Chemical* 2012, *166*, 135–140.

42. Shen, Z.; Wang, J.; Qiu, Z.; Jin, M.; Wang, X.; Chen, Z.; Li, J.; Cao, F., QCM immunosensor detection of *Escherichia coli* O157:H7 based on beacon immunomagnetic nanoparticles and catalytic growth of colloidal gold. *Biosensors and Bioelectronics* 2011, *26* (7), 3376–3381.

43. Li, D.; Wang, J.; Wang, R.; Li, Y.; Abi-Ghanem, D.; Berghman, L.; Hargis, B.; Lu, H., A nanobeads amplified QCM immunosensor for the detection of avian influenza virus H5N1. *Biosensors and Bioelectronics* 2011, *26* (10), 4146–4154.

44. Yuan, Y.; Zhang, J.; Zhang, H.; Yang, X., Label-free colorimetric immunoassay for the simple and sensitive detection of neurogenin3 using gold nanoparticles. *Biosensors and Bioelectronics* 2011, *26* (10), 4245–4248.

45. Wu, W.; Bian, Z.; Wang, W.; Zhu, J., PDMS gold nanoparticle composite film-based silver enhanced colorimetric detection of cardiac troponin I.*Sensors and Actuators B: Chemical* 2010, *147* (1), 298–303.

46. Bera, D.; Qian, L.; Tseng, T.; Holloway, P., Quantum dots and their multimodal applications: a review. *Materials* 2010, *3* (4), 2260–2345.

47. Gill, R.; Zayats, M.; Willner, I., Semiconductor quantum dots for bioanalysis. *Angewandte Chemie-International Edition* 2008, *47* (40), 7602–7625.

48. Medintz, I.; Uyeda, H.; Goldman, E.; Mattoussi, H., Quantum dot bioconjugates for imaging, labelling and sensing. *Nature Materials* 2005, *4* (6), 435–446.

49. Chan, W.; Nie, S., Quantum dot bioconjugates for ultrasensitive nonisotopic detection. *Science* 1998, *281* (5385), 2016–2018.

50. Somers, R.; Bawendi, M.; Nocera, D., CdSe nanocrystal based chem-/bio-sensors. *Chemical Society Reviews* 2007, *36* (4), 579–591.

51. Algar, W.; Susumu, K.; Delehanty, J.; Medintz, I., Semiconductor quantum dots in bioanalysis: crossing the valley of death. *Analytical Chemistry*2011, *83* (23), 8826–8837.

52. Colvin, V.; Goldstein, A.; Alivisatos, A., Semiconductor nanocrystals covalently bound to metal-surfaces with self-assembled monolayers. *Journal of the American Chemical Society* 1992, *114* (13), 5221–5230.

53. Lidke, D.; Nagy, P.; Heintzmann, R.; Arndt-Jovin, D.; Post, J.; Grecco, H.; Jares-Erijman, E.; Jovin, T., Quantum dot ligands provide new insights into erbB/HER receptor-mediated signal transduction. *Nature Biotechnology* 2004, *22* (2), 198–203.

54. Palaniappan, K.; Xue, C.; Arumugam, G.; Hackney, S.; Liu, J., Water-soluble, cyclodextrin-modified CdSe-CdS core-shell structured quantum dots.*Chemistry of Materials* 2006, *18* (5), 1275–1280.

55. Mattoussi, H.; Mauro, J.; Goldman, E.; Anderson, G.; Sundar, V.; Mikulec, F.; Bawendi, M., Self-assembly of CdSe-ZnS quantum dot bioconjugates using an engineered recombinant protein. *Journal of the American Chemical Society* 2000, *122* (49), 12142–12150.

56. Xing, Y.; Chaudry, Q.; Shen, C.; Kong, K.; Zhau, H.; Chung, L.; Petros, J.; O'Regan, R.; Yezhelyev, M.; Simons, J.; Wang, M.; Nie, S., Bioconjugated quantum dots for multiplexed and quantitative immunohistochemistry. *Nature Protocols* 2007, *2* (5), 1152–1165.

57. Liu, G.; Wang, J.; Kim, J.; Jan, M.; Collins, G., Electrochemical coding for multiplexed immunoassays of proteins. *Analytical Chemistry* 2004, *76*(23), 7126–7130.

58. Li, L.; Liu, K.; Yang, G.; Wang, C.; Zhang, J.; Zhu, J., Fabrication of graphene-quantum dots composites for sensitive electrogenerated chemiluminescence immunosensing. *Advanced Functional Materials* 2011, *21* (5), 869–878.

59. Aoyagi, S.; Kudo, M., Observation of fluorescence-labeled protein A on a biosensor surface by means of TOF-SIMS. *Sensors and Actuators B-Chemical* 2005, *108* (1–2), 708–712.

60. Lingerfelt, B.; Mattoussi, H.; Goldman, E.; Mauro, J.; Anderson, G., Preparation of quantum dot-biotin conjugates and their use in immunochromatography assays. *Analytical Chemistry* 2003, *75* (16), 4043–4049.

61. Sun, B.; Xie, W.; Yi, G.; Chen, D.; Zhou, Y.; Cheng, J., Microminiaturized immunoassays using quantum dots as fluorescent label by laser confocal scanning fluorescence detection. *Journal of Immunological Methods* 2001, *249* (1–2), 85–89.

62. Goldman, E.; Balighian, E.; Mattoussi, H.; Kuno, M.; Mauro, J.; Tran, P.; Anderson, G., Avidin: a natural bridge for quantum dot-antibody conjugates.*Journal of the American Chemical Society* 2002, *124* (22), 6378–6382.

63. Kerman, K.; Endo, T.; Tsukamoto, M.; Chikae, M.; Takamura, Y.; Tamiya, E., Quantum dot-based immunosensor for the detection of

prostate-specific antigen using fluorescence microscopy. *Talanta* 2007, *71* (4), 1494–1499.

64. Lao, U.; Mulchandani, A.; Chen, W., Simple conjugation and purification of quantum dot-antibody complexes using a thermally responsive elastin-protein L scaffold as immunofluorescent agents. *Journal of the American Chemical Society* 2006, *128* (46), 14756–14757.

65. Susha, A.; Javier, A.; Parak, W.; Rogach, A., Luminescent CdTe nanocrystals as ion probes and pH sensors in aqueous solutions. *Colloids and Surfaces A: Physicochemical and Engineering Aspects* 2006, *281* (1–3), 40–43.

66. Liu, Y.; Sun, Y.; Vernier, P.; Liang, C.; Chong, S.; Gundersen, M., pH-sensitive photoluminescence of CdSe/ZnSe/ZnS quantum dots in human ovarian cancer cells. *Journal of Physical Chemistry C* 2007, *111* (7), 2872–2878.

67. Herrero-Latorre, C.; Alvarez-Mendez, J.; Barciela-Garcia, J.; Garcia-Martin, S.; Pena-Crecente, R. M., Characterization of carbon nanotubes and analytical methods for their determination in environmental and biological samples: a review. *Analytica Chimica Acta* 2015, *853*, 77–94.

68. Vashist, S.; Zheng, D.; Al-Rubeaan, K.; Luong, J.; Sheu, F., Advances in carbon nanotube based electrochemical sensors for bioanalytical applications.*Biotechnology Advances* 2011, *29* (2), 169–188.

69. Li, J.; Pandey, G. P., Advanced physical chemistry of carbon nanotubes. *Annual Review of Physical Chemistry* 2015.

70. Yang, N.; Chen, X. P.; Ren, T. L.; Zhang, P.; Yang, D. G., Carbon nanotube based biosensors. *Sensors and Actuators B: Chemical* 2015, *207*, 690–715.

71. Girifalco, L.; Hodak, M.; Lee, R., Carbon nanotubes, buckyballs, ropes, and a universal graphitic potential. *Physical Review B* 2000, *62* (19), 13104–13110.

72. Peng-Cheng Maa, N. A. S., Gad Maromb, Jang-Kyo Kima, Dispersion and functionalization of carbon nanotubes for polymer-based nanocomposites: a review. 2010, *41* (10), 1345–1367.

73. Di Crescenzo, A.; Ettorre, V.; Fontana, A., Non-covalent and reversible functionalization of carbon nanotubes. *Beilstein Journal of Nanotechnology*2014, *5*, 1675–1690.

74. Gao, C.; Guo, Z.; Liu, J.; Huang, X., The new age of carbon nanotubes: an updated review of functionalized carbon nanotubes in electrochemical sensors. *Nanoscale* 2012, *4* (6), 1948–1963.

75. Tuncel, D., Non-covalent interactions between carbon nanotubes and conjugated polymers. *Nanoscale* 2011, *3* (9), 3545–3554.

76. Sanchez, S.; Roldan, M.; Perez, S.; Fabregas, E., Toward a fast, easy, and versatile immobilization of biomolecules into carbon nanotube/polysulfone-based biosensors for the detection of hCG hormone. *Analytical Chemistry* 2008, *80* (17), 6508–6514.

77. Viswanathan, S.; Rani, C.; Anand, A.; Ho, J., Disposable electrochemical immunosensor for carcinoembryonic antigen using ferrocene liposomes and MWCNT screen-printed electrode. *Biosensors and Bioelectronics* 2009, *24* (7), 1984–1989.

78. Malhotra, R.; Patel, V.; Vaque, J.; Gutkind, J.; Rusling, J., Ultrasensitive electrochemical immunosensor for oral cancer biomarker IL-6 using carbon nanotube forest electrodes and multilabel amplification. *Analytical Chemistry* 2010, *82* (8), 3118–3123.

79. Munge, B.; Fisher, J.; Millord, L.; Krause, C.; Dowd, R.; Rusling, J., Sensitive electrochemical immunosensor for matrix metalloproteinase-3 based on single-wall carbon nanotubes. *Analyst* 2010, *135* (6), 1345–1350.

80. Chen, Y.; Huang, J., Electrochemical sensing of bovine serum albumin at self-assembled SWCNTs on gold. *Diamond and Related Materials* 2009, *18*(2–3), 516–519.

81. Su, H.; Yuan, R.; Chai, Y.; Zhuo, Y.; Hong, C.; Liu, Z.; Yang, X., Multilayer structured amperometric immunosensor built by self-assembly of a redox multi-wall carbon nanotube composite. *Electrochimica Acta* 2009, *54* (17), 4149–4154.

82. Jacobs, C.; Peairs, M.; Venton, B., Review: carbon nanotube based electrochemical sensors for biomolecules. *Analytica Chimica Acta* 2010, *662* (2), 105–127.

83. Freitas, T.; Mattos, A.; Silva, B.; Dutra, R., Amino-functionalization of carbon nanotubes by using a factorial design: human cardiac troponin T immunosensing application. *Biomed Research International* 2014.

84. Yao, J.; Sun, Y.; Yang, M.; Duan, Y. X., Chemistry, physics and biology of graphene-based nanomaterials: new horizons for sensing, imaging and medicine. *Journal of Materials Chemistry* 2012, *22* (29), 14313–14329.

85. Novoselov, K. S.; Geim, A. K.; Morozov, S. V.; Jiang, D.; Zhang, Y.; Dubonos, S. V.; Grigorieva, I. V.; Firsov, A. A., Electric field effect in atomically thin carbon films. *Science* 2004, *306* (5696), 666–669.

86. Zhang, L.; Zhou, R.; Zhao, X., Graphene-based materials as supercapacitor electrodes. *Journal of Materials Chemistry* 2010, *20* (29), 5983–5992.

87. Lawal, A. T., Synthesis and utilisation of graphene for fabrication of electrochemical sensors. *Talanta* 2015, *131*, 424–443.

88. Zhu, Y.; Murali, S.; Cai, W.; Li, X.; Suk, J.; Potts, J.; Ruoff, R., Graphene and graphene oxide: synthesis, properties, and applications. *Advanced Materials* 2010, *22* (46), 5226–5226.

89. Zhu, Y.; Murali, S.; Cai, W.; Li, X.; Suk, J.; Potts, J.; Ruoff, R., Graphene and graphene oxide: synthesis, properties, and applications. *Advanced Materials* 2010, *22* (35), 3906–3924.

90. Chung, C.; Kim, Y.; Shin, D.; Ryoo, S.; Hong, B.; Min, D., Biomedical applications of graphene and graphene oxide. *Accounts of Chemical Research*2013, *46* (10), 2211–2224.

91. Chen, D.; Feng, H. B.; Li, J. H., Graphene oxide: preparation, functionalization, and electrochemical applications. *Chemical Reviews* 2012, *112* (11), 6027–6053.

92. Dreyer, D.; Park, S.; Bielawski, C.; Ruoff, R., The chemistry of graphene oxide. *Chemical Society Reviews* 2010, *39* (1), 228–240.

93. Shan, C.; Yang, H.; Han, D.; Zhang, Q.; Ivaska, A.; Niu, L., Graphene/AuNPs/chitosan nanocomposites film for glucose biosensing. *Biosensors and Bioelectronics* 2010, *25* (5), 1070–1074.

94. Luo, J.; Jiang, S. S.; Zhang, H. Y.; Jiang, J. Q.; Liu, X. Y., A novel non-enzymatic glucose sensor based on Cu nanoparticle modified graphene sheets electrode. *Analytica Chimica Acta* 2012, *709*, 47–53.

95. Mao, S.; Lu, G.; Yu, K.; Bo, Z.; Chen, J., Specific protein detection using thermally reduced graphene oxide sheet decorated with gold nanoparticle-antibody conjugates. *Advanced Materials* 2010, *22* (32), 3521–+.

96. Zhong, Z.; Wu, W.; Wang, D.; Wang, D.; Shan, J.; Qing, Y.; Zhang, Z., Nanogold-enwrapped graphene nanocomposites as trace labels for sensitivity enhancement of electrochemical immunosensors in clinical immunoassays: carcinoembryonic antigen as a model. *Biosensors and Bioelectronics*2010, *25* (10), 2379–2383.

97. Huang, J. L.; Tian, J. N.; Zhao, Y. C.; Zhao, S. L., Ag/Au nanoparticles coated graphene electrochemical sensor for ultrasensitive analysis of

carcinoembryonic antigen in clinical immunoassay. *Sensors and Actuators B: Chemical* 2015, *206*, 570–576.

98. Tang, J.; Tang, D.; Niessner, R.; Chen, G.; Knopp, D., Magneto-controlled graphene immunosensing platform for simultaneous multiplexed electrochemical immunoassay using distinguishable signal tags. *Analytical Chemistry* 2011, *83* (13), 5407–5414.

99. Teixeira, S.; Burwell, G.; Castaing, A.; Gonzalez, D.; Conlan, R. S.; Guy, O. J., Epitaxial graphene immunosensor for human chorionic gonadotropin.*Sensors and Actuators B-Chemical* 2014, *190*, 723–729.

100. Xu, S.; Liu, Y.; Wang, T.; Li, J., Positive potential operation of a cathodic electrogenerated chemiluminescence immunosensor based on luminol and graphene for cancer biomarker detection. *Analytical Chemistry* 2011, *83* (10), 3817–3823.

101. Li, L. L.; Liu, K. P.; Yang, G. H.; Wang, C. M.; Zhang, J. R.; Zhu, J. J., Fabrication of graphene-quantum dots composites for sensitive electrogenerated chemiluminescence immunosensing. *Advanced Functional Materials* 2011, *21* (5), 869–878.

Chapter 7

APPLICATION OF NANOPARTICLES IN QUARTZ CRYSTAL MICROBALANCE BIOSENSORS

Somayeh Heydari*, **Gholam Hossein Haghayegh**

Faculty of Agriculture, Higher Education Complex of Torbat-e Jam, Torbat-e Jam, Iran

ABSTRACT

Nanoparticles are playing an increasingly important role in the development of biosensors. The sensitivity and performance of biosensors are being improved by using Nanoparticles for their construction. The use of these Nanoparticles has allowed the introduction of many new signal transduction technologies in biosensors. In this report, a comprehensive review of application of nanoparticles in Quartz Crystal Microbalance biosensors is presented. The main advantages of QCM in sensing fields include high sensitivity, high stability, fast response and low cost. In addition, it provides label-free detection capability for bio-sensing applications. Firstly, basic QCM's design and characterization are described. Next, QCM biosensors based on modification of quartz substrate structure and their applications are digested. Nanoparticles and their utilizationin analysis are then illustrated. These include Nanoparticles in bio applications that cover Nanoparticles in Quartz Crystal Microbalance biosensors.

Keywords: Nanoparticles, Quartz Crystal Microbalance, QCM, Biosensors, Review

1. INTRODUCTION

Nanoparticle analysis is a rapidly growing research area with applications in a variety of fields including energy, nanocomposite materials, the environment and healthcare. Nanoparticles are being incorporated into materials such as solar panels, nanomachines, and a variety of thin films and coatings to generate more desirable properties. Between the different applications, the use of NPs for biosensing is showing an increased interest for several areas such as clinical analysis [1][2] , environmental monitoring [3] -[5] as well as safety and security. NPs involvements in DNA, protein and even cell sensing systems have recently been the hottest topics in nanobiotechnology. Although their use and benefits are unquestioned, there are still potential risks in their widespread use to health and the environment. Therefore, there is a need for new techniques for characterizing nanoparticles including their interactions with surfaces, each other and their stability in the environment. Quartz Crystal Microbalance (QCM) is an analytical technique that provides in-situ analysis of interactions taking place at surfaces. The main advantages of QCM in sensing fields include high sensitivity, high stability, fast response and low cost. In addition, it provides label-free detection capability for bio-sensing applications. Being labeled free, it dispenses with the time and cost demanding labeling step, and also eliminates any possible interference of the "true" binding process due to the presence of the labels. Given the special importance of bio analysis, we intend in this review to overview some of the major advances and milestones in the field of Quartz Crystal Microbalance biosensors based on NPs underlying the different approaches reported so far. Since the nanoparticles and biomolecules typically meet at the same nanometer length scale, this interdisciplinary approach will contribute to the establishment of a novel field, descriptively termed immunosensing nanotechnology or nanoimmunosensing. NP based immunosensing systems are proving to assist in the development of new and versatile protein detection methods.

In this report, firstly, basic QCM's design and characterization are described. Next, QCM biosensors based on modification of quartz substrate structure and their applications are discussed. Nanoparticles and their applications in analysis are then illustrated. These include nanoparticles in bio applications that cover nanoparticles in Quartz Crystal Microbalance biosensors.

2. QUARTZ CRYSTAL MICROBALANCE

Piezoelectricity is defined as electric polarization produced by mechanical strain in certain crystals, the polarization being proportional to the strain [6] . The Curies first observed piezoelectricity in 1880 as a potential difference generated across two surfaces of a quartz crystal under strain [7] . The converse piezoelectric effect, the deformation of apiezoelectric material by an applied electric field, was predicted by Lippman [8] . Thus, when a thin wafer from a piezoelectric crystal such as quartz is placed in an alternating electric field of the right frequency it will oscillate in a mechanically resonant mode of the wafer. The resonance frequency depends upon the angles with respect to the opticalaxis at which the wafer was cut from a single crystal and inversely on the crystal thickness. Quartz Crystal Microbalance (QCM) devices derive from piezoelectric quartz crystal resonators that resonate electromechanically in a thickness-shear mode (TSM) (Figure 1, Figure 2). Owing to piezoelectric effects, the resonant properties of crystal can be influenced by external physical loading, which may be classified into two main types, gravimetric and viscoelastic loading. In the first case, gravimetric force due to mass rigidly deposited on the crystal surface is balanced by a force originating from the shear gradient inside the crystal, leading to the resonance frequency shift (Δf_g), which is governed by Sauerbrey's equation [10] [11] :

$$\Delta f_g = -\frac{2 f_0^2}{\sqrt{\rho_q \mu_q}} \frac{\Delta M}{A}$$

$$f_0 = \frac{v}{2t_q} = \frac{\sqrt{(\mu_q / \rho_q)}}{2t_q}$$

where f_0 is the fundamental resonance frequency, ΔM is the mass deposited on QCM surface, A is the active electrode area, v is the acoustic wave velocity, ρ_q is the density of crystal (2.648 g·cm^{-3}for quartz), μ_q is the shear modulus of the cut face (2.947 × 10^{11} g·cm^{-1}·s^2 for AT-cut quartz) and t_q is the crystal thickness. With this loading effect, QCM has been routinely used for thickness monitoring in physical vapor deposition processes.

For the viscoelastic loading, resonance frequency is changed due to acoustic-fluid damping interaction when QCM operates in viscoelastic medium such as liquid [12] . Unlike gravimetric loading, there are large changes of dissipation factor and resonant resistance due to energy losses of shear wave that travels through non-rigidly adsorbed layers [13] -[16] . The viscoselastic resonance frequency shift (Δf_v), dissipation factor change (ΔD) and resonant resistance change (ΔR) are given by:

$$\Delta f_v = -f_0^{3/2}\sqrt{\frac{\eta_L \rho_L}{\pi \mu_q \rho_q}}$$

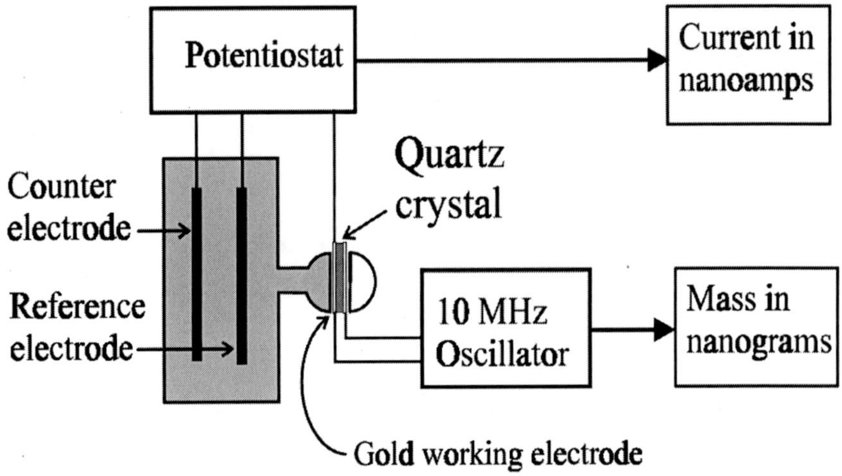

Figure 1. Schematic diagram of Quartz Crystal Microbalance instrumentation. (Crystal is held between two o-rings) Reproduced with permission from Ref. [7], Copyright 2008 Elsevier.

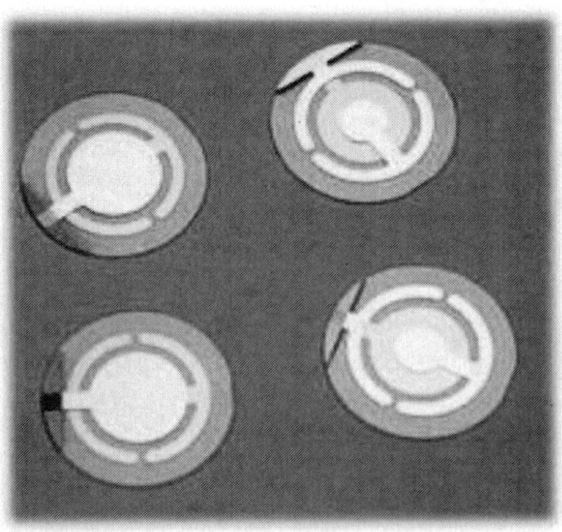

Figure 2. Crystals used in a QCM. Reproduced with permission from Ref. [9], Copyright 2004 Elsevier.

$$\Delta D = 2 f_0^{1/2} \sqrt{\frac{\eta_L \rho_L}{\pi \mu_q \rho_q}}$$

$$\Delta R = \frac{(2\pi \eta_L \rho_L f_0)^{1/2} A}{k^2}$$

where ρ_L, γ_L and k are viscosity, density and electromechanical coupling factor of the liquid medium, respectively. Both dissipation factor and resonant resistance can be directly measured experimentally. The combined measurements of frequency shift and changes in the dissipation factor and resonant resistance have been termed, QCM-D and QCM-R, respectively [17] . With the combined mass and viscosity-density effects, QCM can be applied for bio/chemical sensing by coating with a chemically or biologically sensitive film to sense very small amount of adsorbed gases such as volatile organic compounds (VOCs) [18] [19] and environmental pollutants [20] -[22] and liquid phase biochemicals.

A quartz crystal microbalance (QCM) measures a mass per unit area by measuring the change in frequency of a quartz crystal resonator. The resonance is disturbed by the addition or removal of a small mass due to oxide growth/decay or film deposition at the surface of the acoustic resonator. The QCM can be used under vacuum, in gas phase ("gas sensor", first use described by King) [23] and more recently in liquid environments. It is useful for monitoring the rate of deposition in thin film deposition systems under vacuum. In liquid, it is highly effective at determining the affinity of molecules (proteins, in particular) to surfaces functionalized with 3 to investigate interactions between biomolecules [24] -[28] .

There has been a rapid increase in the number of publications related to QCM sensing applications during the last few decades (Figure 3). QCMs have been used for multifarious applications in various disciplines of science and technology for the detection of metals in vacuum, vapors, chemical analytes, environmental pollutants, biomolecules, disease biomarkers, cells, and pathogens. They are also employed for deposition rate monitoring, electrochemical deposition, corrosion studies, detection in gas chromatography, and controlling thickness and composition of sputtered materials. Presently, there are several companies that are dealing with QCM-based systems.QCM provides one of the most promising sensor technology based on its low cost, rapid response, portability, nonhazardous label-free real-time procedure, and high sensitivity, which is ideal for the sensitive online detection of analytes and holds a great promise for the next-generation sensors.

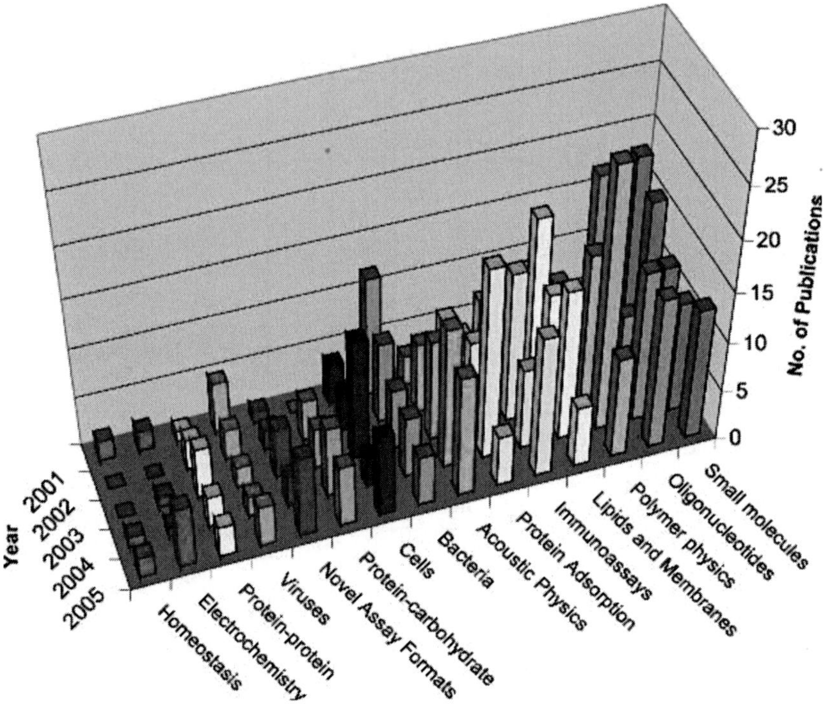

Figure 3. A summary of the number of papers (for specific analytes by class) referencing "QCM" in the title, Keywords or Abstract [29]. Reproduced with permission from Ref. [29], Copyright 2007 John Wiley & Sons, Ltd.

3. QCM AS A BIOSENSOR

Nowadays the applications of sensors are a very promising research field since they find application in many different areas [30] -[35] . Biosensors are an increasingly important technology in the detection of compounds ranging from pesticides to biological weapons. Typically, biosensors consist of a biological macromolecule that is immobilized on the surface of a signal transducer, as shown in Figure 4. As the macromolecule binds specifically to the ligand being detected, the signal transducer can measure a physical change due to the binding event. The transducer usually detects a change in resistance, pH, heat, light, or mass and then converts that data to an electrical signal to be collected and processed. One promising type of detector is the Quartz Crystal Microbalance (QCM). The QCM is a piezoelectric mass-sensing device. A QCM device works by sending an electrical signal through a gold-plated quartz crystal, which causes a vibration at some resonant frequency. The QCM then measures the frequency of oscillation in the crystal. When used as a biosensor, the QCM can detect changes in frequency of the crystal due to changes in mass on the surface of the crystal.

The sensitivity and specificity of QCM-based biosensors is dependent on the immobilization of recognitionlayer. For example various strategies have been employed for the immobilization of antibodies on the crystal surface. Passive adsorption of antibodies has been the most widely used method. However, it leads to random immobilization of antibodies on the surface, which may not be in functionally active orientation. The crystallizable fragment (Fc)-binding proteins such as protein A, protein G and protein A/G are employed for oriented immobilization of antibodies on the gold-coated QCM surface such that their antigen binding sites (Fab region) are completely free for binding antigens. Another strategy employs the initial generation of amino groups on the QCM surface by treatment with 3-aminopropyltriethoxysilane followed by the activation of amine-functionalized surface with glutaraldehyde to generate aldehyde groups, which bind to the antibody through its amino groups.

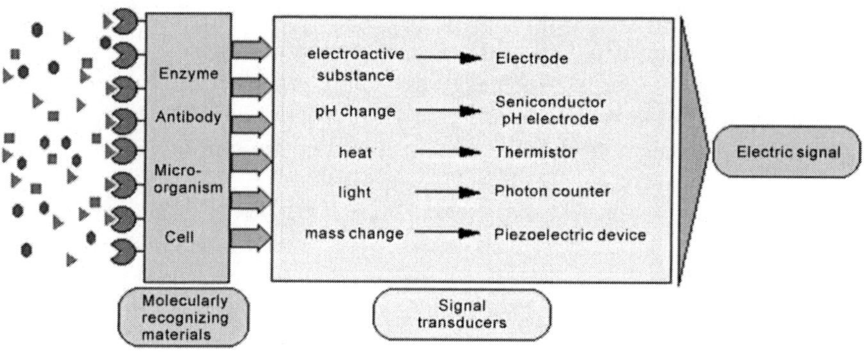

Figure 4. Schematic for generic biosensor [36].

The interactions between thiols and gold are very strong and have also been exploited for antibody binding by employing self-assembled layers of thiols and sulphides. The avidin-biotin interactions have also been employed, where the avidin/streptavid in-coated QCM-gold-surface is bound to the biotinylated antibody. Other approaches based on the use of polyelectrolytes, polymers, and Langmuir-Blodgett films have also been used.

QCM biosensors have been employed in the label-free detection of an incredibly broad range of analytes; from interfacial chemistries and lipid membranes to small molecules and whole cells [37] . They provide a unique method for observing in situ events involving "soft matter", in which changes in contact mechanics, interfacial dynamics, surface roughness, viscoelasticity, density and mass can be monitored in real time [38] -[44] . This level of information content goes beyond that produced by many more widely adopted optical techniques that are used to analyse molecular changes occurring in biochemical processes. Acoustic sensor technology is thus highly interdisciplinary and has encompassed advances and improvements from electrical engineers to cell biologists. Braunhut et al. adapted the Quartz Crystal Microbalance (QCM) biosensor technique to study responses of human mammary epithelial tumor cells to taxanes [45] . Taxanes are used for the treatment of many human cancers, as firstand second-line chemotherapeutics.

The ultimate goal in any biosensor development project is its use for actual sample detection. Recently, there has been an interest in biosensors with aptamers as bio-recognition elements, but reported examples all deal with standards, not human serum. In comparison to antibodies, aptamer receptors have a number of advantages. The main advantage is the avoidance of the use of animals for their production. Moreover, the aptamer select ion process can be manipulated to obtain aptamers that bind to a specific region of the target and with specific binding properties under different binding conditions [46] - [50] . In order to verify the differences of aptamer-based biosensor and antibody-based biosensor in clinical detection, Yao et al. carried out a comparison of the performance of aptamer-based and antibody-based Quartz Crystal Microbalance (QCM) biosensors for the detection of immunoglobulin E (IgE) in human serum [51] . Wu et al. developed a DNA piezoelectric biosensing method for real-time detection of Escherichia coli O157:H7 in a circulating-flow system [52] . The QCM system was also applied in the detection of PCR-amplified DNA from real samples of E. coli O157:H7. The resultant H% of the PCR-amplified double-strand DNA was comparable to that of the synthetic target T-104AS, a single-strand DNA. The piezoelectric biosensing system has potential for further applications. This approach lays the groundwork for incorporating the method into an integrated system for rapid PCR-based DNA analysis.

4. NANOPARTICLES

One of the most important functions of nanoparticles is catalysis, especially with noble metal nanoparticles, which have high catalytic activity for many chemical reactions. Because nanomaterials also have good biocompatibility, they are used to immobilize biomolecules for the fabrication of biosensors. Sensitivity and other attributes of biosensors can be improved by using nanomaterials in their construction. Nanomaterials, or matrices with at least one of their dimensions ranging in scale from 1 to 100 nm, display unique physical and chemical features because of effects such as the quantum size effect, mini size effect, surface effect and macro-quantum tunnel effect.

4.1. Nanoparticles in Bio Applications

Nanoparticles have numerous possible applications in biosensors. For example, functional nanoparticles (electronic, optical and magnetic) bound to biological molecules (e.g. peptides, proteins, nucleic acids) have been developed for use in biosensors to detect and amplify various signals. Some of the nanoparticle-based sensors include the acoustic wave biosensors, optical biosensors, magnetic and electrochemical biosensors. The structure of IgG, the most used antibody in immunosensig assays, has been determined by X-ray crystallography (Figure 5(a), upper part) which shows a Y-shape form consisting of three equal-sized portions, loosely connected by a flexible tether [53] . The antibodies are constituted by two heavy and two light polypeptide chains linked between themas shown in Figure 5(a) (lower part). The C regions determine the isotype of the antibody whereas the variable regions of one heavy and one light chain constitute an Antigen Binding Site (ABS). The digestion of antibody with papain (Figure 5(b)upper part) and pepsin (Figure 5(b) lower part) proteases produces several smaller fragments that may be used instead of the original antibody for immunosensing applications. Immunoassays are based on the interaction between the antibody and the antigen, in particular between the ABS and the epitope. In fact this interaction gives high specificity and sensitivity to the immunoassay. In order to functionalize the Ig, the connection of labels through three main groups: $-NH_2$, $-COOH$ and $-SH$ have been performed. Figure 5(c) (upper part) shows two examples of conjugation of mono-sulfo-NHS-AuNPs with an antibody through the amino group. A conjugation of monomaleimido Au NPs (Figure 5(c), lower part) with an antibody, through the $-SH$ group, is also shown. The use of the thiol group for the functionalization is a good way to control the direction of the bond between the label and the antibody, and prevent the involvement of the ABS. Figure 5(d)is an example of conjugation through the carboxyl group reported by Ahirwal et al. [54] . They connected the carboxi-term of the antibody with an AuNP by glutathione used as aspacer. The C-term region is a good point to attach the label, since it s far from the ABS and should allow the molecule interaction with the antigen. The AuNP modification with protein can even be observed by TEM (Figure 5(e)) [55] .

Nanoparticles in Quartz Crystal Microbalance Biosensors

Quartz crystal microbalances have emerged in the last few years as versatile biosensors demonstrating remarkable achievements such as high sensitivity and label free detection [56] -[58] . In the quartz crystal microbalance, the bioreaction generates a change in the mass, reordering the charges in the surface of the piezoelectric material and giving rise to a change in the resonant frequency of the microbalance. The use of NP labels in sandwich assays can increase both the surface stress and the mass of the immune complex, allowing for the increase in sensitivity of these electromechanical assays. This was the case in work reported for the detection of human IgG[59] and aflatoxin B1 [60] at clinical relevant levels using AuNPs as amplification agents in adynamic mode quartz crystal microbalance biosensor.

Quartz crystal microbalance biosensors were developed to greatly improve the sensitivity and limits of detection [61] . In the mass-amplified quartz crystal microbalance assay variant of this technology, antibody modified sol particles indirectly bind to an electrode surface by complexing to an analyte that has been itself captured by an antibody immobilized on the electrode surface. The large mass of the bound sol particles greatly affects the vibrational frequency of the quartz crystal and this is used as the basis for detection. The assay can be carried out in the competitive mode. The preferred diameter of sol particles is in the range of 5 - 100 nm. Other high-density particles (e.g. Au, Pt, CdS, TiO_2, polymers) may be also suitable [62] [63] .

Recently Chu et al. developed a label-free gliadinimmuno sensor by applying 25 nm gold nanoparticles (AuNPs) to the surface of a bare QCM electrode [64] . Also a rapid and sensitive Quartz Crystal Microbalance (QCM) biosensor combined with gold nanoparticle amplification for diagnosis Mycobacterium tuberculosis was developed by Kaewphinit et al. [65] . The label free QCM was successfully developed with high sensitivity, specificity, low cost and convenience, this technique may prove to be a powerful tool for an early diagnosis of M. tuberculosis.

DNA-conjugated nanoparticles can be used to enhance the signal produced upon hybridization to a surface-bound single-stranded template [66] -[73] . They can also be used to increase the surface area available to immobilize the DNA template on a sensor [74] -[76] . Detection sensitivity is generally in the region of 10^{-15} - 10^{-16} M, with one report of zeptomolar (10^{-21} M) quantities of specific nucleic acid [77] . Ma et al. [78] used a mixture of $HAuCl_4$ and NH_2OH to enlarge the gold particles. However, it has been shown that 20 nm is the optimal particle size for maximally increasing the sensitivity as the larger particles suffer from weaker binding to the surface and are therefore lost on exposure to flow [71] . A quartz crystal microbalance (QCM) DNA sensor, based on the nanoparticle amplification method, was developed for detection of Escherichia coli O157:H7 [79] . A thiolated single-stranded DNA (ssDNA) probe specific to E. coli O157: H7 was immobilized onto the QCM sensor surface through self-assembly (Figure 6). The hybridization was induced by exposing the ssDNA probe to the complementary target DNA, and resulted in the mass change and therefore frequency change of the QCM.

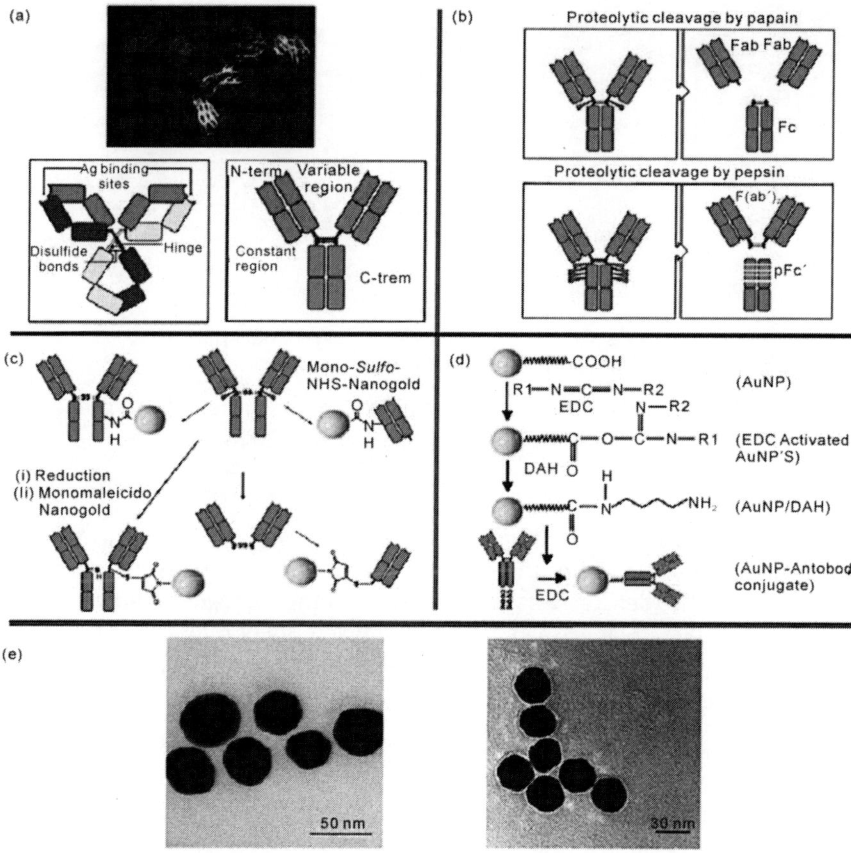

Figure 5. (a) X-ray crystallographic structure of an IgG antibody (upper part); A schematic representation of IgG and a simplified scheme of the antibody's structure (lower part); (b) The papain (upper part) and the the pepsin (lower part) cleavages; (c) Functionalization of a whole IgG or a Fab by a Mono-Sulfo-NHS-AuNP viafree amino groups (upper part); monomaleimidoAuNP binds the S atom after the reduction of the disulfide bond starting from the whole IgG (lower part, left) or from F(ab')$_2$ (lower part, right) [53] ; (d) An activated by EDC AuNP coated with gluthatione, as spacer arm, reacts with an IgG, also activated by EDC [54] ; (e) TEM images ofAuNPs (left part) and AuNPs conjugated with IgG (right part). Image shows a thin white layer, called "halo" effect, surrounding the surface of the AuNPs indicatingcoating with protein. Reproduced with permission from Ref. [55] , Copyright 2010 Elsevier.

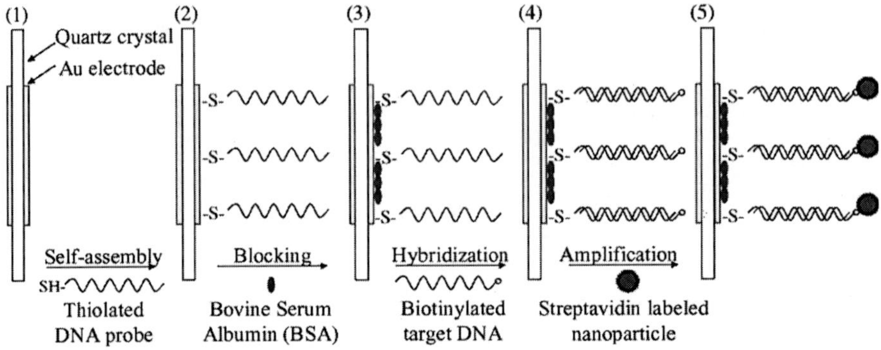

Figure 6. Llustration of the sensor fabrication and detection procedure reproduced with permission from Ref. [79], Copyright 2006 Elsevier.

The enhancement of a single strain DNA probe linked to the sensor surface is of crucial importance in DNA molecule recognition. By means of nanogold modification of the sensor surface in addition to the nanogold amplifier, DNA detection sensitivity higher than 10^{-16} mol/L was obtained in a Quartz Crystal microbalance (QCM) system, much higher than the ordinary QCM sensor without surface modification by nanogold (Figure 7) [63] .

Recently a DNA probe functionalized quartz crystal microbalance (QCM) biosensor was developed to detect B. anthracis based on the recognition of its specific DNA sequences, i.e., the 168 bp fragment of the Ba813 gene in chromosomes and the 340 bp fragment of the paggene in plasmid pXO1 [80] . Willner et al. described three different methods for the amplified detection of a single-base mismatch in DNA using microgravimetric quartzcrystal-microbalance as transduction means (Figure 8) [81] . All methods involve the primary incorporation of a biotinylated base complementary to the mutation site in the analyzed double-stranded primer/DNA assembly. The double-stranded assembly is formed between 25 complementary bases of the probe DNA assembled on the Au-quartz crystal and the target DNA. One method of amplification includes the association of avidinand biotin-labeled liposomes to the sensing interface. The second method of amplified detection of the base mismatch includes the association of an Au-nanoparticle-avidin conjugate to the sensing interface, and the secondary Au-nanoparticle-catalyzed deposition of gold on the particles. The third amplification route includes the binding of the avidin-alkaline phosphatase biocatalytic conjugate to the double-stranded surface followed by the oxidative hydrolysis of 5-bromo-4-chloro-3-indolyl phosphate to the insoluble product indigo derivative that precipitates on the transducer. Comparison of the three amplification routes reveals that the catalytic deposition of gold on the Au-nanoparticle/avidin conjugate is the most sensitive method, and the single-base mismatch in the analyzed DNA is detected with a sensitivity that corresponds to 3×10^{-16} M.

A novel amplification route for DNA detection based on the deposition of gold on a 10 nm Au-colloid/avidin conjugate label acting as a "seeding" catalyst is described by Weizmann et al. (Figure 9) [66] . Microgravimetric quartz-crystal-microbalance measurements are employed to transduce the catalyzed deposition of gold on the piezoelectric crystals. Three different DNA detection schemes are described: 1) analysis of a 27-base nucleic acid fragment; 2) analysis of the entire M13f DNA (7229 bases); and 3) detection of a single-base mismatch in a DNA. Ultrasensitive detection of DNA is accomplished by the catalyzed deposition of gold, detection limit $\approx 1 \times 10^{-15}$ M.

A rapid and sensitive detection method for the early diagnosis of infectious dengue virus (DENV) urgently needs to be developed by Chen et al. [82] . In this study, a circulating-flow Quartz Crystal Microbalance (QCM) biosensing method combining oligonucleotide-functionalized gold nanoparticles (i.e. AuNP probes) used to detect DENV has been established. The sensitivity and specificity of the present DNA-QCM method with nanoparticle technology showed it to be comparable to the fluorescent real-time PCR methods. Moreover, the method described herein was shown to not require expensive equipment, was label-free and highly sensitive. Scodeller et al. reported a fully integrated core-shell nanoparticle system responsive to glucose [83] . The system is comprised of self-assembled glucose oxidase and an osmium molecular wire on core-shell Au nanoparticles.

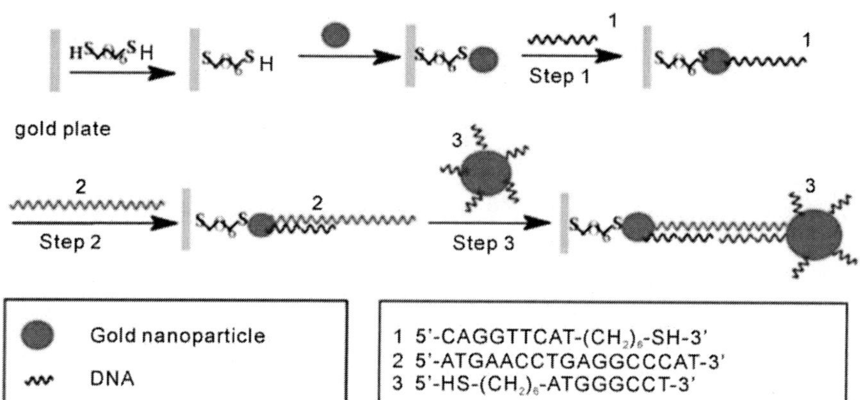

Figure 7. Schematic illustration of the sensing process of the amplifying system based on Au nanoparticle-covered QCM surface. Reproduced with permission from Ref. [63] , Copyright 2004 Elsevier.

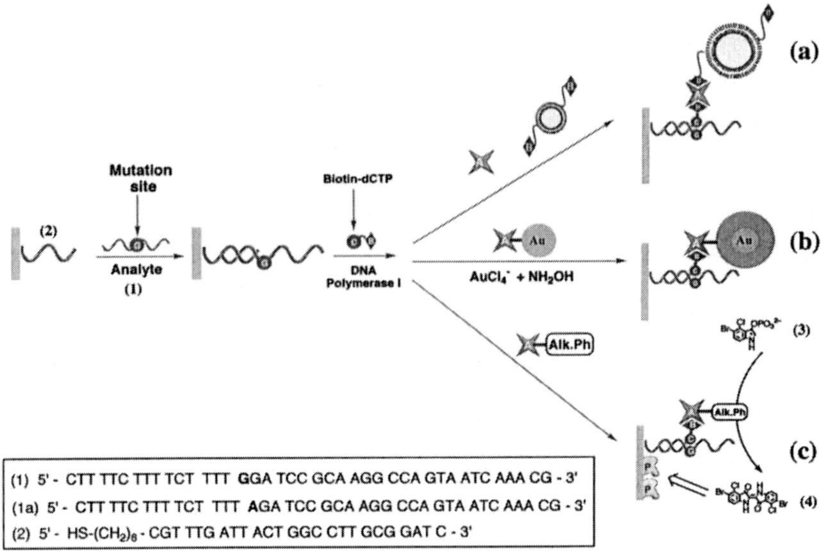

Figure 8. Amplified detection paths of a single-base mismatch in nucleic acids: (a) using avidin and biotin-labeled liposomes; (b) Using an avidin-Au-nanoparticle conjugate and the catalyzed deposition of gold; (c) Using an avidin-alkaline phosphatase bioconjugate and the biocatalyzed precipitation of the insoluble product 4 Reproduced with permission from Ref. [81] , Copyright 2002 Elsevier.

In an interesting work Krozer et al., investigated a new and generic approach for integration of molecularly imprinted nanoparticles into sensing systems [84] . Nanoparticles can be easily immobilized on flat transducer surfaces through a supporting polymer film using spin-coating, and the thickness of the nanoparticle coating can be easily controlled by changing the nanoparticle/polymer ratio and the solid content of the nanoparticle suspension. Kong et al. was developed a molecularly imprinted Quartz Crystal Microbalance (QCM) sensor for ractopamine (RAC) detection by electrodepositing a poly-o-aminothiophenol membrane on an Au electrode surface modified by self-assembled Au nanoparticles (AuNPs) [85] .

Recently a simple and sensitive sensor method for cancer biomarkers [prostate specific antigen (PSA) and PSA-alpha 1 antichymotrypsin (ACT) complex] analysis was developed, to be applied directly with human serum (75%) by using antibody modified quartz crystal microbalance sensor and nanoparticles amplification system [86] . A QCM sensor chip consisting of two sensing array enabling the measurement of an active and control binding events simultaneously on the sensor surface was used in this work. With the achieved detection limit in serum samples, the developed QCM assay show a promising technology for cancer biomarker analysis in patient samples.

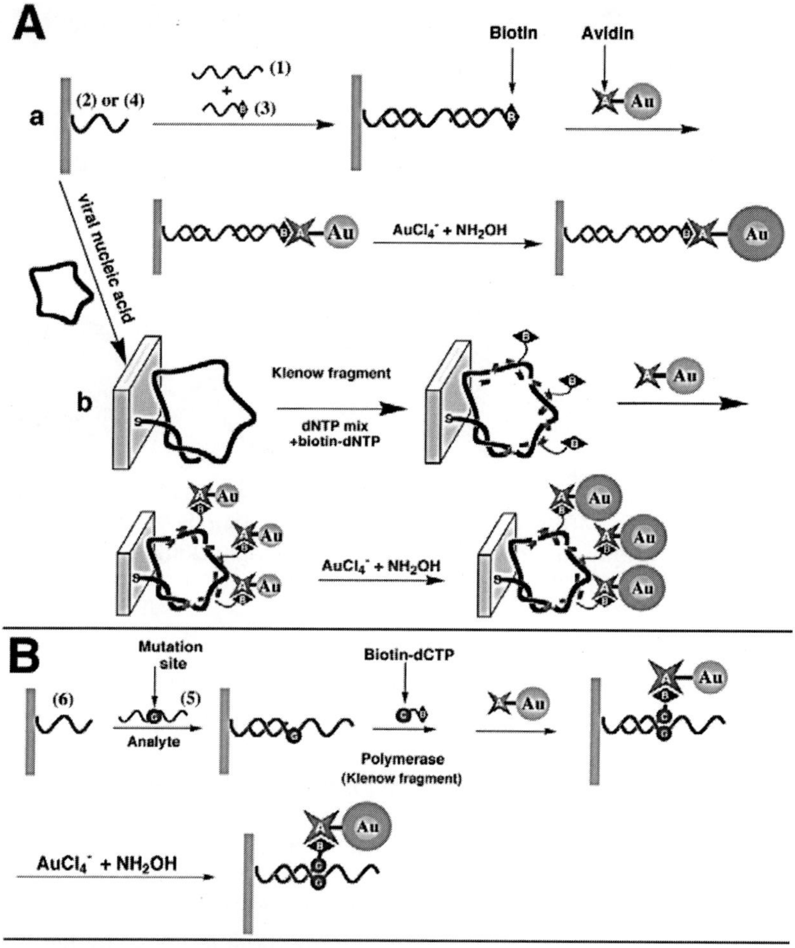

(1) 5'-AGCGTAGGATAGATATACGGTTCGCGC-3'
(1a) 5'-AGCGCTCCAGTGATATACGGTTCGCGC-3'
(2) 5'-TCTATCCTACGCT-(CH_2)-SH-3'
(3) 5'-biotin-GCGCGAACCGTATA-3'
(4) '5-HS-$(CH_2)_6$-CCCCCACGTTGTAAAACGACGGCCAGT-3'
(5) 5'-CTT TTC TTT TCT TTT GGA TCC GCA AGG CCA GTA ATC AAA CG-3'
(5a) 5'-CTT TTC TTT TCT TTT AGA TCC GCA AGG CCA GTA ATC AAA CG-3'
(6) 5'-HS-$(CH_2)_6$- CGT TTG ATT ACT GGC CTT GCG GAT C-3'

Figure 9. (A) Amplified detection of DNA by: (a) association of an Au-avidin conjugate and catalytic deposition of Au; (b) polymerase-induced replication of DNA followed by the electroless deposition of gold on an Au-avidin conjugate linked to the double-stranded assembly. (B) Amplified detection of a singlebase mismatch in DNA by the catalytic deposition of Au on an Au-avidin conjugate linked to the labeled double-stranded assembly. Reproduced with permission from Ref. [66] , Copyright 2001 Royal Society of Chemistry.

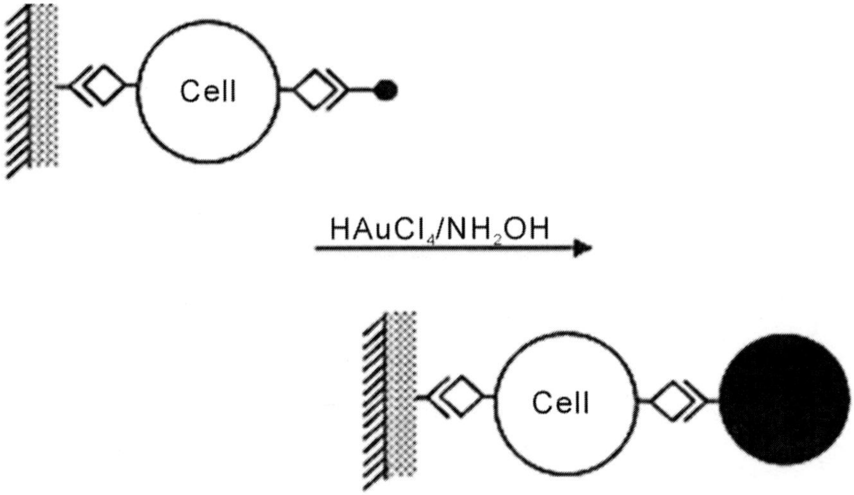

Figure 10. The gold-coated quartz crystal was modified by polystyrene filmsReproduced with permission from Ref. [87], Copyright 2002 Royal Society of Chemistry.

Ma et al. reported a new amplification route for human lung carcinoma cell detection based on the detection of gold deposition on an Au-antibody conjugate label acting as a “seeding” catalyst [87] . Quartz crystal microbalance (QCM) was employed to measure the catalyzed deposition of gold on the piezoelectric crystals. The gold-coated quartz crystal was modified by polystyrene films, which provided a substrate for antibody immobilization by physical adsorption (Figure 10). The polystyrene films significantly improve the surface smoothness of the electrode. Compared to the conventional Enzymelinked Immunosorbent Assay (ELISA), the method described herein is a less time consuming procedure and gives comparable detection sensitivity of human lung carcinoma cells, ca. 100 cells mL^{-1}. This will be of great significance in the diagnosis of this disease.

A highly sensitive quartz crystal microbalance with dissipation monitoring (QCM-D) biosensor for protein was developed using aptamer-functionalized gold nanoparticles (Apt-GNPs) for amplification [88] . Human α-thrombin, an important physiological protease found in blood, was chosen as the target protein. Captured by immobilized aptamers, thrombin was determined on-line using Apt-GNPs to enhance both frequency and dissipation signals. The fabricated sandwich of aptamer/thrombin/Apt-GNPs on chip surface was confirmed by atomic force microscopy (AFM). Compared to direct assay, the detection limit for thrombin was down to 0.1 nM, yielding about 2 orders of magnitude improvement in sensitivity. For the first time, the dual-signal enhancement of Apt-GNPs on QCM-D sensing was demonstrated, and such design could provide a promising detection strategy for proteins with two binding sites. Also an aptamer-based quartz crystal microbalance (QCM) biosensor was developed for the selective and sensitive detection of leukemia cells [89]. In this strategy, aminophenylboronic acid-modified gold nanoparticles (APBA-AuNPs) which could bind to cell membrane were used for the labeling of cells followed by silver enhancement, through which significant signal amplification was achieved.

In an interesting work Kaufman et al., described the adsorption characteristics of three proteins [bovine serum albumin (BSA), myoglobin (Mb), and cytochrome c (CytC)] onto self-assembled monolayers of mercaptoundecanoic acid (MUA) on both gold nanoparticles (AuNP) and gold surfaces (Au) [90]. The combination of quartz crystal microbalance measurements with dissipation (QCM-D) and pH titrations of the ú-potential provide information on layer structure, surface coverage, and potential. A new quartz crystal microbalance immunoassay method based on a novel transparent immunoaffinity reactor was developed for clinical immunoassay by Tang et al. [91]. To construct such an affinity reactor, resonators with a frequency of 10 MHz were fabricated by affinity binding of functionalized gold nanoparticles (nanogold) to quartz crystal with immobilized specific ligand for the label-free analysis of the affinity reaction between a ligand and its receptor. [Recombinant human tumor markers, carcinoembryonic antigen (CEA) was chosen as a model ligand.]. Compared with the conventional enzyme-linked immunosorbent assay, the proposed immunoassay system was simple and rapid without multiple labeling and separation steps. Importantly, the proposed immunoassay system could be further developed for the immobilization of other antigens or biocompounds.

Su et al. fabricated novel low-humidity sensors by the in situ photopolymerization of polypyrrole/Ag/TiO_2 nanoparticles (PPy/Ag/TiO_2 NPs) composite thin films on a quartz-crystal microbalance (QCM) [92]. The sensitivity increased with the doping amount of TiO_2 NPs. Additionally, based on the dynamic analysis of adsorption, the association constants of water vapor molecules with PPy/Ag and PPy/Ag/50 wt% TiO_2 NPs composite thin films were estimated to be 81.6 and 227.9 M^{-1}, respectively, explaining the effect of adding 50 wt% TiO_2 NPs to PPy/Ag; the sensitivity to low humidity increased as the association constant increased. Also a sensitive method for ammonia gas analysis was developed with conducting polymer poly (3,4-ethylenedioxythiophene) (PEDOT) nanoparticles by reverse micelle method, which could be well dispersed in water or alcoholic solvent with ultrasonic treatment [93]. Compare to electrical conductivity of conventional PEDOT particles (ca. 0.5 S/cm), the as-prepared palletized nanoparticles have higher conductivity (ca. 10.2 S/cm) and the conductivity varies with the doped and de-doped states of PEDOT. As for sensing property, it has been found that QCM device coated with nanoparticles shows faster response and recovery to 20 ppm ammonia gas (ca. 50 s) than that of conventional PEDOT particles (Figure 11). The PEDOT nanoparticle covered device exhibits almost linear relation to lower NH_3 gas concentration and shows a saturate tendency of gas sensitivity with the increase of gas concentration over 500 ppm.

Real-time and sensitive detection of Salmonella typhimurium using an automated quartz crystal microbalance (QCM) instrument with gold nanoparticles amplification was described by Salam et al. [94] . Also Guo et al. described a piezoelectric immunosensor for specific capture and enrichment of viable pathogens by quartz crystal microbalance sensor, followed by detection with antibody-functionalized gold nanoparticles [95] . The unique advantages of the system developed offer great potential in the microbial analysis of food samples in routine settings.

One of the applications of QCM technique is study nanoparticle self-assembly processes, nanocomposite materials and track nanoparticle interactions in the environment. Self-assembly is a process where similar particles spontaneously bind to one another to form new materials, coatings or films with a variety of desirable properties (Figure 12). These type of mass sensitive technique is highly useful to study monolayer formation, enzyme immobilization, selective sensing of small organic or large biomolecules. For example, recently Crooks and coworkers have shown that hyperbranch polymer monolayers and poly (amidoamine) dendrimer monolayers can interact with small organic molecules [97] [98] . Similarly Kepley and coworkers have used copper (II) ions bound on SAM to selectively and reversibly detected organophosphonate [99] . Host-guest interactions can be monitored by highly ordered cavitated monolayers as shown using calyx [4]resorcinarene [100] . These cavited monolayers have a high affinity towards perchloroethylene [100] . The QCM technology enables one to measure the kinetics of nanoparticle self-assembly processes; measure the resulting thickness, mass and structural properties of the self-assembled film; and compare the self-assembly process under different conditions such as temperature, solvent type or concentration, flow or static environment and particle type.

Modifying polymer films with nanoparticles to form nanocomposite materials has become a useful method to modulate the structural properties and other characteristics of thin films to have more desirable properties and better stability (Figure 13). The QCM-D technology can be used to study changes in film thickness and rigidity due to the additional of nanoparticles. Properties such as thickness, viscosity, storage and loss modulus of nanocomposite materials can be measured, in real time, by QCM. There is concern that nanoparticles stable in the environment could cause significant health issues. The QCM technology can also be used to track nanoparticle stability and interactions under environmental conditions with a variety of real-life surfaces such as glass, metals and natural organic matter (Figure 14). The technique can track the kinetics, extent and stability of nanoparticle deposition onto different surfaces and in different environmental conditions.

Chen et al. described the fabrication of gold nanoparticle surface thin films by taking advantage of the strong complexation interactions between transition metal ions and pyridine functional groups (Figure 15) [101] . The thickness (layers) of the particle thin films was readily controlled by the repetition of the alternate dipping cycles, as monitored by Quartz Crystal Microbalance (QCM).

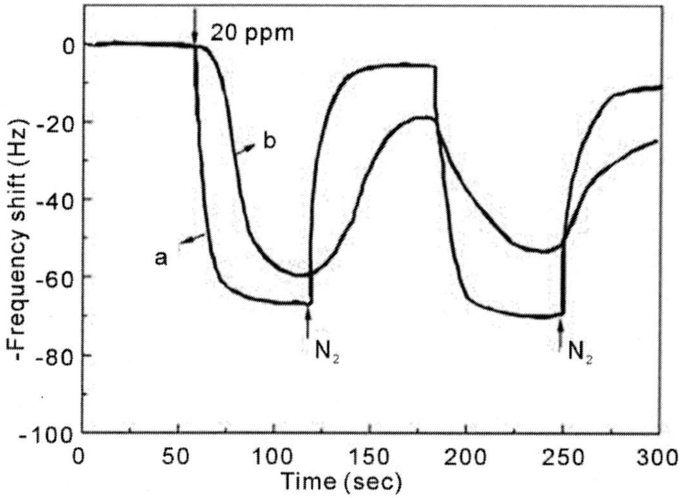

Figure 11. Typical response of PEDOT particles to NH_3 gas (20 ppm) (a) nanoparticles and (b) conventional particles reproduced with permission from Ref. [93] , Copyright 2007 Elsevier.

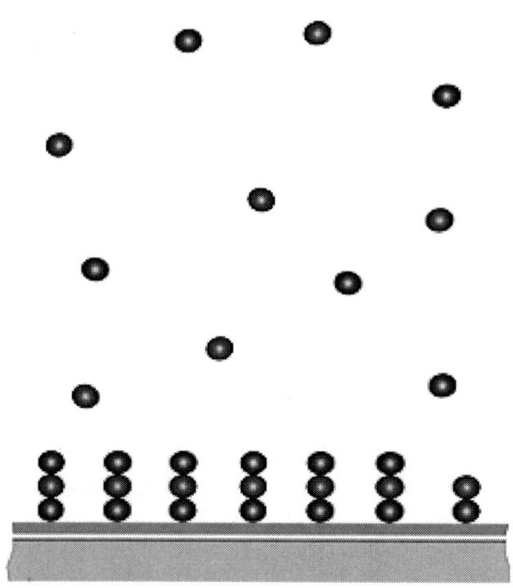

Figure 12. Self assembly processes [96].

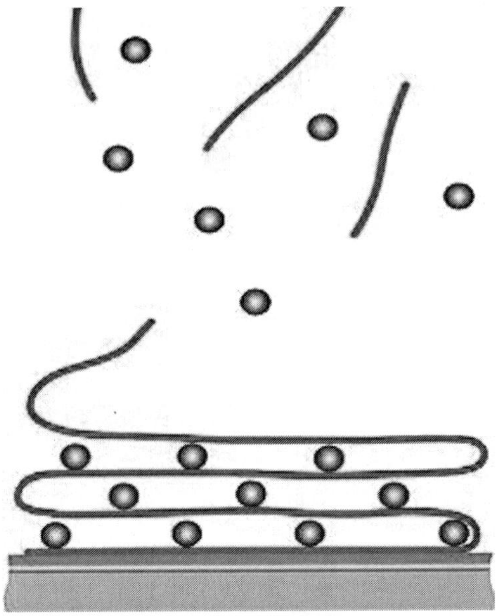

Figure 13. Nanocomposite materials analysis [96].

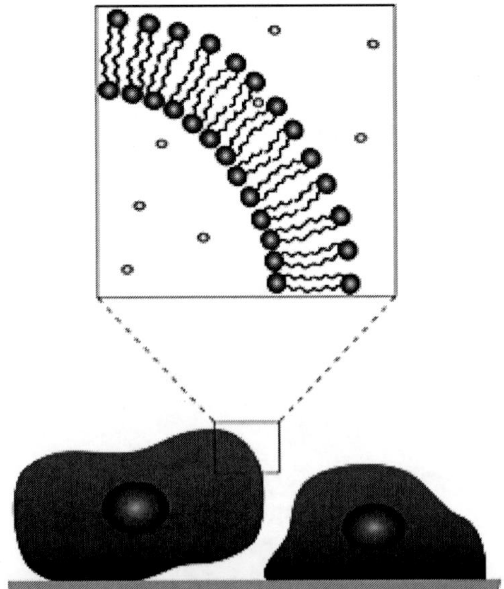

Figure 14. Nanoparticle interactions in the environment [96].

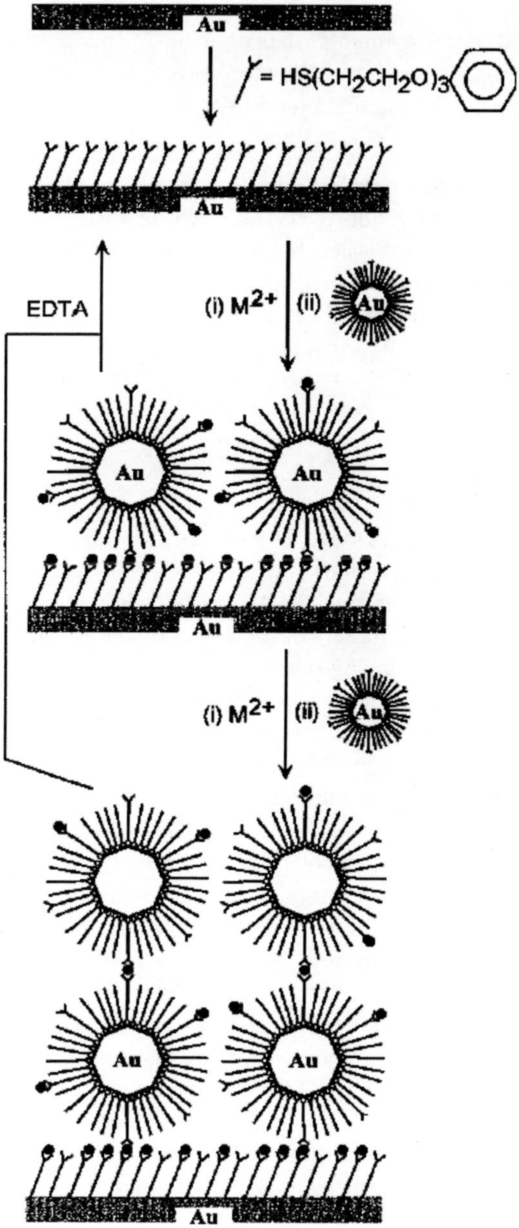

Figure 15. Procedure for nanoparticle assembling by the chelating interactions between divalent (transition) metal ions and pyridine moieties [101] reproduced with permission from Ref. [101] , Copyright 2002 American Chemical Society.

Shen et al. investigated a means of producing Au nanoparticle network-type thin films that are derived by a one-step mixed-assembly cross-linking route, can avoid nonspecific interactions, and provide maximum binding to a specific target (Figure16) [102] .

Liu and Hu showed interaction between myoglobin and hyaluronic acid in their layer-by-layer assembly by quartz crystal microbalance and cyclic voltammetry studies [103] . Quartz Crystal Microbalance (QCM) and CV were used to confirm the film growth and characterize the films. Also the deposition of silica nanoparticles on alumina coated surfaces in aqueous electrolytes was studied by optical reflectometry (OR) and a quartz crystal microbalance with dissipation (QCM-D) [104] .

5. CONCLUSION

Quartz crystal microbalances have emerged in the last few years as versatile biosensors demonstrating remarkable achievements such as high sensitivity and label free detection. In the quartz crystal microbalance, the bioreaction generates a change in the mass, reordering the charges in the surface of the piezoelectric material and giving rise to a change in the resonant frequency of the microbalance. Its detection range and sensitivity are not quite as good as some other label-free sensor technologies (e.g. surface plasmon resonance). The use of NP labels in sandwich assays can increase both the surface stress and the mass of the immune complex, allowing the increase in sensitivity of these electromechanical assays. This progress will require an increased understanding of both theoretical and experimental aspects of the relation among changes in frequency, energy dissipation, and NPs. These advancements will have a tremendous impact on the versatility and effectiveness of the QCM in future applications. In this review, application of nanoparticles in Quartz Crystal Microbalance biosensors has been extensively reviewed. The theoretical/experimental analysis, design and applications of quartz crystal microbalance biosensors combined with nanoparticles amplification have been discussed. These biosensors present as a detection option that can be effectively employed to specific pathogen related requirements. The prospects are low detection limit, less time consuming, economical, biocompatible, precise and accurate like features. New nanoparticles need to be explored for use in these biosensors. Therefore, QCM is a relatively new and promising research area worthy for further exploration.

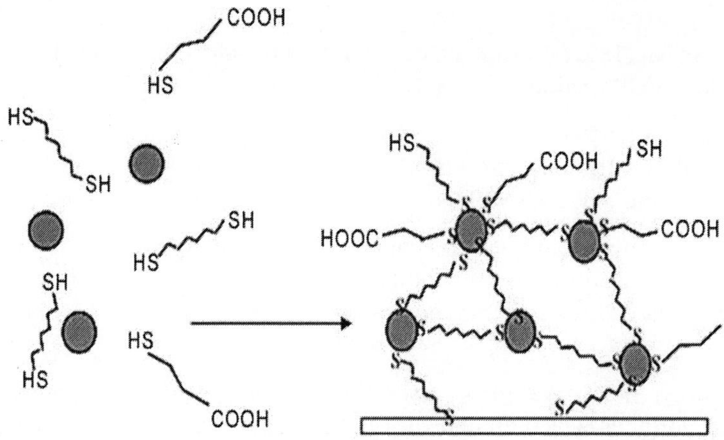

Figure 16. Schematic of the mixed-assembly cross-linking route for the preparation of Au nanoparticle network-type thin film. Reproduced with permission from Ref. [102] , Copyright 2007 Elsevier.

REFERENCES

1. Agasti, S.S., Rana, S., Park, M.H., Kim, C.K., You, C.C. and Rotello, V.M. (2010) Nanoparticles for Detection and Diagnosis. Advanced Drug Delivery Reviews, 62, 316-328.

2. Durner, J. (2009) Clinical Chemistry: Challenges for Analytical Chemistry and the Nanosciences from Medicine. Angewandte Chemie International Edition, 496, 1026-1051.

3. Palchetti, I. and Mascini, M. (2008) Nucleic Acid Biosensors for Environmental Pollution Monitoring. Analyst, 133, 846-854.

4. Arbab Zavar, M.H., Heydari, S., Rounaghi, Gh., Eshghi, H. and Azizi-Toupkanloo, H. (2012) Electrochemical Behavior of Para-Nitroaniline at a New Synthetic Crownether-Silver Nanoparticle Modified Carbon Paste Electrode. Analytical Methods, 4, 953-958.

5. Rounaghi, G.H., Mohamadzadehkakhki, R. and Azizi-Toupkanloo, H. (2012) Voltammetric Determination of 4-Nitro-phenol Using a Modified Carbon Paste Electrode Based on a New Synthetic Crown Ether/Silver Nanoparticles. Material Science and Engineering C, 32, 172-177.

6. Cady, W.G. (1946) Piezoelectricity. McGraw-Hill, New York and London.

7. Curie, J. and Curie, P. (1880) Mull Historical Society, Paris, 1880, 90.

8. Lippman, G. (1881) Principe de conservation de lelectricite. Annales de Chimie et de Physique, 24, 145-178.

9. Dwyer, M.A. and Hellinga, H.W. (2004) Periplasmic Binding Proteins: A Versatile Superfamily for Protein Engineering. Current Opinion in Structural Biology, 14, 495-504.

10. Rabe, J., Buttgenbach, S., Schroder, J. and Hauptmann, P. (2003) Monolithic Miniaturized Quartz Microbalance Array and Its Application to Chemical Sensor Systems for Liquids. IEEE Sensors Journal, 3, 361-368.

11. Sauerbrey, G. (1959) Verwendung von Schwingquarzenzur Wagungdunner Schichten und zur Mikrowagung. Zeitschrift für Physik, 155, 206-222.

12. Kanazawa, K. and Gordon Ii, J.G. (1985) The Oscillation Frequency of a Quartz Resonator in Contact with Liquid. Analytica Chimica Acta, 175, 99-105.

13. Ward, M.D. and Delawski, E.J. (1991) Radial Mass Sensitivity of the Quartz Crystal Microbalance in Liquid Media. Analytical Chemistry, 63, 886-890.

14. Lin, Z., Yip, C.M., Scott Joseph, I. and Ward, M.D. (1993) Operation of an Ultrasensitive 30 MHz Quartz Crystal Microbalance in Liquids. Analytical Chemistry, 65, 1546-1551.

15. Rodahl, M. and Kasemo, B. (1996) A Simple Setup to Simultaneously Measure Liquid Deposits on a QCM Electrode. Sensors and Actuators B, 37, 111-116.

16. Muramatsu, H., Tamiya, E. and Karube, I. (1988) Computation of Equivalent Circuit Parameters of Quartz Crystals in Contact with Liquid and Study of Liquid Properties. Analytical Chemistry, 60, 2142-2146.

17. Rodahl, M., Hook, F., Fredriksson, C., Keller, C.A., Krozer, A., Brzezinski, P., Voinova, M. and Kasemo, B. (1997) Simultaneous Frequency and Dissipation Factor QCM Measurements of Biomolecular Adsorption and Cell Adhesion. Faraday Discussions, 107, 229-246.

18. Penza, M., Cassano, G., Aversa, P., Antolini, F., Cusano, A., Cutolo, A., Giordano, M. and Nicolais, L. (2004) Alcohol Detection Using Carbon Nanotubes Acoustic and Optical Sensors. Applied Physics Letters, 85, 2379-2381.

19. Matsuguchi, M. and Uno, T. (2006) Molecular Imprinting Strategy for Solvent Molecules and Its Application for QCM-Based VOC Vapor Sensing. Sensors and Actuators B, 113, 94-99.

20. Brousseau III, L.C., Aurentz, D.J., Benesi, A.J. and Mallouk, T.E. (1997) Molecular Design of Intercalation-Based Sensors. 2. Sensing of Carbon Dioxide in Functionalized Thin Films of Copper Octanediylbis(phosphonate). Analytical Chemistry, 69, 679-687.

21. Ding, B., Kim, J., Miyazaki, Y. and Shiratori, S. (2004) Electrospun Nanofibrous Membranes Coated Quartz as Gas Sensor for NH3 Detection. Sensors and Actuators B, 101, 373-380.

22. Wang, X.H., Zhang, J., Zhu, Z.Q. and Zhu, J.Z. (2007) Humidity Sensing Properties of Pd2+-Doped ZnO Nanotetrapods. Applied Surface Science, 253, 3168-3173.

23. King, W.H. (1964) Piezoelectric Sorption Detector. Analytical Chemistry, 36, 1735-1739.

24. Bottom, V.E. (1982) Introduction to Quartz Crystal Unit Design. Van Nostrand Reinhold, New York.

25. Bruckenstein, S., Fensore, A., Li, Z.F. and Hillman, A.R. (1994) Dual Quartz Crystal Microbalance Compensation Using a Submerged Reference Crystal. Effect of Surface Roughness and Liquid Properties. Journal of Electroanalytical Chemistry, 370, 189-195.

26. Bruckenstein, S., Michalskl, M., Fensore, A., Zhufen, L.I. and Hillman, A.R. (1994) Dual Quartz Crystal Oscillator Circuit. Minimizing Effects Due to Liquid Viscosity, Density, and Temperature. Analytical Chemistry, 66, 1847-1852.

27. Dunham, G.C., Benson, N.H., Petelenz, D. and Janata, J. (1995) Dual Quartz Crystal Microbalance. Analytical Chemistry, 67, 267-272.

28. Uno, T. (1996) Frequency Control Symposium, 50th, Proceedings of the 1996 IEEE International, 526-531.

29. Cooper, M.A. and Singleton, V.T. (2007) A survey of the 2001 to 2005 Quartz Crystal Microbalance Biosensor Literature: Applications of Acoustic Physics to the Analysis of Biomolecular Interactions. Journal of Molecular Recognition, 20, 154-184.

30. ArbabZavar, M.H., Heydari, S., Rounaghi, G. and Ashraf, N. (2011) Graphite Disk Lanthanum(III)-Selective Electrode Based on Benzo-15-Crown-5. Journal of the Electrochemical Society, 158, F142-F146.

31. Rounaghi, G.H. and Adzadeh Kakhki, R.M. (2011) Highly Selective and Sensitive Coated-Wire Yttrium (III) Cation Selective Electrode Based on Kryptofix-22DD. Journal of the Electrochemical Society, 158, F121-F125.

32. Arbab Zavar, M.H., Heydari, S. and Rounaghi, G. (2013) Electrochemical Determination of Salicylic Acid at a New Biosensor

Based on Polypyrrole-Banana Tissue Composite. Arabian Journal for Science and Engineering, 38, 29-36.

33. Zadeh kakhki, R.M., Ronagi, G.H. and Sadeghian, H. (2011) A New Cerium (III) Ion Selective Electrode Based on 2,9-Dihydroxy-1,10-Diphenoxy-4,7-Dithia Decane, a Novel Synthetic Ligand. Electrochimica Acta, 56, 9756-9761.

34. Arbab Zavar, M.H., Heydari, S., Rounaghi, G.H., Eshghi, H. and Sadeghian, H. (2012) Nano-Level Monitoring of Yttrium by a Novel PVC-Membrane Sensor Based on 2,9-dihydroxy-1,10-diphenoxy-4,7-dithiadecane. Croatica Chemica Acta, 85, 131-137.

35. Zadeh Kakhki, R.M. and Rounaghi, G. (2011) Selective Uranyl Cation Detection by Polymeric Ion Selective Electrode Based on Benzo-15-Crown-5. Materials Science & Engineering: C, 31, 1637-1642.

36. Carmon, K. (2004) Development of a QCM Biosensor Using Receptor Proteins to Detect the Binding of Small Ligands. Thesis. Clarkson University, Potsdam.

37. Marx, K.A. (2003) Quartz Crystal Microbalance: A Useful Tool for Studying Thin Polymer Films and Complex Biomolecular Systems at the Solution-Surface Interface. Biomacromolecules, 4, 1099-1120.

38. Thompson, M. and Hayward, G.L. (1997) Mass Response of the Thickness-Shear Mode Acoustic Wave Sensor in Liquids as a Central Misleading Dogma. IEEE International Frequency Control Symposium, Orlando, 28-30 May 1997, 114-119.

39. Janshoff, A., Galla, H.J. and Steinem, C. (2000) Piezoelectric Mass-Sensing Devices as Biosensors. An Alternative to Optical Biosensors. Angewandte Chemie International Edition, 39, 4004-4032.

40. Cooper, M.A. (2003) Biosensor Profiling of Molecular Interactions in Pharmacology. Current Opinion in Pharmacology, 3, 557-562.

41. Cooper, M.A. (2006) Resonant Acoustic Profiling (RAPTM) and Rupture Event Scanning (REVSTM). In: Steinem, C. and Janshoff, A., Eds., Piezoelectric Sensors, Vol. 5, Springer-Verlag, Berlin, Heidelberg, 449-481.

42. Cote, G.L., Lec, R.M. and Pishko, M.V. (2003) Emerging Biomedical Sensing Technologies and Their Applications. IEEE Sensors Journal, 3, 251-266.

43. Hook, F. and Kasemo, B. (2006) In the QCM-D Technique. In: Steinem, C. and Janshoff, A., Eds., Piezoelectric Sensors, Vol. 5, Springer-Verlag, Berlin, Heidelberg, 425-449.

44. Steinem, C. and Janshoff, A. (2006) In Piezoelectric Sensors. Springer-Verlag, Berlin, Heidelberg.

45. Braunhut, S., McIntosh, D., Vorotnikova, E., Zhou, T. and Marx, K.A. (2005) Detection of Apoptosis and Drug Resistance of Human Breast Cancer Cells to Taxane Treatments Using Quartz Crystal Microbalance Biosensor Technology. ASSAY and Drug Development Technologies, 3, 77-88.

46. Kim, Y.S., Niazi, J.H. and Gu, M.B. (2009) Specific Detection of Oxytetracycline Using DNA Aptamer-Immobilized Interdigitated Array Electrode Chip. Analytica Chimica Acta, 634, 250-254.

47. Proske, D., Blank, M., Buhmann, R. and Resch, A. (2005) Aptamers—Basic Research, Drug Development, and Clinical Applications. Applied Microbiology and Biotechnology, 69, 367-374.

48. Xie, S. and Walton, S.P. (2009) Application and Analysis of Structure-Switching Aptamers for Small Molecule Quantification. Analytica Chimica Acta, 638, 213-219.

49. Fang, L., Lu, Z., Wei, H. and Wang, E. (2008) A Electrochemiluminescence Aptasensor for Detection of Thrombin Incorporating the Capture Aptamer Labeled with Gold Nanoparticles Immobilized onto the Thio-Silanized ITO Electrode. Analytica Chimica Acta, 628, 80-86.

50. Farokhzad, O.C., Jon, S., Khademhosseini, A., Tran, T.N., Lavan, D.A. and Langer, R. (2004) Nanoparticle-Aptamer Bioconjugates: A New Approach for Targeting Prostate Cancer Cells. Cancer Research, 64, 7668-7672.

51. Yao, C.Y., Zhu, T.Y., Qi, Y.Z., Zhao, Y.H., Xia, H. and Fu, W.L. (2010) Development of a Quartz Crystal Microbalance Biosensor with Aptamers as Bio-Recognition Element. Sensors, 10, 5859-5871.

52. Wu, V.C.H., Chen, S.H. and Lin, C.S. (2007) Real-Time Detection of Escherichia coli O157:H7 Sequences Using a Circulating-Flow System of Quartz Crystal Microbalance. Biosensors and Bioelectronics, 22, 2967-2975.

53. Janeway Jr., C.A., Walport, M.J. and Travers, P. (2005) Immunobiology: The Immune System in Health and Disease. 6th Edition, Taylor & Francis Group, New York.

54. Ahirwal, G.K. and Mitra, C.K. (2010) Gold Nanoparticles Based Sandwich Electrochemical Immunosensor. Biosensors and Bioelectronics, 25, 2016-2020.

55. Noble, J., Thobhani, S., Attree, S., Boyd, R., Kumarswami, N., Szymanski, M. and Porter, A. (2010) Bioconjugation and Characterisation of Gold Colloid-Labelled Proteins. Journal of Immunological Methods, 356, 60-69.

56. Fritz, J. (2008) Cantilever Biosensors. Analyst, 133, 855-863.

57. Wu, G.H., Datar, R.H., Hansen, K.M., Thundat, T., Cote, R.J. and Majumdar, A. (2001) Bioassay of Prostate-Specific Antigen (PSA) Using Microcantilevers. Nature Biotechnology, 19, 856-860.

58. Milburn, C., Zhou, J., Bravo, O., Kumar, C. and Soboyejo, W.O. (2005) Sensing Interactions between Vimentin Antibodies and Antigens for Early Cancer Detection. Journal of Biomedical Nanotechnology, 1, 30-38.

59. Chu, X., Zhao, Z.L., Shen, G.L. and Yu, R.Q. (2006) Quartz Crystal Microbalance Immunoassay with Dendritic Amplification Using Colloidal Gold Immunocomplex. Sensors and Actuators B, 114, 696-704.

60. Jin, X., Jin, X., Chen, L., Jiang, J., Shen, G. and Yu, R. (2009) Piezoelectric Immunosensor with Gold Nanoparticles Enhanced Competitive Immunoreaction Technique for Quantification of Aflatoxin B1. Biosensors & Bioelectronics, 24, 2580-2585.

61. Ward, M.D. and Ebersole, R.C. (1996) Piezoelectric Specific Binding Assay with Mass Amplified Reagents. US Patent No. 5501986.

62. Su, X.D., Chew, F.T. and Li, S.F.Y. (2000) Design and Application of Piezoelectric Quartz Crystal-Based Immunoassay. Analytical Sciences, 16, 107-114.

63. Liu, T., Tang, J. and Jiang, L. (2004) The Enhancement Effect of Gold Nanoparticles as a Surface Modifier on DNA Sensor Sensitivity. Biochemical and Biophysical Research Communications, 313, 3-7.

64. Chu, P.T., Lin, C.S., Chen, W.J., Chen, C.F. and Wen, H.W. (2012) Detection of Gliadin in Foods Using a Quartz Crystal Microbalance Biosensor that Incorporates Gold Nanoparticles. Journal of Agricultural and Food Chemistry, 60, 6483-6492.

65. Kaewphinit, T., Santiwatanakul, S. and Chansiri, K. (2012) Gold Nanoparticle Amplification Combined with Quartz Crystal Microbalance DNA Based Biosensor for Detection of Mycobacterium Tuberculosis. Sensors & Transducers Journal, 146, 156-163.

66. Weizmann, Y., Patolsky, F. and Willner, I. (2001) Amplified Detection of DNA and Analysis of Single-Base Mismatches by the Catalyzed Deposition of Gold on Au-Nanoparticles. Analyst, 126, 1502-1504.

67. Zhao, H.Q., Lin, L., Tang, J., Duan, M.X. and Jiang, L. (2001) Enhancement of the Immobilization and Discrimination of DNA Probe on a Biosensor Using Gold Nanoparticles. Chinese Science Bulletin, 46, 1074-1077.

68. Willner, I., Patolsky, F., Weimann, Y. and Willner, B. (2002) Amplified Detection of Single-Base Mismatches in DNA Using Microgravimetric Quartz-Crystal-Microbalance Transduction. Talanta, 56, 847-856.

69. Liu, T., Tang, J. and Jiang, L. (2002) Sensitivity Enhancement of DNA Sensors by Nanogold Surface Modification. Biochemical and Biophysical Research Communications, 295, 14-16.

70. Fritzsche, W. and Taton, T.A. (2003) Metal Nanoparticles as Labels for Heterogeneous, Chip-Based DNA Detection. Nanotechnology, 14, R63-R73.

71. Liu, T., Tang, J., Han, M.M. and Jiang, L. (2003) Surface Modification of Nanogold Particles in DNA Detection with Quartz Crystal Microbalance. Chinese Science Bulletin, 48, 873-875.

72. Ha, T.H., Kim, S., Lim, G. and Kim, K. (2004) Influence of Liquid Medium and Surface Morphology on the Response of QCM during Immobilization and Hybridization of Short Oligonucleotides. Biosensors and Bioelectronics, 20, 378-389.

73. He, F.J. and Liu, S.Q. (2004) Detection of P. Aeruginosa Using Nano-Structured Electrode-Separated Piezoelectric DNA Biosensor. Talanta, 62, 271-277.

74. Liu, T., Tang, J., Zhao, H.Q., Deng, Y.P. and Jiang, L. (2002) Particle Size Effect of the DNA Sensor Amplified with Gold Nanoparticles. Langmuir, 18, 5624-5626.

75. Liu, S.F., Li, J.R. and Jiang, L. (2005) Surface Modification of Platinum Quartz Crystal Microbalance by Controlled Electroless Deposition of Gold Nanoparticles and Its Enhancing Effect on the HS-DNA Immobilization. Colloids and Surfaces A, 257-258, 57-62.

76. Zhang, R.Y., Pang, D.W., Zhang, Z.L., Yan, J.W., Yao, J.L., Tian, Z.Q., Mao, B.W. and Sun, S.G. (2002) Investigation of Ordered ds-DNA Monolayers on Gold Electrodes. Journal of Physical Chemistry B, 106, 11233-11239.

77. Mo, Z.H., Wang, H.L., Liang, Y.L., Liu, F.W. and Xue, Y.X. (2005) Highly Reproducible Hybridization Assay of Zeptomole DNA Based on Adsorption of Nanoparticle-Bioconjugate. Analyst, 130, 1589-1594.

78. Ma, Z.F., Li, J.R., Jiang, L., Yang, M.S. and Sui, S.F. (2002) High Sensitive DNA Detection Amplified by Enlarging Au-Nanoparticles in Situ. Chemistry Letters, 31, 328-329.

79. Mao, X., Yang, L., Su, X.L. and Li, Y.B. (2006) A Nanoparticle Amplification Based Quartz Crystal Microbalance DNA Sensor for Detection of Escherichia coli O157:H7. Biosensors and Bioelectronics, 21, 1178-1185.

80. Hao, R.Z., Song, H.B., Zuo, G.M., Yang, R.F., Wei, H.P., Wang, D.B., Cui, Z.Q., Zhang, Z., Cheng, Z.X. and Zhang, X.E. (2011) DNA Probe Functionalized QCM Biosensor Based on Gold Nanoparticle Amplification for Bacillus anthracis Detection. Biosensors and Bioelectronics, 26, 3398-3404.

81. Willner, I., Patolsky, F., Weizmann, Y. and Willner, B. (2002) Amplified Detection of Single-Base Mismatches in DNA Using Microgravimetric Quartz-Crystal-Microbalance Transduction. Talanta, 56, 847-856.

82. Chen, S.H., Chuang, Y.C., Lu, Y.C., Lin, H.C., Yang, Y.L. and Lin, C.S. (2009) A Method of Layer-by-Layer Gold Nanoparticle Hybridization in a Quartz Crystal Microbalance DNA Sensing System Used to Detect Dengue Virus. Nanotechnology, 20, Article ID: 215501.

83. Scodeller, P., Flexer, V., Szamocki, R., Calvo, E.J., Tognalli, N., Troiani, H. and Fainstein, A. (2008) Wired-Enzyme Core-Shell Au Nanoparticle Biosensor. Journal of the American Chemical Society, 130, 12690-12697.

84. Reimhult, K., Yoshimatsu, K., Risveden, K., Chen, S., Ye, L. and Krozer, A. (2008) Characterization of QCM Sensor Surfaces Coated with Molecularly Imprinted Nanoparticles. Biosensors and Bioelectronics, 23, 1908-1914.

85. Kong, L.J., Pan, M.F., Fang, G.Z., He, X.L., Yang, Y.K., Dai, J. and Wang, S. (2014) Molecularly Imprinted Quartz Crystal Microbalance Sensor Based on Poly(o-Aminothiophenol) Membrane and Au Nanoparticles for Ractopamine Determination. Biosensors and Bioelectronics, 51, 286-292.

86. Uludag, Y. and Tothill, I.E. (2010) Development of a Sensitive Detection Method of Cancer Biomarkers in Human Serum (75%) Using a Quartz Crystal Microbalance Sensor and Nanoparticles Amplification System. Talanta, 82, 277-282.

87. Ma, Z.F., Wu, J.L., Zhou, T.H., Chen, Z.H., Dong, Y., Tang, J.T. and Sui, S.F. (2002) Detection of Human Lung Carcinoma Cell Using Quartz Crystal Microbalance Amplified by Enlarging Au Nanoparticles in Vitro. New Journal of Chemistry, 26, 1795-1798.

88. Chen, Q., Tang, W., Wang, D., Wu, X., Li, N. and Liu, F. (2010) Amplified QCM-D Biosensor for Protein Based on Aptamer-Functionalized Gold Nanoparticles. Biosensors and Bioelectronics, 26, 575-579.

89. Shan, W.Q., Pan, Y.L., Fang, H.T., Guo, M.L., Nie, Z., Huang, Y. and Yao, S.Z. (2014) An Aptamer-Based Quartz Crystal Microbalance Biosensor for Sensitive and Selective Detection of Leukemia Cells Using Silver-Enhanced Gold Nanoparticle Label. Talanta, 126, 130-135.

90. Kaufman, E.D., Belyea, J., Johnson, M.C., Nicholson, Z.M., Ricks, J.L., Pavak, K., Shah, P.K., Bayless, M., Pettersson, T., Feldoto, Z., Blomberg, E., Claesson, P. and Franzen, S. (2007) Probing Protein Adsorption onto Mercaptoundecanoic Acid Stabilized Gold Nanoparticles and Surfaces by Quartz Crystal Microbalance and Potential Measurements. Langmuir, 23, 6053-6062.

91. Tang, D.Q., Zhang, D.J., Tang, D.Y. and Ai, H. (2006) Amplification of the Antigen-Antibody Interaction from Quartz Crystal Microbalance Immunosensors via Back-Filling Immobilization of Nanogold on Biorecognition Surface. Journal of Immunological Methods, 316, 144-152.

92. Su, P.G. and Chang, Y.P. (2008) Low-Humidity Sensor Based on a Quartz-Crystal Microbalance Coated with Polypyrrole/Ag/TiO2 Nanoparticles Composite Thin Films. Sensors and Actuators B, 129, 915-920.

93. Yang, Y.J., Jiang, Y.D., Xu, J.H. and Yu, J.S. (2007) Conducting Polymeric Nanoparticles Synthesized in Reverse Micelles and Their Gas Sensitivity Based on Quartz Crystal Microbalance. Polymer, 48, 4459-4465.

94. Salam, F., Uludag, Y. and Tothill, I.E. (2013) Real-Time and Sensitive Detection of Salmonella Typhimurium Using an Automated Quartz Crystal Microbalance (QCM) Instrument with Nanoparticles Amplification. Talanta, 115, 761- 767.

95. Guo, X., Lin, C.S., Chen, S.H., Ye, R. and Wu, V.C.H. (2012) A Piezoelectric Immunosensor for Specific Capture and Enrichment of Viable Pathogens by Quartz Crystal Microbalance Sensor, Followed by Detection with Antibody-Functionalized Gold Nanoparticles. Biosensors and Bioelectronics, 38, 177-183.

96. www.q-sense.com/file/qcm-d-and-nanoparticles-application-overview.pdf

97. Crooks, R.M. and Ricco, A.J. (1998) New Organic Materials Suitable for Use in Chemical Sensor Arrays. Accounts of Chemical Research, 31, 219-227.

98. Ricco, A.J., Crooks, R.M. and Osbourn, G.C. (1998) Surface Acoustic Wave Chemical Sensor Arrays: New Chemically Sensitive Interfaces Combined with Novel Cluster Analysis to Detect Volatile Organic Compounds and Mixtures. Accounts of Chemical Research, 31, 289-296.

99. Kepley, L.J., Crooks, R.M. and Ricco, A.J. (1992) A Selective SAW-Based Organophosphonate Chemical Sensor Employing a Self-Assembled, Composite Monolayer: A New Paradigm for Sensor Design. Analytical Chemistry, 64, 3191-3193.

100. Schonherr, H., Vancso, G.J., Huisman, B.H., Van Veggel, F.C.J.M. and Reinhoudt, D.N. (1997) An Atomic Force Microscopy Study of Self-Assembled Monolayers of Calix[4]resorcinarene Adsorbates on Au(111). Langmuir, 13, 1567-1570.

101. Chen, S., Pei, R., Zhao, T. and Dyer, D.J. (2002) Gold Nanoparticle Assemblies by Metal Ion-Pyridine Complexation and Their Rectified Quantized Charging in Aqueous Solutions. Journal of Physical Chemistry B, 106, 1903-1908.

102. Shen, G., Wang, H., Shen, G. and Yu, R.Q. (2007) Au Nanoparticle Network-Type Thin Films Formed via Mixed Assembling and Cross-Linking Route for Biosensor Application: Quartz Crystal Microbalance Study. Analytical Biochemistry, 365, 1-6.

103. Liu, H. and Hu, N.F. (2006) Interaction between Myoglobin and Hyaluronic Acid in Their Layer-by-Layer Assembly:Quartz Crystal Microbalance and Cyclic Voltammetry Studies. Journal of Physical Chemistry B, 110, 14494-14502.

104. Guleryuz, H., Kaus, I., Buron, C.C., Filiatre, C., Hedin, N., Bergstrom, L. and Einarsrud, M.A. (2014) Deposition of Silica Nanoparticles onto Alumina Measured by Optical Reflectometry and Quartz Crystal Microbalance with Dissipation Techniques. Colloids and Surfaces A: Physicochemical and Engineering Aspects, 443, 384-390.

Chapter 8

Nano biosensors for neurochemical monitoring

M. Meyyappan

NASA Ames Research Center

ABSTRACT

Neurochemicals such as dopamine (DA) and serotonin (S-HT) are linked to disorders such as Parkinson's disease, epilepsy, addiction and many others. Detection of and monitoring these neurochemicals *in vivo* and *in vitro* has become important in treating various disorders. The electroactive nature of DA and S-HT has enabled employing electrochemical techniques to detect them at low concentrations, and a variety of electrodes and approaches have been reported. The use of nanomaterials such as carbon nanotubes, graphene and nanowires has been advocated in recent years for the sensitive detection of neurochemicals. This article reviews the advances in nano biosensors for this application and discusses the future outlook and challenges.

Keywords: Biosensors; Nanomaterials Neurochemical detection; Carbon nanotubes; Graphene; Neural disorders

1 INTRODUCTION

Research activities on nanomaterials such as carbon nanotubes (CNTs), graphene, various inorganic nanowires and nanoparticles have grown exponentially over the last two decades. The nanomaterials offer interesting and unique electrical, mechanical, optical and other properties compared to their bulk counterparts. Exploiting these differences to construct useful devices, systems and architecture is expected to benefit all economic sectors, and the research activities span all areas of science and engineering [1–4]. Sensor technology is one of the areas to reap the benefits of nanoscience and technology, offering tremendous potential to develop sensors for early warning of infectious diseases, cancer diagnostics, and lab-on-a-chip for rapid and inexpensive routine health monitoring involving blood and urine analysis [5]. A subset of medical diagnostics to benefit from nanosensors includes sensitive detection of neurochemicals for the treatment of various neural disorders such as Parkinson's disease, epilepsy, etc. [6–8]. This article first provides a brief background on neurochemicals, biosensors and nanomaterials and then reviews the sensor development efforts to date. This is followed by a detailed discussion of carbon nanofiber electrodes developed in our laboratory for this purpose and finally, future outlook and challenges are summarized.

1.1 Neurochemicals and the need for monitoring

Neurotransmitters are chemicals in the brain that transmit signals from the neurons to the target cells. Dopamine (DA) and Serotonin (5-HT) are two of these chemicals, which have been associated with a number of neural disorders [9–15]. Dopamine concentration is associated with Parkinson's disease, epilepsy and addiction. Insufficiency of dopamine is thought to cause Parkinson's disease whereas excess release of DA leads to schizophrenia. Similarly, serotonin levels are linked to disorders such as addiction and depression. Both DA and HT-5 are present in very small quantities in a sea of ascorbic acid (AA) with 100 – 1000 times higher concentration. Most of the detection techniques used in the literature and reviewed here, especially the electrochemical techniques, have the advantage of fast response time, but the real challenge is the selectivity between DA and HT-5 in AA. The two neurochemicals have oxidation potentials close to each other, separated only by 150 mV. Therefore, many conventional electrodes have difficulty yielding very distinct peaks for the two targets especially when both are present together in AA at realistic concentration levels. This has prompted the investigation of nanomaterial based electrodes or modification of conventional electrodes with coatings of nanomaterials.

In addition to recording the concentrations of neurochemicals, electrodes are desired for stimulation as well [9, 10] since some of the neural disorders are treated by deep brain stimulation (DBS). The stimulation is believed to produce evoked release of the neurochemicals. Therefore, monitoring the released chemicals adds to the importance of developing sensitive techniques to provide feedback for the DBS procedure and improve the outcome. Figure 1 shows a schematic of the electrode inserted in the brain for DBS, which is typically 1.3 mm in diameter with four 1.5 mm contacts. This electrode is used for stimulation only at present without any feedback from monitoring the concentrations of neurochemicals. Targeted activation of specific and precise location with a much smaller electrode along with guidance/feedback from measured neurotransmitter levels is highly desirable. Nanoelectrodes have the potential to meet this challenge, and the requirements would include ultrahigh sensitivity ~1 nM, fast speed of 10 ms resolution and reliability and stability for long term implantation.

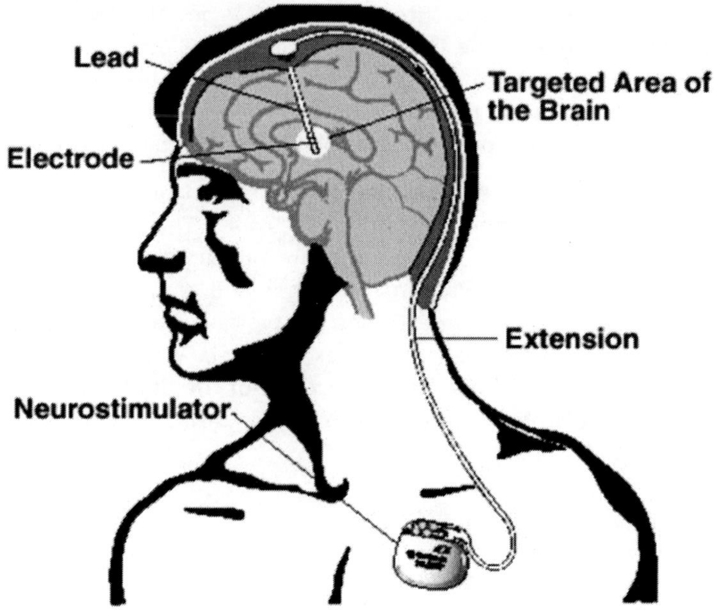

Figure 1. Schematic of the electrode in deep brain stimulation. Image courtesy of Jun Li

1.2 Biosensors

Biosensors are devices that provide the surface to host the probe-target interaction and translate this interaction into some form of a measurable signal. The signal output can be in many forms: optical, deflection of a cantilever, electrical or electrochemical signals [5, 16–19]. The International Union of Pure and Applied Chemistry (IUPAC) defines a biosensor as "a device that uses specific biochemical reactions mediated by isolated enzymes, immunosystems, tissues, organelles, or whole cell to detect chemical compounds usually by electrical, thermal or optical signals" [5]. Of these, optical methods have been well developed and shown capable of detecting down to a single molecule. They are commonly fluorescence based wherein the bio-recognition molecules or probes are labeled with fluorescent tags (example: dye) and the probe-target interaction results in a fluorescent signal. The intensity relates to the concentration of the target molecules and capture efficiency. Though labeling is laborious, label-free optical techniques have been developed as well.

Electrical biosensors may take the form of bio field effect transistor (BioFET) wherein the conventional transistor gate is replaced with a liquid gate and a reference electrode [18, 19]. When a probe is attached to the gate, the current–voltage (I-V) characteristics would shift from the behavior of the unmodified gate. Further probe-target interaction would bring about additional shift of the I-V curve proportional to the concentration of the target molecules in the solution. BioFETs have been extensively investigated for the detection of various biomarkers related to healthcare and environmental monitoring.

Another form of electrical transduction involves electrochemical techniques that use metal (Pt for example) or carbon electrodes. The latter can be in the form of graphite, glassy carbon, carbon paste or diamond, which have long been in use; more recently, carbon nanotubes and graphene have become popular as well [1, 3, 5]. These electrodes are typically used in amperometric, potentiometric or impedimetric modes wherein the electrodes are functionalized with probes that have selectivity or affinity for the targets of interest [16, 17]. The probes in all cases can be DNA, RNA, antibodies, aptamers etc.

1.3 Nanomaterials

Since their discovery in the early 1990s, carbon nanotubes have commanded significant attention in terms of growth, characterization of their interesting properties and application development [1]. The CNTs are seamless cylinders arising from the roll-up of graphene sheets. In the case of single-walled carbon nanotubes (SWCNTs), a chiral vector (n, m) defines their intrinsic properties. When (n-m)/3 is an integer, the resulting structure is metallic; otherwise, it is semiconducting. Multiwalled carbon nanotubes (MWCNTs) are concentric cylinders with a hollow core, resembling rolled-up graphite sheets. Both SWCNTs and MWCNTs are grown by techniques such as arc synthesis, laser ablation, chemical vapor deposition (CVD) and plasma enhanced chemical vapor deposition (PECVD). While all of these approaches can produce CNTs in bulk quantities, CVD and PECVD are suited to grow CNTs on substrates with controlled thickness and even vertical orientation, suitable for device fabrication. The properties of CNTs can be modified by functionalizing them with other atomic or molecular groups. For example, fluorination makes SWCNTs insulating [20]. Attaching bio probes of DNA, RNA, aptamers, antibodies etc. enables development of biosensors for selective capturing of biotargets of interest [1, 5]. CNTs have been investigated for a wide variety of applications including transistors, memory devices, chem, bio and radiation sensors, nanoelectromechanical devices, high strength light weight composites, batteries, supercapacitors, filters and membranes, thermal interface materials, electrical interconnects in integrated circuits, electrical cables, transparent electrodes, shielding of electromagnetic interference, field emission devices, catalyst support and many others [1].

Graphene is a one atom thick quasi two-dimensional material made up of sp^2-bonded carbon atoms, which are packed in a honeycomb-like lattice. It is a semi-metal, that is, a zero-gap semiconductor. Graphene exhibits a very low resistivity of $\sim 10^{-6}$ ohm-cm and a high mobility of ~15000 $cm^2/v.s$ [3]. The Young's modulus of graphene is estimated to be ~1 TPa. Graphene is synthesized by exfoliation of graphite or reduced graphite oxide. CVD also is a popular method to produce graphene films on nickel and copper substrates. PECVD can provide multilayer graphene vertically oriented to the substrate surface. Graphene has been investigated for almost every application listed above for CNTs.

Thin films of various inorganic materials such as Si, Ge, GaAs, InP, oxides, nitrides etc. have helped with advances in microelectronics, optoelectronics, microelectromechanical systems and others. In recent years, most of these materials have been grown in the form of one-dimensional (1D) nanowires. When the radius of a semiconducting nanowire is smaller than its Bohr radius, the bandgap increases from the corresponding value for the thin film. Nanowires offer interesting electrical, optical, thermal and other properties relative to their bulk counterparts and therefore, have received much attention in the development of electronic, photonic, sensor and other devices [2].

2 REVIEW

2.1 Sensors for neurochemical monitoring

All of the nanomaterials mentioned above have been used for the detection of neurochemicals [21–59]. A review of selected works is presented here. Baldrich et al. [21] prepared SWCNT electrodes through entrapment of nanotubes on the surface of protein-coated magnetic particles. This electrode was able to distinguish the peaks of DA and uric acid (UA) at 200 μM concentrations. Li et al. [22] modified glassy carbon electrodes (GCEs) with SWCNTs to obtain a DA detection limit of 50 nM with peak current linearly varying between 5 and 100 μM. A graphite electrode modified with MWCNTs and pretreated for DA detection enabled Moreno et al. [23] to obtain a linear response in the range of

50 nm – 1 μM in the presence of AA. Electrodes using a nanocomposite of MWCNTs, sol gel silica and gold nanoparticles were tried by Komathi et al. [24] for the sensitive detection of DA. The sensitivity was found to be nearly four times smaller without the gold nanoparticles. The pores in the silica network host MWCNTs which are hydrophobic and silica with $-NH_2$ group which are hydrophilic. The latter attract the ascorbate ions and reject the DA molecules, which reside on the surface of MWCNTs and exhibit strong electron transfer reactions. In addition, the presence of gold nanoparticles enhances the electrocatalytic oxidation of DA. The linear range for DA detection reported in this work is 0.1 to 30 nM. Yang et al. [25] prepared a composite of nano-sized copper oxide / MWCNTs / nafion and used it to modify a GCE which showed a detection limit of 0.4 μM. Aravind and Ramaprabhu [26] created a nanocomposite of Pt nanoparticles and MWCNTs and immobilized it with SS-DNA to prevent nanotube agglomeration. The SS-DNA also facilitated electron transfer reaction with dopamine. This hybrid electrode was shown to have a DA detection limit of 0.8 μM. Kumaraswamy and Venton [27] modified a carbon fiber microelectrode with SWCNTs and used it for the co-detection of DA and ST *in vivo* in the striatum of anesthetized rat after administration of a serotonin synthetic precursor. The measured concentrations 24 min. after the administration were 250 nM for DA and 130 nM for seratonin.

Kim et al. [28] modified a glassy carbon electrode (GCE) with graphene to reduce the capacitive background current and increase the current intensity of DA. This allowed them to obtain a complete peak separation between DA and AA with a linear range for DA detection between 4 and 100 μM. Alwarappan et al. [29] also used a graphene-modified GCE but reported selective detection results only for 1 mM each of DA, AA and serotonin. Sun et al. [30] modified a glassy carbon electrode with graphene/Pt to obtain distinct peaks in DPV curves for AA, DA and UA. The 1.7 nm Pt particles self-assembled on graphene showed better performance than GC and GC/graphene electrodes. Modification of GCE with a nanocomposite of β-cyclodextrin and graphene sheet enabled Tan et al. [31] to obtain a linear current response in the range of 9 nM to 12.7 μM DA in PBS. The linear range obtained in the amperometric mode was from 0.9 to 200 μM. In both cases, the nanocomposite-modified GCE did better than both GC and GC/graphene electrodes; this was attributed to the mass transfer control of the electrochemical reaction of dopamine on the nanocomposite in contrast to the usual adsorption-limited process. Han et al. [32] modified a GCE with a graphene-chitosan composite and compared it against a GC-chitosan electrode for the detection of DA, AA and UA. The addition of graphene was shown to increase the electrocatalytic activity towards the oxidation of all three chemicals. Gao et al. [33] modified a GCE with graphene oxide and showed a detection limit of 0.27 μM DA in the presence of AA and the oxidation peak current showed a linear relation with concentration in the range of 1.0–15.0 μM. Thomas et al. [34] modified a carbon paste electrode with graphene oxide to obtain a DA detection limit of

15 nM in the presence of 1000-fold concentration of AA and UA. Sun et al. [35] prepared a nanocomposite of graphene and tin oxide nanosheets and used it to modify a carbon ionic liquid electrode. This modified electrode provided peak currents proportional to DA concentration in the range of 0.5 to 500 μM with a detection limit of 0.13 μM. Tsai et al. [36] coated GCEs with Te nanowires and the modified electrode was further coated with nafion to improve the stability and selectivity.

Chandrashekar et al. [38] electropolymerized L-arginine on a carbon paste electrode to form a biopolymer and utilized it for the detection of DA, AA and UA. The peak current in the CV measurement showed a linear range for dopamine concentration between 50 μM and 0.1 mM with a detection limit of 0.5 μM. In addition to CNTs and graphene, diamond based electrodes have also been explored. For example, Raina et al. [39] used a nitrogen-incorporated nano-diamond based ultramicroelectrode array and produced steady state CV curves for DA concentrations in the range of 100–800 μM in 0.1 M PBS.

Figure 2. Vertical carbon nanofibers grown by plasma enhanced chemical vapor deposition. The nanofibers have a diameter of ~50 nm and height of 3–4 μm. Image courtesy of Prabhu Arumugam

Besides electrochemical approaches such as the ones discussed above, BioFETs have been used as well for detecting dopamine. For example, Li et al. [40]

fabricated an open gate ion sensitive field effect transistor and measured the I-V characteristics for increasing DA concentrations from 1 fM to 1 nM. However, no data was given on how the device would perform when DA is present in mixtures of AA and/or UA.

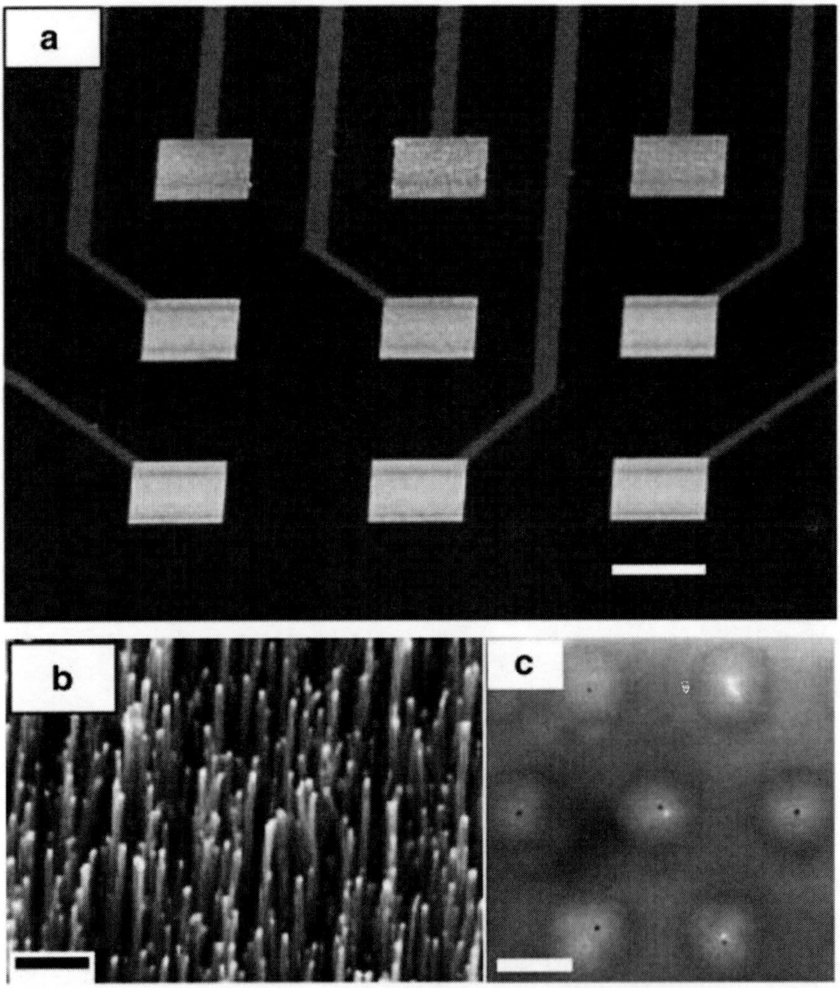

Figure 3. (**a**) A 3 × 3 electrode array, (**b**) stimulating electrode consisting of CNFs with large area (scale bar: 1 μm) and (**c**) recording electrode embedded in SiO_2 with high sensitivity (scale bar: 1 μm). Image courtesy of Jun Li

2.2 Carbon nanofiber electrodes for neurochemical monitoring

Our group has been developing carbon nanofiber based nanoelectrodes for the detection of neurochemicals and also for stimulation in collaboration with Mayo Clinic [56–59]. The CNFs are ~50 nm in diameter, grown by plasma enhanced chemical vapor deposition (PECVD). The as-grown structures are vertical and free-standing, as shown in Fig. 2, due to the electric field perpendicular to the wafer. The broken walls of CNFs along the outside of the structure as evidenced by TEM analysis (data not shown here) and the resulting defects are ideal for good electron transfer. The fabrication of the electrodes has been described previously [60–64] and a brief description is provided below with the fabricated electrodes in Fig. 3.

A silicon wafer with a 500 nm thick thermal oxide is used as substrate and a thin layer of Pt or Cr is deposited on the substrate to provide electrical contact between the CNFs and the measurement circuit. This is followed by sputtering of 10–30 nm nickel as catalyst layer. Acetylene is used to grow CNFs at a pressure of 1–3 Torr and temperature of 600-700 °C. The catalyst film breaks up into tiny droplets of random sizes on the wafer and growth on these catalysts particles yields nanofibers shown in Fig. 3 (b). These uncoated CNFs can be used as stimulating electrode of desirable size, even down to a single CNF electrode of 50 nm. Patterning of the catalyst instead of a blanket sputtered catalyst layer can provide CNFs of specified diameter at desired locations. The gap between the nanofibers is filled with SiO_2 to provide mechanical stability. Then the top surface of the wafer with CNFs is subjected to chemical mechanical polishing which results in a smooth oxide surface with few nm of CNFs protruding out as seen in Fig. 3 (c). This embedded electrode is suitable for recording neurochemical concentrations.

Figure 4 shows a color plot of the response of the electrode shown in Fig. 3 (c) to injected dopamine (2.5 μM) in real time. The response is seen 25 s after injection, ending around 60 s [56]. Fig. 5 compares the performance of this electrode with that of a conventional GCE for a mixture containing 1 mM of AA and 10 μM each of DA and 5-HT. The GCE is not able to resolve the peaks whereas the CNF nanoelectrodes shows very distinct peaks in the differential pulse voltammetry plots [58].

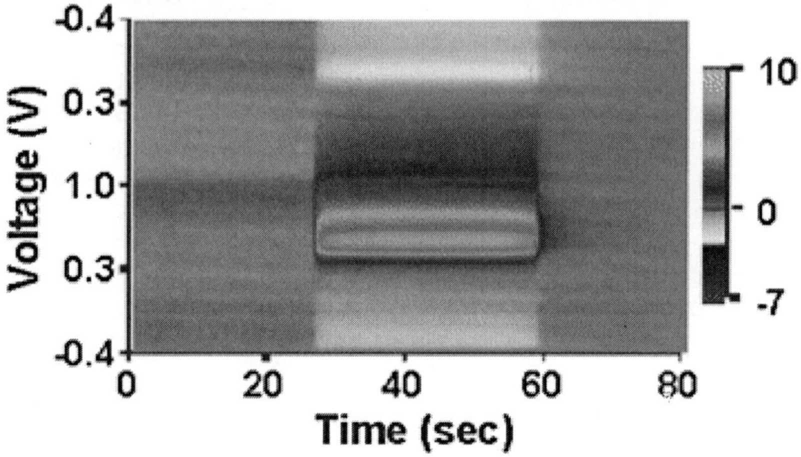

Figure 4. Electrode response to injected dopamine (2.5 μM) in real time [56]. Image courtesy of Jessica Koehne

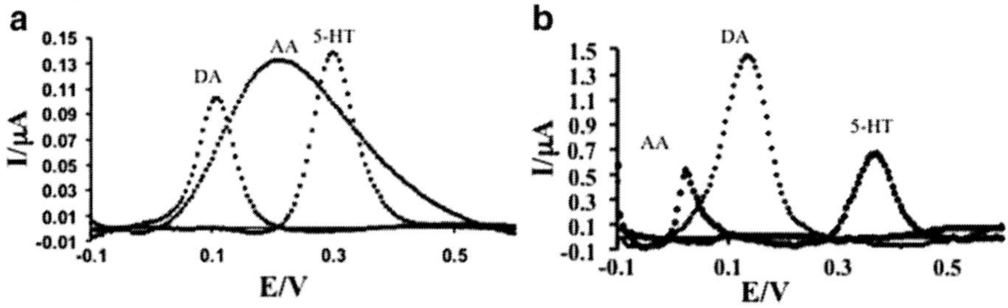

Figure 5. Differential pulse voltammetry results for 1 mM AA, 10 μM DA and 10 μM 5-HT [58]. (**a**) glassy carbon electrode and (**b**) carbon nanofiber electrode. Image courtesy of Jessica Koehne

3 CONCLUSIONS

A review of the current status of sensors for the detection and monitoring of neurochemicals has been presented. Most electrodes do not have any difficulty in detecting down to few nM concentrations of dopamine by itself in buffer solutions. However, sensitive detection of DA in mixtures of 5-HT, ascorbic acid and others is challenging. Nanoelectrodes using CNTs, graphene and CNFs have been reported to meet these challenges at present. More studies demonstrating the utility of these techniques *in vivo* are needed. Nanoelectrodes for stimulation show much potential but no significant demonstrations have appeared yet. This is an area that requires further investigation.

ACKNOWLEDGEMENTS

The author acknowledges Jessica Koehne, the Biosensor Group Leader at NASA Ames Center for Nanotechnology who is responsible for Technology development and other past and present members of this team Russell Andrews, Jun Li, Prabhu Arumugam, Adai Periyakaruppan, Emily Rand, Ramprasad Ghandiraman and collaborators from Mayo Clinic Kendall Lee, Mike Marsh and Kevin Bennett.

REFERENCES

1. M Meyyappan (ed.), *Carbon Nanotubes: Science and Applications* (Boca Raton, FL, CRC Press, 2004)
2. M Meyyappan, M Sunkara, *Inorganic Nanowires: Applications, Properties and Characterization*, (Boca Raton, FL, CRC Press, 2010)
3. W B Choi, J W Lee (eds.), *Graphene Synthesis and Application* (Boca Raton, FL CRC Press, 2012)
4. C Altavilla, E Ciliberto (eds.), *Inorganic Nanoparticles Synthesis, Applications and Perspectives* (Boca Raton, FL, CRC Press, 2011)
5. J Li, N Wu (eds.), *Biosensors Based on Nanomaterials and Nanodevices* (Boca Raton, FL, CRC Press, 2014)
6. RJ Andrews, Neuromodulation: Deep Brain Stimulation, Sensory Neuroprostheses and the Neural-Electrical Interface, in *Progress in Brain Research*, ed. by HS Sharma, 2009. 180, 127
7. RJ Andrews, Nanotechniques for Neurosurgery, Chapter 28, in *Current Progress in Neurosurgery, Vol 1*, ed. by BK Misra, AH Kaye, ER Laws (Kothari Medical Publishers, Mumbai, 2014), pp. 454–474
8. RR Llinas, KD Walton, M Nakao, I Hunter, PA Anquetil, J. Nanopart. Res. **7**, 111 (2005)
9. F Agnesi, SJ Tye, JM Bledsoe, CJ Griessenuer, CJ Kimble, GC Sieck, KE Bennett, PA Garris, CD Blaha, KH Lee, J. Neurosurg. **111**, 701 (2009)
10. KH Lee, SY Chang, DW Roberts, U Kim, J. Neurosurg. **101**, 511 (2004)
11. S Kapur, G Remington, Am. J. Psychiatry **153**, 466 (1996)
12. Y Ni, EB Malarkey, V Parpura, J. Neurochem. **103**, 1273 (2007)
13. RM Wightman, JM Leslie, AC Michael, Anal. Chem. **60**, 769A (1988)
14. DL Robinson, A Hermans, AT Seipel, RM Wightman, Chem. Rev. **108**, 2554 (2008)

15. TJ Webster, MC Waid, JL McKenzie, RL Price, JU Ejiofor, Nanotechnol. **48**, 48 (2004)

16. JS Daniels, N Pourmond, Electroanal. **19**, 1239 (2007)

17. J Gooding, Electroanal. **14**, 1149 (2002)

18. T Rim, KY Kim, CH Baek, YH Jeong, JS Lee, M Meyyappan, J. Nanosci. Nanotechnol. **14**, 273 (2014)

19. F G Banica, *Chemical Sensors and Biosensors*, (Wiley, New York, 2012)

20. ET Michelson, CB Huffman, AG Rinzler, RE Smalley, RH Hauge, JL Margrave, Chem. Phys. Lett. **296**, 188 (1998)

21. E Baddrich, R Gomez, G Gabriel, IX Munoz, Biosens. Bioelec. **5**, 1876 (2011)

22. Y Li, J Du, J Yang, D Liu, X Lu, Colloid Surf. B. Bioint. **97**, 32 (2012)

23. M Moreno, AS Arribas, E Berijo, M Chicharro, A Zapardiel, MC Rodriguez, Y Jalit, GA Rivas, Talanta **80**, 2149 (2010)

24. S Komathi, AI Gopalan, KP Lee, Analyst **135**, 397 (2010)

25. S Yang, G Li, Y Yin, R Yang, J Li, L Qu, J. Electroanal. Chem. **703**, 45 (2013)

26. SSJ Aravind, S Ramaprabhu, Sens. Actuat. B. Chem. **155**, 679 (2011)

27. BE Kumaraswamy, BJ Venton, Analyst **132**, 876 (2007)

28. YR Kim, SY Bong, YJ Kang, YT Yang, RK Mahajan, JS Kim, H Kim, Biosens. Bioelec. **25**, 2366 (2010)

29. S Alwarappan, A Erdem, C Liu, CZ Li, J. Phys. Chem. C **113**, 8853 (2009)

30. CL Sun, HH Lee, JM Yang, CC Wu, Biosens. Bioelec. **26**, 3450 (2011)

31. L Tan, KG Zhou, YH Zhang, HX Wang, XD Wang, YF Guo, HL Zhang, Electrochem. Commun. **12**, 557 (2010)

32. D Han, T Han, C Shan, A Ivaska, L Niu, Electroanal. **22**, 2001 (2010)

33. F Gao, X Cai, X Wang, C Cao, S Liu, F Gao, QX Wang, Sens. Actuat. B. Chem. **186**, 380 (2013)

34. T Thomas, RJ Mascarenhas, C Nethravathi, M Rajamathi, BE Kumaraswamy, J. Electroanal. Chem. **659**, 113 (2011)

35. W Sun, X Wang, Y Wang, X Ju, L Xu, G Li, Z Sun, Electrochem. Acta. **87**, 317 (2013)

36. HY Tsai, ZH Lin, HT Chang, Biosens. Bioelec. **35**, 479 (2012)

37. S Minnikanti, MGAG Pereira, S Jaraiedi, K Jackson, CM Costa-Neto, Q Li, N Peixoto, J. Neural Eng. **7**, 016002 (2010)

38. BN Chandrashekar, BE Kumaraswamy, M Pandurangachar, TV Sathisha, BS Sherigara, Colloid Surf. B. Bioint.**88**, 413 (2011)

39. S Raina, WP Kang, JL Davidson, Diamond Rel. Mater. **19**, 256 (2010)

40. DC Li, PH Yang, MSC Lu, IEEE Trans. Electron Dev. **57**, 2761 (2010)

41. MPN Bui, CA Li, GH Seong, Biochip. J. **6**, 149 (2012)

42. MS Lu, YC Chen, PC Huang, Biosens. Bioelec. **26**, 1093 (2010)

43. S Sansuk, E Bitziou, MB Joseph, JA Corington, MG Boutelle, PR Unwin, JV MacPherson, Anal. Chem. **85**, 163 (2013)

44. S Yang, G Li, R Yang, M Xia, L Qu, J. Solid State Electrochem. **15**, 1909 (2011)

45. M Zhu, C Zeng, J Ye, Electroanal. **23**, 907 (2011)

46. TV Sathisha, BE Kumaraswamy, S Reddy, BN Chandrashekar, B Eswarappa, J. Mol. Liq. **172**, 53 (2012)

47. X Yu, K Sheng, G Shi, Analyst **139**, 4525 (2014)

48. H Kim, JH Lee, EJ Chae, JW Choi, BK Oh, J. Nanosci. Nanotech. **11**, 7516 (2011)

49. D Sun, Q Zhao, F Tans, X Wang, J Gao, Anal. Methods. **4**, 3253 (2012)

50. JH Jin, E Cho, SH Jung, Biotechnol. Lett. **32**, 413 (2010)

51. AN Patel, SY Tan, TS Miller, JV MacPherson, PR Unwin, Anal. Chem. **85**, 11755 (2013)

52. M Hasanzadeh, X Shadjou, E Omidinia, J. Neurosci. Methods **219**, 53 (2013)

53. P Veerakumar, R Madhu, SM Chen, CT Hung, PH Tang, CB Wang, SB Liu, Analyst **139**, 4994 (2014)

54. TDB Nguyen-Vu, H Chen, AM Cassell, R Andrews, M Meyyappan, J Li, Small **2**, 89 (2006)

55. TDB Nguyen-Vu, H Chen, AM Cassell, RJ Andrews, M Meyyappan, J Li, IEEE Trans. Biomed. Eng. **54**, 1121 (2007)

56. JE Koehne, M Marsh, A Boakye, B Douglas, IY Kim, SY Chang, DP Jang, KW Bennet, C Kimble, R Andrews, M Meyyappan, KH Lee, Analyst **136**, 1802 (2011)

57. MP Marsh, JE Koehne, RJ Andrews, M Meyyappan, KE Bennett, KH Lee, Biomed. Eng. Lett. **2**, 271 (2012)

58. E Rand, A Periyakaruppan, Z Tanaka, DA Zhang, MP Marsh, RJ Andrews, KH Lee, M Meyyappan, JE Koehne, Biosens. Bioelect. **42**, 434 (2013)

59. DA Zhang, E Rand, M Marsh, RJ Andrews, KH Lee, M Meyyappan, J Koehne, Mol. Neurobiol. **48**, 380 (2013)

60. JE Koehne, H Chen, AM Cassell, Q Ye, J Han, M Meyyappan, J Li, Clinical Chem. **50**, 1886 (2004)

61. PU Arumugam, H Chen, S Siddiqui, JAP Weinrich, A Jejelowo, J Li, M Meyyappan, Biosens. Bioelect. **24**, 2818 (2009)

62. PU Arumugam, E Yu, R Riverie, M Meyyappan, Chem. Phys. Lett. **499**, 241 (2010)

63. A Periyakaruppan, RP Gandhiraman, M Meyyappan, JE Koehne, Anal. Chem. **85**, 3858 (2013)

64. RK Gupta, A Periyakaruppan, JE Koehne, M Meyyappan, Biosens. Bioelect. **59**, 112 (2014)

Chapter 9

Microfabricated Electrochemical Cell-Based Biosensors for Analysis of Living Cells *In Vitro*

Jun Wang, Chengxiong Wu, Ning Hu, Jie Zhou, Liping Du and Ping Wang *

Biosensor National Special Lab, Key Lab for Biomedical Engineering of Ministry of Education, Department of Biomedical Engineering, Zheda Road No. 38, Zhejiang University, Hangzhou 310027, China

ABSTRACT

Cellular biochemical parameters can be used to reveal the physiological and functional information of various cells. Due to demonstrated high accuracy and non-invasiveness, electrochemical detection methods have been used for cell-based investigation. When combined with improved biosensor design and advanced measurement systems, the on-line biochemical analysis of living cells *in vitro* has been applied for biological mechanism study, drug screening and even environmental monitoring. In recent decades, new types of miniaturized electrochemical biosensor are emerging with the development of microfabrication technology. This review aims to give an overview of the microfabricated electrochemical cell-based biosensors, such as microelectrode arrays (MEA), the electric cell-substrate impedance sensing (ECIS) technique, and the light addressable potentiometric sensor (LAPS). The details in their working principles, measurement systems, and applications in cell monitoring are covered. Driven by the need for high throughput and multi-parameter detection proposed by biomedicine, the development trends of electrochemical cell-based biosensors are also introduced, including newly developed integrated biosensors, and the application of nanotechnology and microfluidic technology.

Keywords: electrochemical biosensor; cell-based biosensor; microfabrication; *in vitro*monitoring

1. INTRODUCTION

A living cell can be properly described as an electrochemical dynamic system [1]. Due to various reduction-oxidation (redox) reactions and changes of ionic composition and concentration [2] in biological processes, cellular life activities are accompanied with electron generation and charge transfer, which can be exploited using electrochemical methods to reveal information about changes in cell function as well as cell growth and development. In this case, cell biochemical parameters, such as concentrations of inorganic ions (H^+, K^+, Na^+, Ca^{2+}, Cl^-, *etc.*), morphological change, membrane potentials and redox potentials, can be detected electrochemically.

Cell-based biosensors on the basis of electrochemical detection can realize cell analysis and evaluation through measurement of electrochemical signals such as current, potential, impedance, conductivity, capacitance, *etc.*, and have become a focus of research in the field of biosensor. In recent years, with the development of nanotechnology and enhancement of interface between sensors and cells, a series of electrochemical cell-based biosensors is used in the study of cell type, concentration, viability, proliferation, apoptosis and molecule distribution inside and outside cells [3,4,5,6,7,8,9,10,11,12,13,14]. From another point of view, cell-based biosensors can employ immobilized living cells as sensing elements, and the specific reaction between stimulus and cells can reveal the intracellular and extracellular microenvironment condition, which can be applied in drug screening and even environmental monitoring [15].

In recent decades, the development of microfabrication technology and micro-electromechanical systems (MEMS) have helped to materialize the miniaturization of cell-based biosensors. The most typical is the development of MEA, which has become a powerful tool in research on bioelectric potential [15] since first reported by Thomas [16] in 1972. Recently, there have been new developments of using microfabricated electrode array systems and stimuli-responsive biomaterials to immobilize viable cells *in situ* [17,18]. The ECIS technique [19], which has matured in cell morphology study [20], is greatly promoted by the microfabrication technology, and thus diversification of electrode design is facilitated [21,22,23]. Semiconductor technology stimulates the development of new cell-semiconductor hybrid biosensor systems, such as the ion-selective field effect transistor (ISFET) [24] based on the properties of the electrolyte insulator semiconductor (EIS) system, and another type of promising field effect transistor utilizing the electrolyte-semiconductor interface for achieving biosensing [25,26]. Among these, LAPS [27], based on the photovoltage technique, received extensive attention because of its good sensitivity, stability and high signal-to-noise ratio. Using LAPS, the response of cells to chemical substances is studied by monitoring the acidification of living cells [28] and changes in concentration of other inorganic ions [29]. These miniaturized cell-based biosensor systems are capable of real-time, noninvasive, label-free measurements, which guarantees the potential in on-line biochemical analysis of living cells *in vitro* and facilitates the development of new analytical instruments based on these biosensors.

Here, we start with the presentation of principles of biochemical cell-based biosensors, including MEA, ECIS and LAPS. Then, their applications in biochemical monitoring of living cells are introduced combined with descriptions of MEMS technology and photovoltage technology. Finally, we survey the developing trends of biochemical cell-based biosensors, including the integration and multifunction requirements, combined with hot topics about microfluidic technology and nanotechnology.

2. PRINCIPLES OF ELECTROCHEMICAL CELL-BASED BIOSENSORS

2.1. Theory and Structure of Microelectrode Array

MEA is an electrochemical biosensor developed to detect the action potential (AP) in the extracellular microenvironment of cells. On an MEA, a thin metallic film is fabricated between a substrate of glass or silicon and a passivation layer with several electrode sites exposed for sensing the extracellular field potential changes generated by the objective cells.

When spreading on the microelectrodes, cultured cells adhere to the substrate. But there is still a minute volume of electrolyte between the cells and the microelectrodes; thus, a solid-liquid interface on the electrode surfaces is formed. The electrochemical properties of the interface are the basis of the sensing mechanism of MEA.

According to the electric double layer (EDL) theory, when a metal is placed into ionic liquid, an equilibrium condition is established once the charge transfer between the metal and the solution is equal. The electric field on the interface generated by electron transfer causes the formation of an inner Helmholtz plane (IHP) and an outer Helmholtz plane (OHP). The net reaction induces the creation of an electric double layer, which is also an electrified interface describing the interphase region at the electrolyte boundary [30].

The equivalent circuit of metal-electrolyte interface can be explained with the Randles model, as shown in Figure 1(a). In the circuit, an interfacial capacitance (C_I) is in parallel with charge transfer resistance (R_t) and diffusion related Warburg element (R_W and C_W). The spreading resistance (R_S) represents the effect of current spreading from the localized electrode to a distant counter electrode.

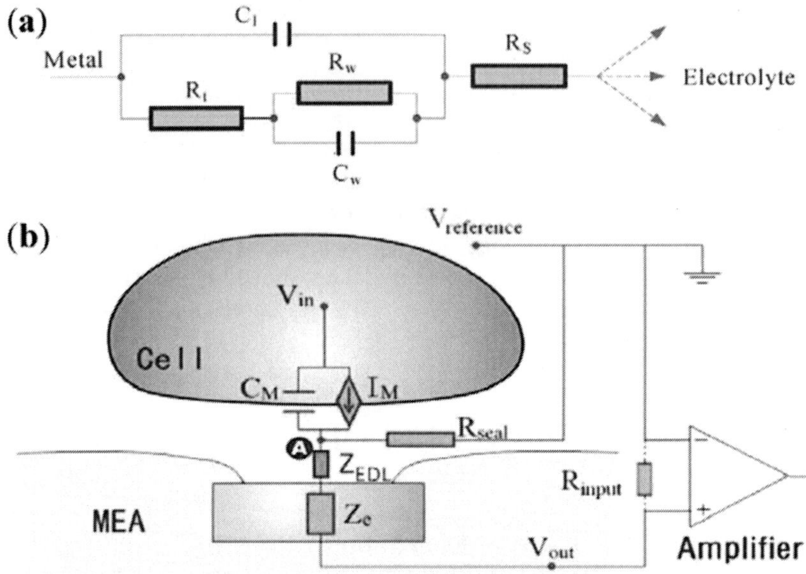

Figure 1. (**a**) The equivalent circuit of metal-electrolyte interface; (**b**) The equivalent circuit of the signal pathway in MEA system. V_{in}: the intracellular potential; C_M: the capacity of the cellular membrane; I_M: the current source of the cellular membrane; A: the junction between the cell and electrode; R_{seal}: the sealing resistance between the cell and chip; R_{input}: the equivalent of the input of preamplifier; V_{out}: the output potential of the microelectrode; $V_{reference}$: the grounded bulk media potential.

An equivalent circuit (Figure 1(b)) of the signaling pathway illustrates how a biological signal is converted into an electrical one [31]. There are mainly three components in the equivalent circuit. Firstly, the transmembrane potential functions as an electric source (V_{in}). Secondly, the parallel circuit of C_M and I_M expresses an overall effect of the cellular double-lipid layer structure, ion channels and cellular signal pathway. Finally, the electrolyte between the cells and the electrodes induces part of the current to leak out to the bulk medium. This effect is expressed as the sealing resistance (R_{seal}). The current flowing through the basal membrane is the sum of the ionic current (I_{ionic}) and the capacitive current (I_M). Thus, the total transmembrane current is given by:

$$I_{\text{total}} = K\frac{dV_{\text{in}}}{dt} = C_{\text{M}}\frac{dV_{\text{in}}}{dt} + I_{\text{ionic}} \quad (1)$$

Accordingly, the voltage at node A is proportional to the second derivative of the transmembrane potential. This voltage is also proportional to the magnitude of R_{seal} [32]. The larger R_{seal} is, the better the recorded signal can reflect the transmembrane potential. In the extreme condition of an infinite seal resistance, the voltage at node A would correspond to the intracellular potential, thereby simulating a whole-cell patch configuration. This trend is shown inFigure 2 [32,33,34].

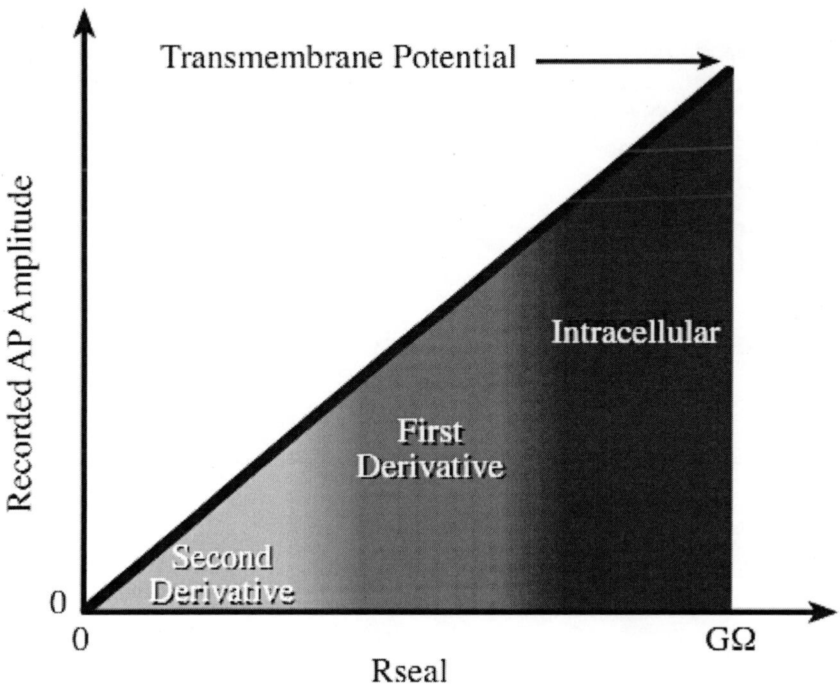

Figure 2. Characteristic AP signal recorded with an extracellular microelectrode as the seal resistance is varied. At low R_{seal}, the amplitude is small and a second derivative behavior is observed (assuming no additional derivative is generated due to the electrode itself). As R_{seal} increases, the amplitude of the recorded AP signal increases and the order of the derivative decreases. For extremely high R_{seal}, a whole-cell patch configuration is approached and the intracellular signal is measured.

In MEA design, the 'sandwich' structure has been employed since it was first pioneered by Thomas in 1972 [16]. Firstly, a thin metallic layer is deposited on an insulative substrate as the sensing layer. Then a passivation layer is coated on it with electrode sites and pads exposed for electric sensing and signal output.

The flow chart in Figure 3 is a general fabrication process of glass-based MEA. In the beginning, Cr (30 nm) and Au (300 nm) are sequentially sputtered or evaporated onto the glass substrate. The use of the Cr layer can enhance the adhesion of the Au layer onto the substrate. Then the electrodes and traces are patterned from this composited metallic layer by conventional lithography and etching techniques. After that, an insulative layer of SiO_2 is deposited onto the chip surface by plasma-enhanced chemical vapor deposition (PECVD). Finally, the electrodes and pads are exposed from the silicon nitride layer by reactive ion etching. The fabrication of 3D MEA shares the design idea of planar MEA [35,36].

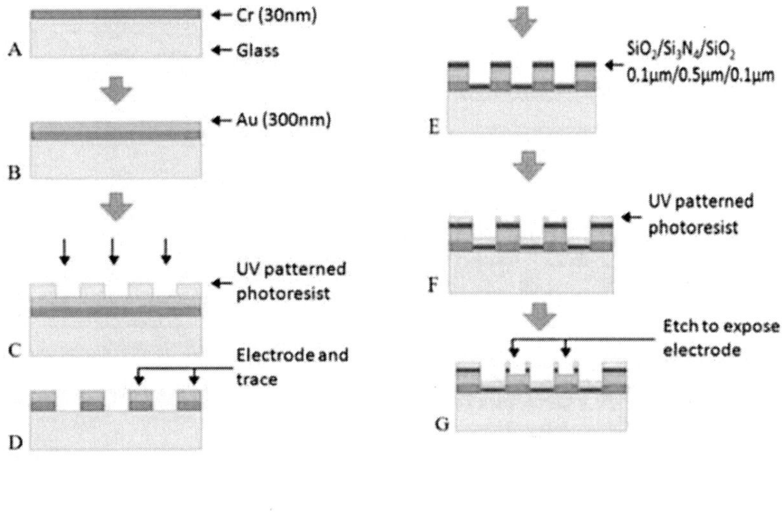

Figure 3. General fabrication process of MEA.

Figure 4 shows a typical fabricated MEA chip. The electrode is designed to match the size of a cell, with a diameter of 10–100 μm. To reduce the electric interference between electrodes, the center-to-center distance is usually more than 100 μm. To lower the impedance and improve the performance of the electrodes, usually a TiN or platinum black layer is deposited on electrodes [37].

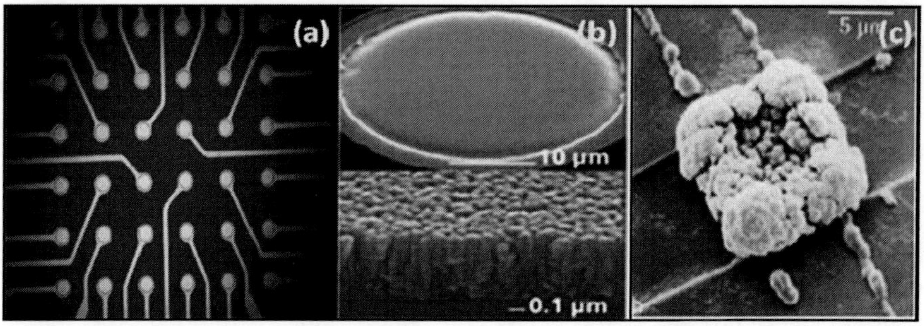

Figure 4. The surface morphology of microelectrodes with different treatment. (**a**) The chip layout of a 6 × 6 microelectrode array; (**b**) A microelectrode sputtered with TiN (http://www.multichannelsystems.com/); (**c**) A microelectrode electroplated with platinum black. (Reprinted from [38]. © 1980, with permission from Elsevier).

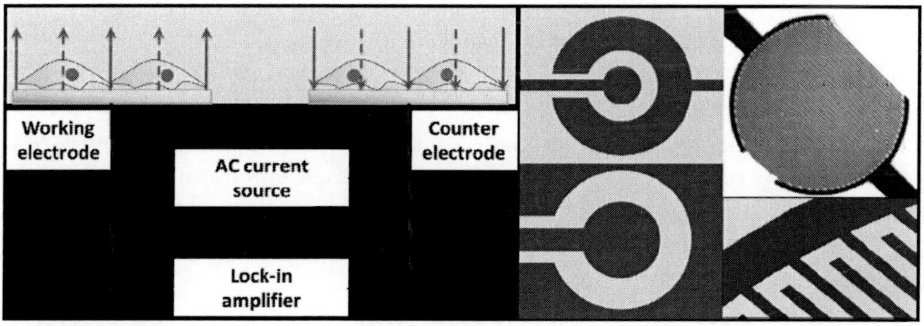

Figure 5. (**a**) Schematic of the measurement setup; (**b**,**c**) Monopolar electrodes: a small electrode is used as the working electrode and a larger electrode as the counter electrode. Since the ratio of the surface area of the two electrodes is over a couple of hundred, the total impedance of the system is dominated by the impedance change at the small measuring electrode; (**d**) Interdigitated electrodes (IDEs): a plurality of independently-operating IDE units connect to a terminal strip to form one branch of an electrode. The other identical one is parallel spaced; (**e**) Interdigitated electrode array with parallel lines.

2.2. Electric Cell-Substrate Impedance Sensor

Before growth and propagation, cell adherence to substrate or surface is always the first step in a cell-based experiment. ECIS is one of the electrochemical techniques that can be applied to monitor cell adhesion, spreading, and motility in real time [39]. Figure 5 shows a schematic overview of the measurement setup pioneered by Giaever and Keese [40]. In this setup, microelectrodes are constructed beneath the cell attachment platform. An alternating current is applied on the electrodes and the voltage is monitored using a lock-in amplifier. If no cells are cultured on the surface, electric current can flow freely from the surface to the electrodes. Cells growing on the electrode will impede the current flow and thus increase the resistance of the system (measured in ohms) because cell membrane acts as an insulator. The magnitude and time course of the increase is found to become more pronounced when the cell density during seeding approaches confluence. Electrodes without any cells will produce a minimal 'base-line' resistance. The addition of analytes that either rupture the cell monolayer or cause disturbances will lead to changes in impedance.

Typically, two structures are used in impedance sensing: the monopolar electrode [41,42] and the IDEs [21,43]. With monopolar structure, the area excluding the working electrode is relatively large, only a few cells on the working electrode contribute to the measured impedance, which results in fluctuations among experiments. Moreover, large counter electrode hinders further miniaturization, which is essential in a high-throughput screening (HTS) system. In contrast, the interdigitated structure shows improved performance as the current flows in the vicinity of the electrodes surface. A much higher sensitivity to impedance change is demonstrated compared with the conventional design. In addition, the IDEs commonly cover up to about 50% of the area of each well, lowering the well-to-well differences in the HTS system. However, no preference can be given to these two prototypes without specific applications.

Generally, the fabrication process of ECIS is the same as MEA, with a metallic layer deposition on the substrate and a following lithography and etching process, as shown in Figure 3.

2.3. Light Addressable Potentiometric Sensor

LAPS is a semiconductor device proposed by Hafeman *et al.* in 1988 [27]. With a structure of $Si/SiO_2/Si_3N_4$, it can be excited by a modulated light source to generate a photocurrent which corresponds to the potential on the Si_3N_4 surface. In principle, any effect that results in the change of surface potential can be detected by LAPS, including the ionic change [44], redox reaction [45],*etc*. The fabrication process of LAPS is easier due to the simple structure, and the encapsulation of LAPS is much less critical. Besides, the extremely flat surface makes it convenient to incorporate into the micro-volume chamber, which facilitates trace measurement. Therefore, LAPS holds promise as a preferred platform for future cell-based biosensors.

LAPS is a potentiometric semiconductor device that could have an EIS structure or an electrolyte-metal-insulator-semiconductor (EMIS) structure, as shown in Figure 6. Generally, it has the layer sequence of an Si/SiO_2/sensitive layer. An external DC bias voltage is applied to form an accumulation layer, a depletion layer or an inversion layer at the interface of insulator (SiO_2) and semiconductor (Si). When an AC modulated light illuminates the LAPS chip, the semiconductor absorbs energy, leads to energy band transition and produces electron-hole pairs. Electron and hole would combine soon and photocurrent cannot be detected by peripheral circuit without light illumination. When LAPS is biased in depletion, an internal electric field is produced across the depletion layer, and the width of the depletion layer is a function of the local surface potential. When a modulated light illuminates the bulk silicon, light induced charge carriers are separated by the internal electric field and thus photocurrent can be detected by the peripheral circuit. The amplitude of the photocurrent depends on the local surface potential. Therefore, by detecting the photocurrent of LAPS, local surface potential can be obtained [27].

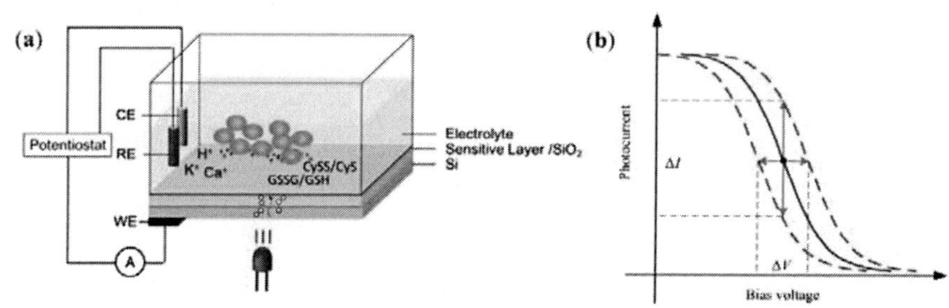

Figure 6. The schematic diagram of LAPS as a cell-based biosensor. (**a**) Working principle of LAPS; (**b**) Characteristic *S*-shaped curves.

When cells change the surface potential of LAPS through the ions (H^+, Na^+, K^+, Ca^{2+}, *etc.*) or redox couples like CySS/CyS (cystine/cysteine) and GSSG/GSH (oxidized/reduced glutathione), the actual bias voltage varies, and fluctuations are generated in the corresponding photocurrent (Figure 6(a)). Therefore, by focusing the light spot on the surface cultured with cells, ionic changes or redox potential can be detected through the local surface potential measurement. Thus LAPS is applied in cellular metabolism detection, cytotoxicity evaluation and drug analysis.

For pH detection, a layer of Si_3N_4 is fabricated on the surface of LAPS. This thin layer of silicon oxynitride is used as the insulating layer that can effectively separate the silicon substrate from the electrolyte. According to the site-binding theory [30,46] which is also demonstrated in the semiconductor sensor based on field-effect [47,48], a potential difference related to H^+concentration of the electrolyte arises on the insulator

(Si_3N_4/SiO_2)-solution interface. The insulating layer interacts with protons in the solution to form groups of silanol (Si-OH) and silamine (Si-NH_2). Therefore, the sensor surface potential is determined by the protons in the solution. As is shown in Figure 7, if a voltage is applied to the sensor, an electric field is formed at the silicon-insulator interface. By illuminating the rear side of the sensor, a photocurrent is generated [49]. This photocurrent corresponds to a hole-electron pair created by radiation absorption at an atomic level. Since a field potential is formed in the chip, holes and electrons move in opposite directions, resulting in the creation of local current which can be output from the backside aluminum layer.

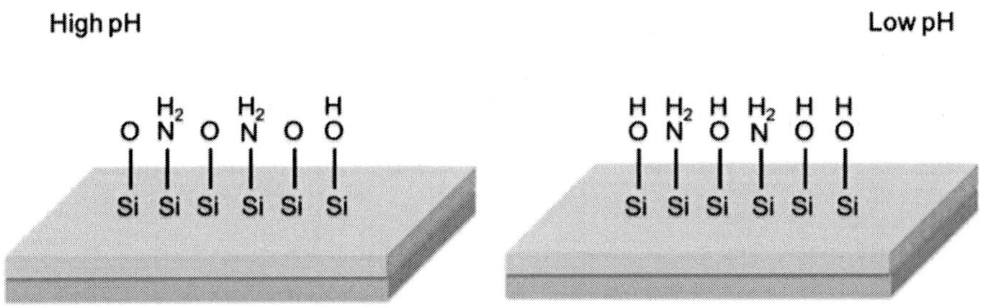

Figure 7. The formation state of silanol (Si-OH) and silamine (Si-NH_2) groups both in high and low pH conditions.

The amplitude of the photocurrent depends on the applied bias voltage and the sensor surface potential determined by the pH value of the electrolyte. The insulator surface is partially covered by silanol and silamine, both of which can be considered as a function of pH. The surface is neutral around pH 3.5. Therefore, under physiological conditions (pH 7.4), the surface becomes negatively charged, which makes the surface potential pH-dependent. Such a dependence is Nernstian and linear from pH 2 to 11 [27]. Generally the characteristic current-voltage curve (I–V curve) is S-shaped (Figure 6(b)). In the continuous measurement, the applied bias voltage is fixed at a constant value, which is selected as the inflection point of the I-V curve. By doing this, the recorded photocurrent change can reflect the pH variation most sensitively. Then the variation of photocurrent ΔI represents the pH change.

Electron exchange accompanies oxidation-reduction reactions, which can be detected by a noble electrode (Pt or Au electrode) through potentiometric measurement. The potential on the electrode (also redox potential) is determined by the ratio of the concentrations of the solution species [50] and the value is stated by the Nernst equation. Redox potential measurements can also be obtained by depositing a metal layer (often Au) on the insulator surface of the EIS structured sensor. When in contact with an aqueous solution which contains a redox pair, the surface potential of the metal layer reflects the ratio of the redox pair's concentration. The Au layer deposited functions as an electron transfer electrode similar to a traditional Au electrode, and the equilibrium potential formed on the surface of the layer is measured by photovoltage technology. As the EMIS structured sensor and EIS structured sensor have similar structure, fabrication and measurement technology, they can share the same measuring system.

It is worth mentioning that measurements of reversible, Nernst potentials are accompanied by a number of requirements, among which the most important is the need of fast electron exchange between the metal layer and the analyte. Slow electron transfer is not propitious to the formation of stable equilibrium potential, and this is responsible for the failure of potentiometric measurements of biological samples. To obtain meaningful analytic information, mediators are usually used so that stable redox potential measurements could be made. The mediators can be redox couples directly added to the samples, like Fe(III)/Fe(II) [51] or electron transport promoter, such as 4-pydidinethiol (4-PySH) and bis(4-pyridyl)disulfide [52] that are to be modified on the metal surface (Figure 8).

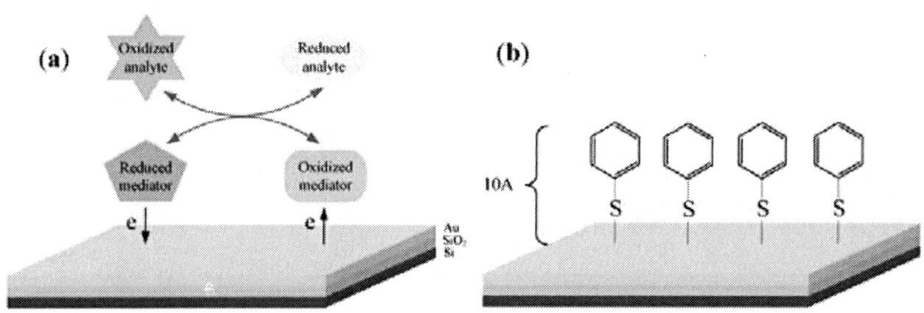

Figure 8. Schematic diagram of the interface between biological analyte and EMIS structured sensor. (**a**) Redox couples functioning as mediators used to be added into samples, which usually can be Fe(III)/Fe(II) or quinhydrone; (**b**) Modified interface using electron transport promoter to accelerate electron transport between analyte and sensor surface, here 4-PySH is taken as an example.

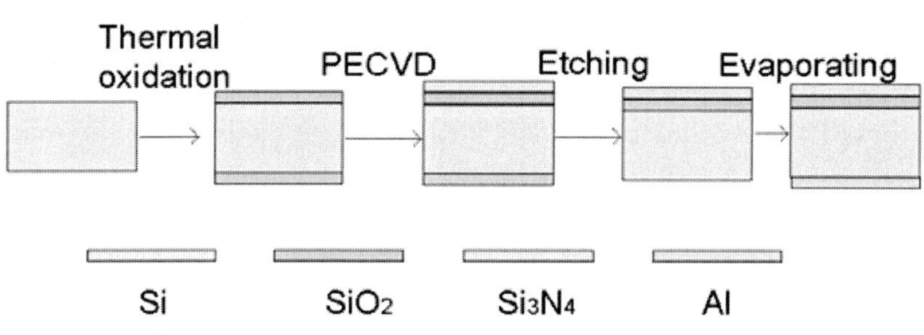

Figure 9. Fabrication process of LAPS chip: thermal oxidation, PECVD, etching, evaporating.

The LAPS device usually consists of the heterostructure of $Si/SiO_2/Si_3N_4$. Fabrication of LAPS is easy and fully compatible with the standard microelectronics facilities (Figure 9). A silicon chip is chosen as the substrate with a resistance of about 5–10 Ω·cm. Then a SiO_2 layer is formed with a thickness of 30–50 nm. A 50–100 nm layer of Si_3N_4 was deposited by PECVD as a sensitive layer on the upper side of the bulk. For the EMIS structured sensor, a deposition step of 30 nm Cr and 100 nm Au on the Si_3N_4 surface is necessary. Finally, an aluminum membrane about 300 nm in thickness is evaporated on the backside of the silicon chip to form an ohmic contact.

3. APPLICATIONS IN BIOCHEMICAL MONITORING OF LIVING CELLS

3.1. Cellular Electrophysiology Detection

MEA is a noninvasive and long-term method enabling stimulation and recording of bioelectricity with high spatial and temporal resolution in cell and tissue cultures. As the MEA technology can be applied to any electrogenic tissue or cell, that is, central and peripheral neurons, cardiomyocytes, and retinas, the MEA biosensor is an ideal *in vitro* system to study the dynamics and physiology of these cells or tissues and to monitor both acute and chronic effects of drugs and toxins.

3.1.1. Cardiomyocyte Detection

The cardiac biomarker sensing [26] using MEA has been reported. In addition, the cultured cardiomyocytes form a confluent contracting layer when mature and they exhibit spontaneous, rhythmic and synchronous excitability. The physiology of cardiomyocytes such as propagation and conduction properties is also learnt by the MEA cardiac biosensor. Natarajan *et al.* [53] developed a modified MEA for cell patterning to study the cardiac physiology. Self-assembled monolayers (SAMs) were modified on the surface of MEA with a photolithography method. An adsorbed fibronectin layer was used as the foreground to support cardiac myocytes attachment and growth, and a poly (ethylene glycol) (PEG) SAM was used as the background to prevent protein adsorption. By doing this, the beating cells were allowed to grow exclusively over specific electrodes. With patterning of the cells on the electrode array, the exact path of a spontaneous excitation wave can be determined and then, using the path length, conduction velocity can be calculated with a high degree of accuracy. The conduction velocity was 0.190–0.025 m/s for spontaneous firing of the patterned cardiomyocytes over eight different MEAs (Figure 10). The conduction velocity at stimulation mode under different frequencies was also measured by this system. This helped to understand the electrophysiological properties of cultured cardiomyocytes. A device for separated and reversible co-culture of cardiomyocytes provides a meaningful platform for studying the AP propagation between cardiomyocytes and skeletal myoblasts through the analysis of conduction velocity across the MEA [54].

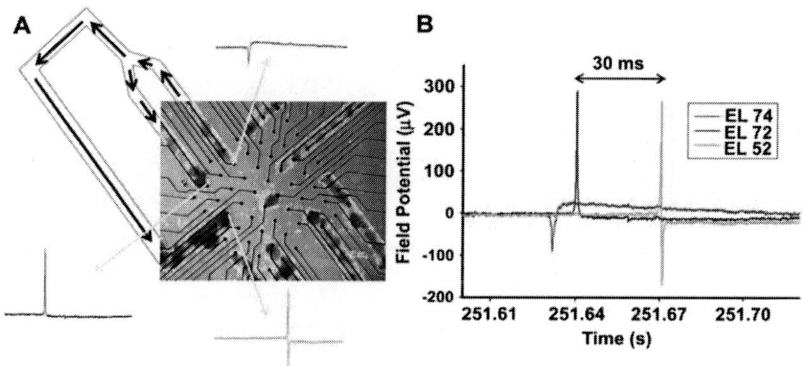

Figure 10. AP conduction in patterned cardiac myocyte monolayers. (**a**) Phase contrast picture of cardiac myocyte patterns on the top of the substrate embedded electrodes (electrode distance is 200 μm) with a sketch of the conduction pathway on the completed pattern and with the recorded extracellular signals on the given electrodes; (**b**) Overlay of the recorded traces showing the temporal relationship of the signals. (Reprinted from [53]. © 2011, with permission from Elsevier).

3.1.2. Brain Slice Detection

Compared with the dissociated cultures, slices hold the closer characteristics to that of the intact tissues. Hippocampal slice is the optimal object to study excitatory and inhibitory transmission as well as synaptic plasticity. Microelectrodes are designed in a tissue-conformal and partially high-density distribution for specific stimulation and recording experiments in acute hippocampal slices [55]. Four custom-designed planar MEAs (cMEAs) in different configurations which conform in design to the slice cytoarchitecture are well suited for specific electrophysiological applications (Figure 11). The high-density provides high spatial resolution for selective stimulation of afferent pathways (Schaffer collaterals; medial *versus* lateral perforant path) and recording of the corresponding responses. It also enables current source density (CSD) analysis of laminar profiles obtained through sequential stimulation along the CA1 pyramidal tree.

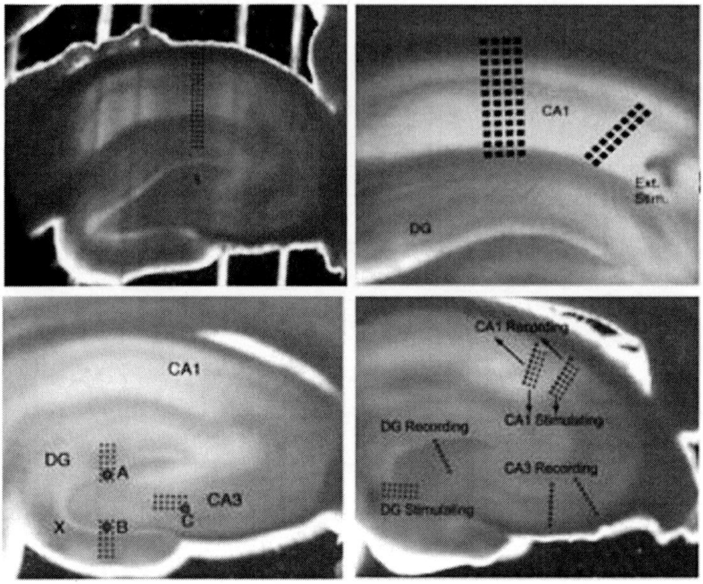

Figure 11. Conformal probes. (**a**) cMEA#1 is a 3 × 20 rectangular array well suited for electrophysiological investigations of the pyramidal and granular cells of hippocampus; (**b**) Using cMEA#2, monosynaptic input/output (IO) curves were recorded in CA1 in response to SchC stimulation; (**c**) Monosynaptic and disynaptic responses were recorded from DG and CA3 respectively by stimulating (perforant path) PP with external electrodes; (**d**) The cMEA#4 was designed to record the monosynaptic response of the DG, disynaptic response in the CA3 area and trisynaptic response in the CA1 when the PP is stimulated. (Reprinted from [55]. © 2006, with permission from Elsevier).

MEAs are also feasible for studying neuronal regeneration between different tissues and the effect of drugs [56,57]. The 3D-structure MEAs can achieve a better tissue-adhesive surface to increase electrical recording and stimulation performance.

High-density three-dimensional carbon nanotube (CNT) can be coated on MEAs to create extremely rough surfaces and this is effective in interfacing with slice or supporting neuronal adhesion. Thus a remarkable signal-to-noise ratio can be obtained [58]. Based on the deep reactive ion etching (DRIE) technique for silicon substrate, Charvet *et al.* [59] developed high-density arrays of 3D microelectrodes, which offered new possibilities for *in vitro* studies of acute brain slice. Rajaraman *et al.* [60] fabricated solid or hollow towers at a height up to 500 μm above a substrate with metal lines defined on them. The towers targeted specifically electrophysiological measurements of slices or large neural network. Figure 12 shows two typical 3D electrodes with cone-shape or cylindrical tips. However, the hard contact between slice or cells and a bed of needles or pillars may have some unpredictable effects on cell membranes and functions.

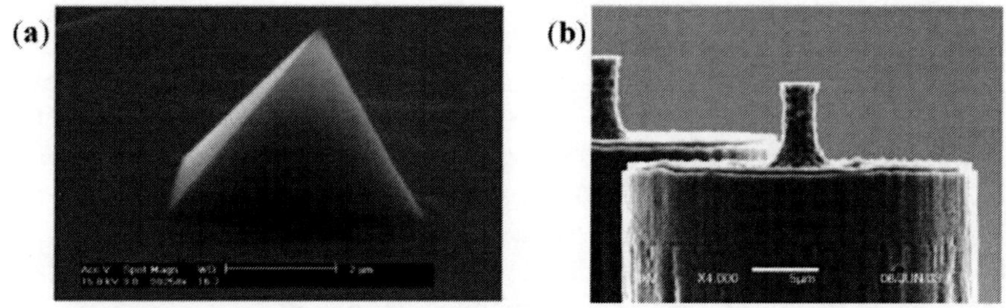

Figure 12. (**a**) SEM of cone-shaped electrodes on glass substrate. (Reprinted from [36]. ©2006, with permission from Elsevier) (**b**) SEM of cylindrical electrodes on silicon substrate. (Reprinted from [61]. © 2006, with permission from Elsevier).

All experiments and results above indicate that MEA is feasible for acute brain slice experiments and effective for neural networks.

3.1.3. Dissociated Neural Network Detection

The dissociated neural network from mammalian tissues is an invaluable model to study learning, memory and information processing in the brain. When cultured on MEA, the spike activity from neurons can be recorded at about 5 to 7 days. A consistent bursting activity is likely to be dominant after 18 to 25 days *in vitro* [62], depending on the cell culture system used. Bursting stems from the regulation of the balance between intrinsic excitation and inhibition within the network. It may be related to calcium wave and sodium currents [63,64]. The synaptic connections and network topology play fundamental roles in the bursting process. Figure 13 shows the spinal cord neural networks grown on an MEA chip (Figure 13(a)) and developmental changes in neuronal activity of the cortex neural network (Figure 13(b)). Thus, excellent experimental conditions are provided for studying structural, functional development of cultured neurons and dynamic neural network.

3D-structure MEAs are also used for strengthening the contact between neural networks and electrodes to obtain a high signal-to-noise ratio. In addition, planar MEAs are designed to guide the neurons' growth onto the defined electrodes or to guide the neural network to form predetermined patterns. Mostly, well-defined electrode arrays were fabricated on silicon substrate to generate dielectrophoretic (DEP) forces, and the desired number of neurons were immobilized onto recording electrodes [65]. Jing *et al.* [66] used SAM to create hydrophobic and hydrophilic areas on MEAs, which would lead a patterned neuronal network on the substrate. Spontaneous electrical signals were detected from the networks of GT1-7 neurons and primary mouse hippocampal neurons. By using a polymer layer such as polydimethylsiloxane (PDMS) and defined chambers or microfluidic channels, the growth of the cultured neuronal network was guided. Thus, the communication between groups of neuron networks or signal transduction in different cells can be mapped [67,68,69].

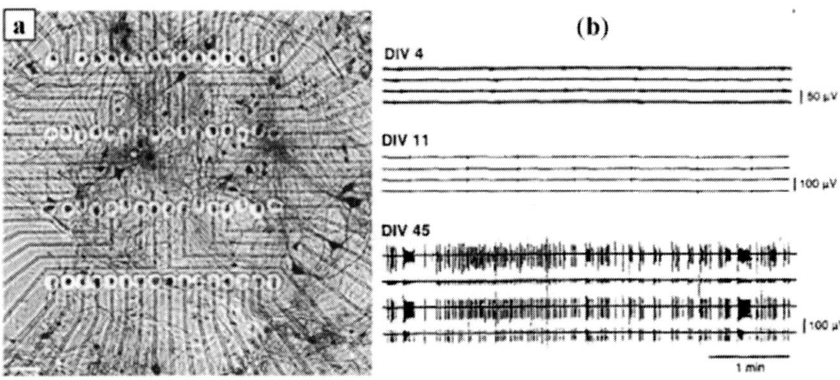

Figure 13. (**a**) Neuronal network derived from murine spinal cord tissue (92 days *in vitro*), grown on the recording matrix of a 64-electrode array plate. (Reprinted from [70]. ©2007, with permission from Elsevier) (**b**) Developmental changes in neuronal activity. Random bursts were observed at DIV4, tightly synchronized activity appeared at DIV11. The mature neuronal activity consisted of a complicated, high order pattern of spike-like firing and bursting at DIV45. (Reprinted from [71]. © 1996, with permission from Elsevier).

3.1.4. Other Electrogenic Tissues or Cells Detection

Embryonic stem cells can be induced to differentiate into different cell types which display similar molecular, structural, and functional properties. It may provide a possible solution for the lack of *in vitro* human cardiac tissue models for a variety of applications. MEA can be used to record the electrophysiology activity of stem cell-derived electrogenic cells such as neurons and cardiomyocytes [72,73]. The combined system could be a high-throughput platform for detecting and analyzing any desired type of toxicant or pathogenic threat.

The vertebrate retina, an easily accessible part of the central nervous system, is a light-sensitive part located in the inner layer of the eye. It is composed of three layers of nerve cell bodies and two layers of synapses. Light stimulation results in a complex signaling of neurons within the layers of the retina. The ganglion cells, a type of neuron located near the inner surface of retina, transmit visual information from retinal photoreceptors to the brain nervous system. When light impulses fall on the retina, trans-retinal voltage change due to the activation of potassium channel on neurons and Müller glia cells, and the changes are measured as the electroretinogram (ERG) [74].

MEA is also a stable platform for detecting electrophysiological activities of olfactory and gustatory system. The biological olfactory system has high sensitivity and specificity to discriminate different odors. In the work of Liu *et al.* [75], the stripped olfactory epithelium was fixed on the surface of a 36-channel MEA. Then, the extracellular potentials of the olfactory receptor neurons in the epithelium were recorded after stimulation of odorants such as acetic acid and butanedione. The olfactory cells generated firing with different modes (Figure 14). Through time-domain and frequency-domain analysis, the firing characteristics were found to be different from that of spontaneous potentials. The detection principle of gustatory cell-based MEA was similar. When exposed to substances with different tastes, such as sodium chloride, quinine and sodium glutamate, the recorded potentials represented different firing modes according to the power spectrum analysis. The multi-channel signal analysis has potentials in revealing some spatial and temporal information of early olfactory/gustatory sensing for bioelectronic nose/tongue.

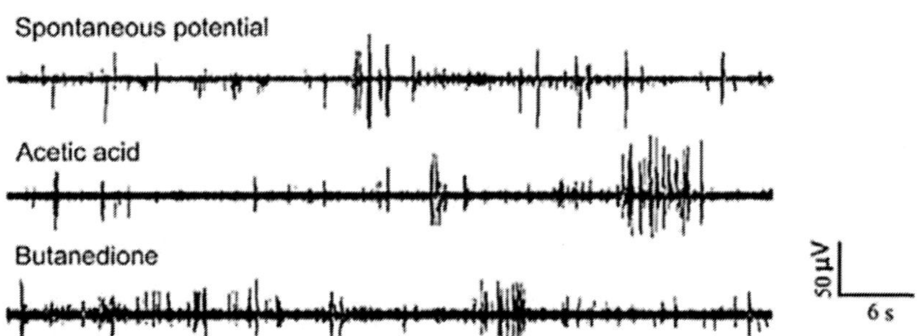

Figure 14. Electrophysiological signal changes of olfactory cells after stimulation of acetic acid and butanedione. (Reprinted from [75]. © 2010, with permission from Elsevier).

3.2. Cellular Morphology Detection

3.2.1. Cell Adhesion and Proliferation

When a cell in suspension encounters a solid surface, a series of cellular morphology changes arise until a complete adhesion is formed between the cell and the substrate [76,77]. The phenomenon of cell adhesion and spreading on the substrate can be detected using ECIS technology; thus, the underlying interactions of cells and substrates can be figured out, which can be used to promote cell culture quality and cell assays *in vitro*.

Cell adhesion and proliferation depend strongly on the interactions between cells and extra-cellular matrix (ECM) substrates. Although the traditional detecting methods such as WST-1 assay, XTT/MTT assay, BrdU assay, fluorescence microscopy and force microscope are capable of exploring cell-substrate interactions quantitatively, they are usually laborious, indirect and not time-resolved. The impedance sensing provides a solution for the revelation of the kinetics of cell adhesion and proliferation. Efforts using ECIS to study cell-ECM interactions are mainly divided into two categories. In the first case a certain type of cells adhere to a specific adhesive substrate, modified by various ECMs, such as fibronectin, collagens, laminins, poly-L-Lysine and vitronectin [78,79]. In the other case, specific reagents are used to block the cell-ECM interaction or to disrupt the cytoskeletal architecture or cell-signaling pathways, including G-protein coupled receptor activation and protein tyrosine kinase activation [80,81,82].

After cell adhesion, ECIS can be effectively used in the study of cell proliferation progression, revelation of mechanisms underlying cell cycle [83,84,85], identification and quality control of cell differentiation [86,87,88], and characterization of tissue cultures [89,90,91,92,93,94]. In comparison with conventional methods such as autoradiography, fluorescent staining and flow cytometry, the advantages of ECIS are real-time, label-free, non-invasive monitoring, and easy-construction. Moreover, an *in vivo* environment can be mimicked in an ECIS system *in vitro*, which is essential in various biomedical applications such as toxicological analysis, high-throughput drug screening, stem cell research, and tissue engineering.

Figure 15 displays a typical investigation of cell adhesion and proliferation on different ECM-coated surfaces. NIH3T3 cells were cultured in the wells of ACEA E-plates, and the chamber coated with fibronectin (FN) or poly-L-lysine (PLL) was used as a control. At the same time, chamber slides were also coated with FN or PLL and the same numbers of cells were added to each well. The dynamic monitoring of cell-substrate impedance change due to cell adhesion and proliferation was completed by RT-CES system (by ACEA Biotech Inc.), while the real state of cells was determined by staining with rhodamine-phalloidin and visualized using an epifluorescent microscope. Figure 15(a) shows the relationship between cell-substrate impedance and the physiological state of cells (mainly cell number and morphology), while Figure 15(b) presents the quality of cultured cells on various ECMs. To determine the concentration-dependent effect of coated FN on the extent of cell adhesion and spreading, E-plates were coated with increasing concentrations of FN ranging from 0 to 20 μg/mL. Further results were presented in the literature [95].

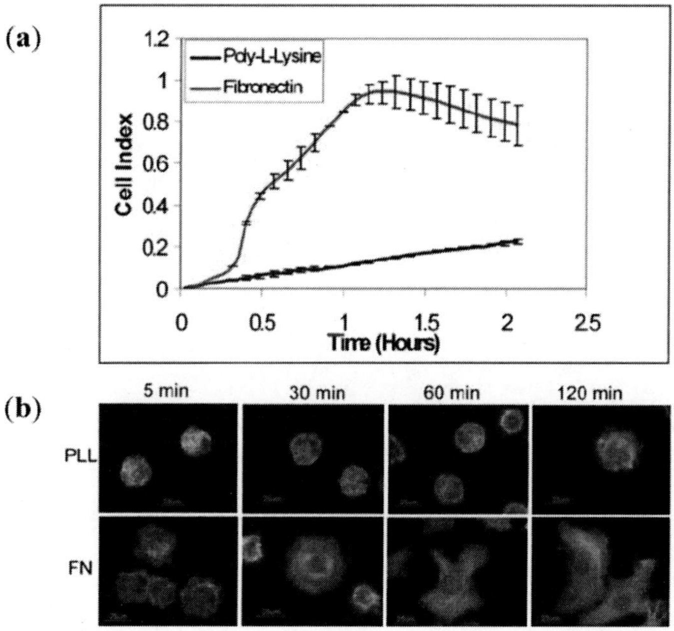

Figure 15. (**a**) Attachment and spreading of NIH3T3 cells on FN and PLL coated ECIS sensors monitored by RT-CES system. The cell index was displayed with an arbitrary unit. Ten thousand cells were seeded per well of E-plate in triplicate. The experiment was carried out at least 6 times; (**b**) Attachment and spreading of NIH3T3 cells on FN- and PLL-coated chamber slides. (Reprinted from [95]. © 2005, with permission from SAGE Publications).

3.2.2. Cell Migration and Metastasis

Cell migration plays an essential role in various complex processes such as embryonic development, homeostasis, immune response, wound healing, and cancer metastasis. Up to now migration has become the focus of many studies [96].

Wound healing assays have been carried out in tissue culture for many years in the study of cell behavior, including appraising the migration and proliferative capacities of different cells under various culture conditions.

The application of impedance sensing in wound-healing investigation was first reported by Noiri *et al.* [97]. They electro-permeated the confluent cell monolayer to generate wounds on the electrode using a DC current. The rate of restitution of monolayer integrity, as judged by the restoration of the electrical impedance, was monitored by an electrical impedance sensor for 20 h. Then, Keese *et al.* [98] improved the wound generation methods by using AC currents in the milliampere range at high frequency, so the wound was restricted to the small electrode and no electrode damage was found. This method provided highly reproducible results comparable to that observed in traditional wound-healing experiments. Wang *et al.* [42] modified the wound healing methods by using SAMs. The SAMs were formed on the electrodes to inhibit cell adhesion, which could effectively mimic wounds in a cell monolayer. After a DC pulse was applied, the SAMs were destroyed and cells began to migrate (Figure 16).

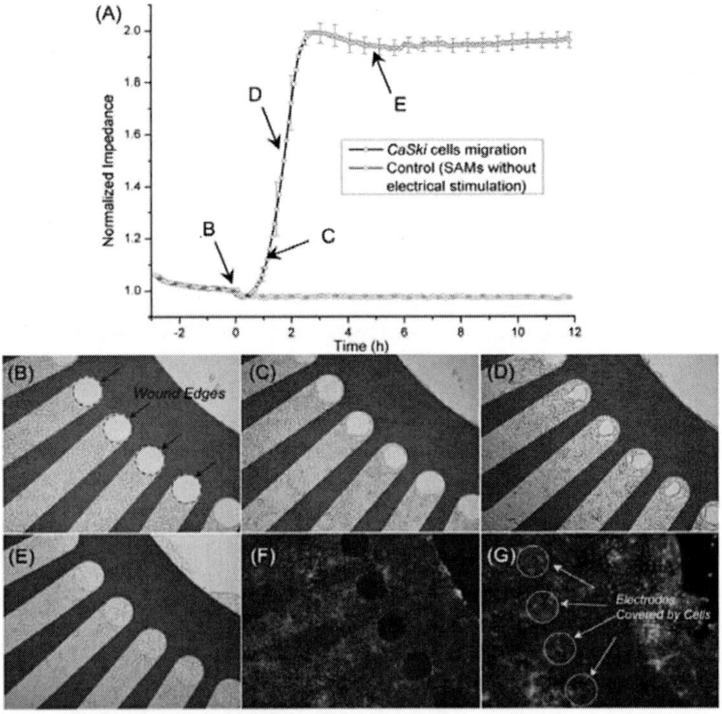

Figure 16. (**a**) The impedance variation for CaSki cells showing the real-time progress of cell migration; (**b–e**) The photographs taken in the progress of cell migration onto the electrodes; (**f,g**) The fluorescence image of the sensor, showing cell viability after modification of the SAMs and application of the DC current (4 times independent repeats for the experimental data). (Reprinted from [42]. © 2012, with permission from the Royal Society of Chemistry).

Migration also occurs during metastasis when some cancer cells migrate out of the initial tumor into the circulation and move to new locations, where they form a secondary tumor. As angiogenesis and invasion of cancer cells from the primary site into the surrounding area are essential for tumor development, assays have been developed to study these processes [99]. The ECIS-based assay used in cancer metastasis was first presented by Keese *et al.* [100], while the previous study was based on microscopic observations, where metastatic cells added over established endothelial cell layers were observed to attach and invade the cell layer. The extent and rate of drop in impedance could be correlated with the metastatic potential of cancer lines tested (Figure 17). For highly metastatic sublines, within an hour after being challenged, the impedance of the confluent human umbilical vein endothelial cells (HUVEC) layer was substantially reduced, while that caused by the weakly metastatic sublines was less pronounced.

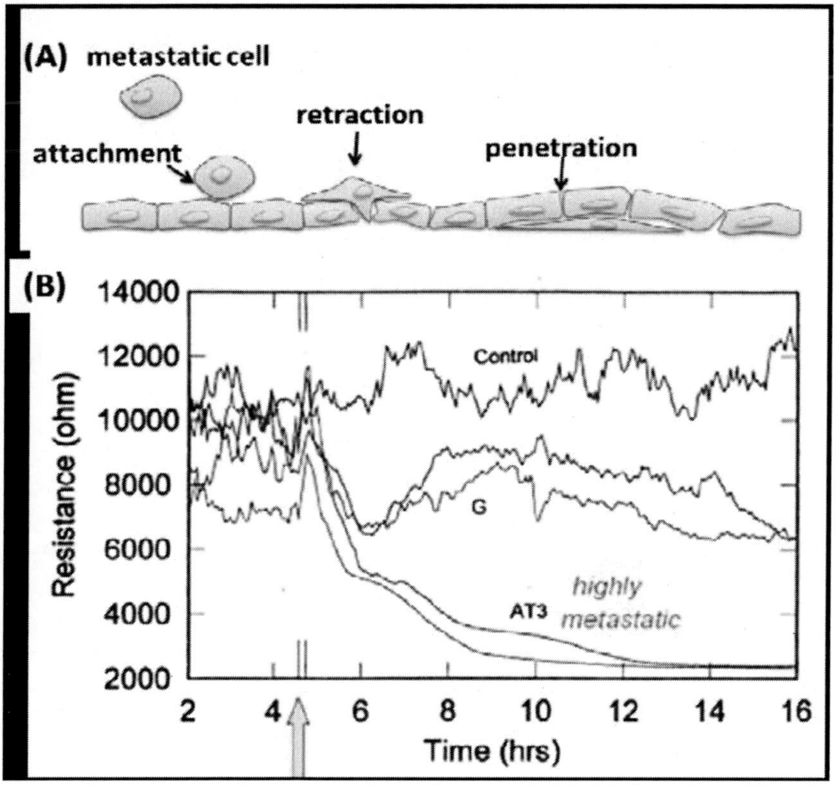

Figure 17. (**a**) Metastatic cells that invade the confluent HUVEC layer and cause damage to the barrier integrity; (**b**) HUVEC layer challenged with the weakly metastatic G subline or the highly metastatic AT3 subline. The loss of resistance was due to the loss of integrity of the endothelial cell layer in response to the activities of the cancer cells.

Tumors undergo a series of changes in the formation of metastases. Many genes and factors have been involved in tumor invasion, including adhesion molecules, cytoskeletal proteins and proteases [101]. Efforts have been made to investigate the function of such factors [102,103,104,105,106]. In these papers, the basic experimental settings were similar to that of Keese's work. The enhancement of migration caused by certain molecules (lysophosphatidic acid (LPA), placenta growth factor (PIGF), *etc.*), receptor (CD44), cytokine (leptin) or proteases (AKT, rho-associated kinase (ROCK), pocal adhesion kinase (FAK), *etc.*) was first observed. As genes in signaling pathways express biochemical substances that can effectively block invasion and metastasis, the effect of pharmacologic inhibitors like JAK/STAT inhibitors and EGFR inhibitors on the signaling pathways was evaluated. Interestingly, the stable expression of certain genes, molecules or proteases was found to have the ability to inhibit tumor metastasis [107,108,109].

In comparison with other chemotaxis assays like Boyden chamber assays, ECIS has several advantages: better performance in mimicking *in vivo* events and high sensitivity and real-time monitoring, which makes it a valuable tool in the exploration of biomarkers or therapeutic targets.

3.3. Cellular Metabolism Detection

As the basic physiological feature of life, heterotrophic cells absorb various nutrients, produce energy and secrete acidic wastes for growth and development. Metabolic energy is generated by carbon sources such as sugars, amino acids and fatty acid. The schematic of cellular metabolism and physiological processes is displayed in Figure 18. Under normal conditions, glucose is taken up by the cells and degraded into energy and acidic products. Under natural aerobic conditions, glucose is converted into CO_2 and energy via glycolysis, the citric acid cycle and oxidative phosphorylation. While under anaerobic conditions, glucose is converted into lactate with energy [110] via glycolysis and combining lactate dehydrogenase. The ATP generation per glucose molecule under aerobic conditions is 19 times higher than that under anaerobic conditions (38 ATP/glucose *versus* 2 ATP/glucose), while the acidic product is much less than anaerobic conditions (0.167 H^+/ATP *versus* 1 H^+/ATP).

Lactic acid and CO_2 are the main products from aerobic and anaerobic glucose degradation. They can be hydrolyzed into lactate/H^+ and HCO_3^-/H^+ respectively, both in the cell and outside the cell after passing through the plasma membrane in the form of an unhydrolyzed state. Lactate is excreted with facilitated transport by monocarboxylate carriers and anion exchange proteins [112]. Meanwhile HCO_3^- can be transported outside the cell by an antiport. During the hydrolysis process, intracellular and extracellular protons are generated. The intracellular ones have to be transported outside by a Na^+/H^+ exchanger. Besides, when the cell receives external stimulation, e.g., exposure to a toxic agent or receptor activation, the metabolic activity or the ionic equilibrium in the cell can be interfered with, and the consumption of cellular ATP may change. Since ATP hydrolysis is closely related to the production of acidic metabolites, it can cause changes of extracellular acidification rate (ECAR).

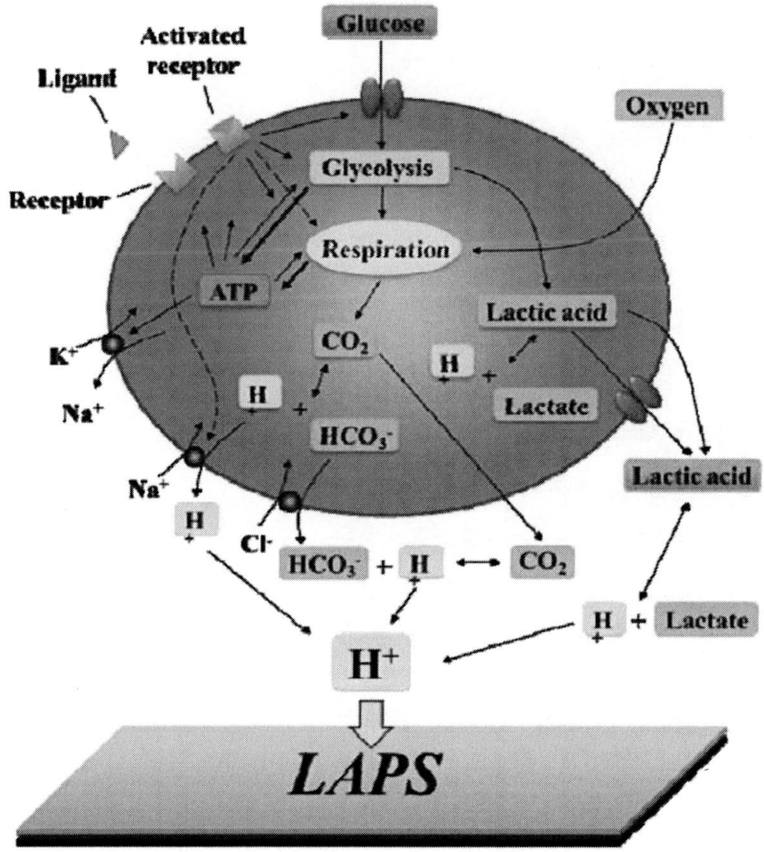

Figure 18. Schematic diagram of extracellular acidification induced by cellular metabolism and physiological processes. With receptor stimulation, the cellular physiological activities will be affected. The corresponding ATP consumption is compensated by the increased uptake and metabolism of glucose, which results in an increased secretion of acidic products. The extracellular acidification can be detected by LAPS. (Reprinted from [111]. © 2012, with permission from World Scientific Publishing Co.).

As mentioned above, LAPS has many advantages for constructing cell-based biosensors. Since the first publication of the Cytosensor™ Microphysiometer, it has been widely used. Several reviews [28,44,49,113,114] have been published to introduce the application of the microphysiometer. Besides, the newly proposed cell-semiconductor LAPS device for extracellular detection is considered as a useful tool for cell electric biology study. With the microphysiometer, functional characteristics of ligand/receptor can be studied, usually by monitoring the time-dependent and dose-dependent response of ligand/receptor binding. The time-dependent response shows the dynamic characteristics of ligand/receptor binding, while the dose-dependent response indicates the activation dose of the receptor.

The identification of specific and functional orphan receptors will facilitate studies on their physiological roles and the search for receptor agonists and antagonists. Due to the lack of a proper radioactive label, screening this type of ligand and receptor with radio receptor assay (RRA) is too difficult. It is also difficult to solve this problem with the biochemical methods, as the details in the mechanism of this type of receptor are not very clear. When using the microphysiometer, a generic characterization of ligand/receptor can be obtained. But it is usually insufficient for the identification of ligand/receptor. Therefore, other methods are usually used along with the microphysiometer to give precise identification of ligand/receptor.

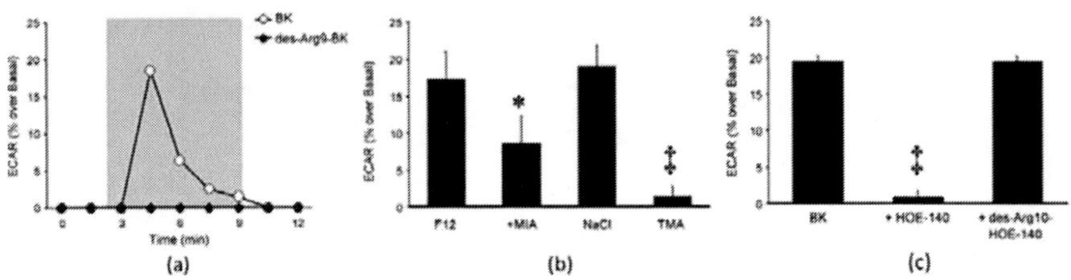

Figure 19. ECAR of HEK293 cells under stimulation of BK. (**a**) ECAR measurements were monitored. BK (white circles) stimulates ECAR, whereas the BK B_1 receptor agonist des-Agr9-BK (dark circles) does not. Cells were exposed to perfusate-containing drug during the time span encompassed by the gray box; (**b**) ECAR stimulated by 10^{-6} M of BK in various buffers, including Ham's F12 medium, without and with 10^{-5} M MIA, a balanced salt solution containing NaCl or TMA substituted mM per mM for sodium. * $P < 0.05$ *vs.* BK alone; zP < 0.01 *vs.* BK in balanced salt solution with NaCl; (**c**) Effects of BK B_2 (HOE-140) and BK B_1 (des-Arg^{10}-HOE-140) receptor antagonists on BK-stimulated ECAR. Antagonists (10^{-5} M) were added 30 min prior to addition of BK. All experiments were performed at least 4 times. zP < 0.01 *vs.* BK alone. Error bars in (b) and (c) represent the S.E.M. (Reprinted from [115]. © 2009, with permission from Elsevier).

ECAR of CHO cells expressing orphan GPCRs was measured [116]. When treated with synthetic compounds, such as bioactive peptides, it was found that CHO cells expressing an orphan GPCR, FM-3 were responsible to neuromedin U in the microphysiometric assay. It was also discovered that bradykinin (BK) activated extracellular signal-regulated protein kinase 1 and 2 (ERK) in the human embryonic kidney (HEK) 293 cells [115]. ECAR was a reflection of the Na^+/H^+ exchange (NHE). When combined with intracellular Ca^{2+} concentration from fluorescent measurements, and ERK activation assessment through western blotting with a phosphor-specific ERK antibody, it was found that signals obtained by exposure of HEK 293 cells to BK, were blocked by HOE-140 (B_2receptor antagonist) but not by des-Arg^{10}-HOE-140 (B_1 receptor antagonist) (Figure 19). Thus it was concluded that HEK 293 cells express endogenous functional BK B_2 receptors.

3.4. Cellular Redox Potential Detection

Cell metabolism is accompanied by various oxidation-reduction reactions, such as hydrogen transfer reactions and electron transfer reactions in the respiratory chain. Maintaining the proper concentrations of oxidants and reductants may be essential for cell health. The excretion of acidic waste from intra-cellular fluid has been used as a parameter to evaluate the metabolism state of cells. Similar to the acid-base mechanism, the redox potential in tissue maintains within a narrow range and injury may be brought about at the cellular, subcellular, and molecular level when the balance breaches. The concept of redox potential balance is not new but has not received the attention of that associated with acid-base chemistry. One of the reasons may be the lack of effective and direct measuring technology for redox potential when involved in biological systems.

The early measuring method is based on the dyes (such as mythylene blue) that indicate the redox potential through their color change [117]. The earliest application of redox potential for the extra-cellular phenomenon was the measurement of oxidation potential of *in vitro* plant oxidase using a gold electrode electrochemically. Then, electrodes (mainly platinum or gold electrode) had been used as the major effective way of redox potential detection in several decades. The cells and tissues measured included yeast, various bacteria [118,119,120], thin slices of liver and muscle suspended in buffer solutions, and tumors *in vivo* [121]. Until now the potentiometric measurement using noble electrodes has been widely used in redox potential monitoring of cell culture [122,123]. Usually, the redox potential electrode is worked as a working electrode and connected to a reference electrode through a potentiometer. Besides, Ward *et al.* [52] developed an universal redox electrode by modifying a Au wire electrode with electron transport promoter 4-pydidinethiol and bis(4-pyridyl)disulfide, which improved the performance of the electrode and avoided biofouling from biological samples. As the electrode technology and manufacturing process approaches maturity, commercial combined redox potential electrodes have come within our range of vision. The combined electrodes are mostly applied in biotechnology or fermentation because of their convenient usage in comparison with the traditional two-electrode system in electrochemical cells.

In biotechnology and microbiology, redox potential is one of the most complex indicators of the physiological state of microbial cultures and its measurement is useful for qualitative and quantitative determination of the microbial contamination [124]. Besides, the significance of redox potential is not limited to functioning as an indicator to reveal related state about cultures, another no less important application is to externally control the redox potential to lead the system developing as premeditated. Zheng *et al.* [125] adopted extracellular oxidation-reduction potential (ORP) of −170 mV and obtained a maximal 176 g/L lactic acid production compared to other different ORPs. Koo *et al.* [126] found that the production of coenzyme Q_{10} can be improved by increasing the $NADH/NAD^+$ ratio in agrobacterium tumefaciens, which is also an example of redox potential working as a control factor. The indicator and controller roles are just like the two sides of a coin, which also can be applied to cell-based research.

For decades, the redox potential has become an important parameter monitored in cell culture medium. The study focuses on cell metabolism detection. Considering the complexity of redox potential in cell culture medium, other parameters like ATP, dissolved oxygen, glucose and glutamine consumption are also monitored to obtain a general analysis of cellular physiological states. Hwang and Sinskey [127] used the redox potential measurement of the medium using a platinum electrode and empirical correlation of the cell concentration with the redox potential for on-line estimation of viable cell concentration. It has been demonstrated that this method is applicable to various cell lines. Later, Eyer and Heinzle [122] used a method based on an ATP balance with ATP steady-state assumption to estimate viable cells in a hybridoma culture and confirmed the validity of viable cell estimation using redox potential measurement. Cell metabolism is always accompanied by substance consumption or depletion, which can be indicated by a decreasing redox potential. Higareda *et al.* [123] studied the relationship between culture redox potential (CRP) and glucose depletion in hybridoma cultures, and found that simultaneous measurement of oxygen uptake rate (OUR) and CRP can permit the discrimination between operational eventualities and real metabolic events.

However, the measured redox potential of cell culture, also extracellular redox potential, is a comprehensive assessment of all redox pairs existing in the culture medium, and the measurement can be easily affected by pH and dissolved oxygen. Recent evidence suggests that the concentrations of specific redox couples may play a role in the regulation of other cellular functions, including gene expression. It is expected that the intracellular concentrations of redox couples like NAD^+/NADH, $NADP^+$/NADPH, CySS/CyS and GSSG/GSH may change selectively with changes in the cellular environment. Biochemical techniques exist for direct quantitation of most import cellular redox couples listed above, such as reverse-phase high performance liquid chromatography (rpHPLC) [128] or cellulose high performance thin-layer chromatography (HPTLC) [129], which have the ability to separate these species and realize measurement of specified species based on the reactivity of the thiol moiety, and a less specific detection method using membrane-permeable compounds which are fluorescent upon thiol conjugation [130]. Nevertheless, these biochemical techniques are invasive for cells and often based on endpoint detection. Electrochemical detection is expected to avoid the drawbacks and realize noninvasive and continuous monitoring of the intracellular redox state. An important difference between the electrochemical and biochemical methods is that the former measures the activity of cellular reactions rather than the steady-state concentration of a particular biochemical species.

The first application of electrochemical-based biosensor for intracellular redox potential detection is reported by Rabinowitz *et al.* [51] who employed an EMIS structured sensor to probe intracellular redox activity in live cells. The sensor structure and measurement system adopted a modified Cytosensor microphysiometer. The electrochemical potential at the gold-electrolyte interface contributes to the reverse-bias potential acting across the photosensitive silicon semiconductor. Illumination of the silicon with a light-emitting diode generates an AC photocurrent in the external circuit, determined by the voltage applied through the controlling electrode. Changes in the redox potential at the gold surface are reflected through this current. In addition, there are no measurable direct current flows through the circuit during these measurements, which enables stable, rapid, low-noise measurements on microliter volumes.

The most essential point in their study lies in measurement of the redox activity of the interior of cells noninvasively from extracellular microenvironment change. A carrier mediator that exchanges electrons rapidly with both the exterior and interior couple and links the extracellular electrolyte solution to intracellular redox pairs was used. As shown in Figure 20, the couple menadione/menadiol was used as an effective carrier mediator, because both of them are lipid soluble. Menadione and menadiol can be oxidized and/or reduced by intracellular enzymes. The major enzyme had been demonstrated to be NADPH, produced by the pentose phosphate pathway. In the meantime, the external medium bathing the cells contains a ferricyanide/ferrocyanide couple which reacts rapidly with menadiol. Thus the monitored redox potential of the ferricyanide/ferrocyanide couple is a measure of intracellular redox activity. Using this method they studied three cell lines (L929, CHO, and CH27), and the effect of reactions of mitochondrial electron transport chain and signal transduction cascade on the cell reduction rate.

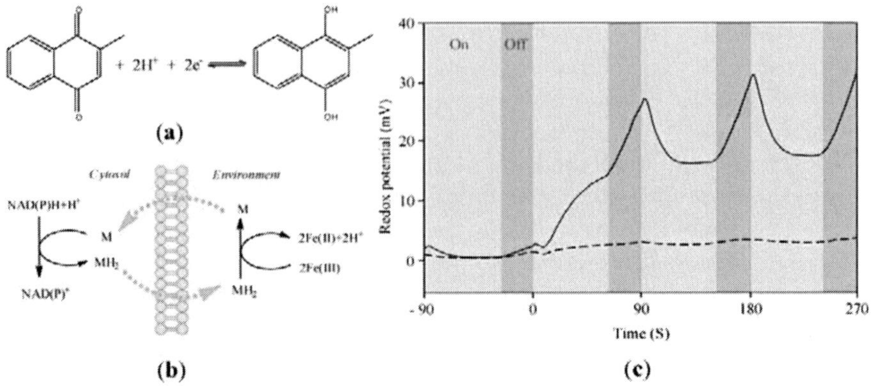

Figure 20. Detection of cellular redox reactions with use of menadione to link intra- and extra-cellular redox pairs. (**a**) Two-electron reduction of menadione to menadiol; (**b**) Hypothetical menadione reaction cycle; (**c**) Potentiometric measurements of menadione mediated ferricyanide reduction by cells. (Reproduced with permission from [51]. © 1998 American Chemical Society).

3.5. Cytotoxicity Assays and Drug Screening

3.5.1. Drug Analysis Using MEA

Cardiomyocytes on MEA have been used in a number of studies to investigate the effect of cardioactive drugs and environmental toxins, such as pesticides [131] which affected the shape of field potentials recorded by MEA.

Drugs can interfere with the normal operation of many ion channels, including sodium, potassium, and calcium ion channels that regulate heart activities. The fast inward Na^+ current is responsible for the depolarization. When treated with Na^+ channel blockers, the amplitude of extracellular field potential greatly decreases and the conduction delay increases [132,133,134]. Ca^{2+} is closely related to cardiac contractility. Drugs like quinidine and nifedipine can cause changes in the shape of extracellular field potentials and beating behavior [134,135]. The cardiotoxicity of different drugs of potassium channel openers, such as angiotensin II [136,137], is also studied. Generally, the drugs caused complicated effects on more than one specific type of ion channel and cellular organ.

An important parameter that cause cardiac arrhythmia is the prolongation of the QT interval of the surface electrocardiogram (ECG), which corresponds to the duration of the field potential recorded by MEA [138]. The QT-Screen is a 96-channel device for high-throughput pharmacological screening, and each channel contains one microelectrode (Φ = 100 μm) [139]. It is specially designed to natively analyze the drug-induced QT-prolongation and cardiac arrhythmia*in vitro*.

Table 1. The study of pharmacology on neuronal networks based on MEAs.

Stimulus and concentration	Cellular component	Description	Reference
N-methyl-D-aspartate (NMDA) (10 μM)	Cortical neurons	NMDA drastically increased the bursting activity. The spike rates and frequencies were increased, but the pattern and shape regularity of bursting neural signal were decreased.	[141]
γ-aminobutyrate (GABA) (30 μM)	Mouse spinal nerve cells	The neural signal bursting was totally inhibited at 30 μM GABA.	[142]
Dopamine (15–100 μM)	Cortical neurons	Dopamine dispersed correlations between individual neuronal activities while preserving the global distribution of correlations at the network level.	[143]
APV (NMDA receptor antagonist) (40–100 μM)	Spinal cord networks	The rate of neural signal bursting was reduced by half.	[144]
Strychnine (0 nM–20 μM)	Spinal cord networks	Neural signal bursting was increased at 5–20 nM and coordinated above 5 μM.	[145]
GABA (10 μM), Bicuculline (10 μM)	Olfactory placode neurons	GABA almost immediately inhibited firing activities. Bicuculline had the facilitatory effect or lack of effect.	[151]
Eserine (10–150 μM)	Cortical cultures	The increase of bursting and spiking at 10 and 25 mM was concentration dependent, but it was inhibited at concentrations above 50 mM.	[147]
Ethanol (40 mM)	Embryonic murine neuronal networks	Spontaneously active frontal cortex cultures showed repeatable, concentration-dependent sensitivities to ethanol, with initial inhibition at 20 mM and a spike rate 50% effective concentration (EC_{50}) of 48.8 ± 5.4 mM.	[148]
Tetrodotoxin (TTX, 100 μM)	Retina	The transient ON-response of ganglion cells was blocked by the sodium channel antagonist TTX.	[150]
Zn	Cortical neurons	The enhanced, coordinated bursting and spiking were increased, followed by irreversible activity decay.	[64]
Botulinum toxin (BoNT, 10 ng/mL)	Cortical networks	The duration and number of spikes were increased.	[149]

Neuronal network grown on MEA responds to neuroactive compound sensitively with changes in their native, spontaneous activity patterns. This altered activity is often substance- and concentration-specific. The influence includes direct metabolic effects, specific synaptic effects, transmission effects that stop AP propagation and generic mebrance effects mediated through non-synaptic Ca^{2+} or K^{+} channels or by the generation of new channels [140]. Some typical neuroactive substances with effects of excitation, inhibition or disinhibition have been studied [64,141,142,143,144,145,146,147,148,149,150], as shown in Table 1.

3.5.2. Drug Analysis Using ECIS

Cell-based assays have been used as a substitutional method for animal experiments in pre-clinical research, development of drugs and toxicological testing [152,153,154]. It has been concluded that *in vitro* cytotoxicity assay can be used as an adjunct to animal testing to improve dose level selection and thus further reduce the number of animals used [155]. ECIS has been proven valuable and reliable in real-time monitoring of dynamic changes induced by cell–toxicant interactions [156].

Cells respond to cytotoxins in the form of loss of adhesion and cell rounding, membrane protrusions or blebbing, formation of apoptotic bodies, and the ultimate engulfment of apoptotic bodies induced by phagocytosis. These apoptotic responses indicate changes in cell adhesion and morphology, which will ultimately induce a decline of cell-substrate impedance in an acute or chronic manner. Such nonlinear dynamic changes depend largely on cell types, compound properties and concentration, and compound exposure duration (Figure 21). Keese *et al.* [157] demonstrated the applicability of the ECIS system for toxicological testing of fibroblasts and epithelial cells. Subsequently, many chemicals, such as antipyrine, trichlorfon, dimethyl formamide and sodium dichromate, as well as some familiar toxins such as sodium arsenite (As(III)), mercury (II) chloride, benzalkonium chloride (BAK), triton X-100, sodium lauryl sulfate, cadmium chloride ($CdCl_2$), 1,3,5-trinitrobenzene (TNB), cycloheximide (CHX), and neutral red solution have been tested for cytotoxicity evaluation with various types of cells [156,158,159]. The effect of cytotoxins on cells is usually presented as time-response and dose-response curves using ECIS technology in a label-free way.

Conventional cell-based assays are often labor-intensive, non-automatic and time-consuming. These major drawbacks are more prominent in drug testing. With the advantages of real-time monitoring and easy operation, ECIS comes to be especially suitable for cell-based HTS. Some corporations have launched their own products for HTS based on ECIS, such as ACEA Biotech Inc., Applied Biophysics Inc., Bionas Inc., and Molecular Devices Inc. In addition to HTS, measurement of an increasing number of cellular parameters in parallel is another promising method in screening applications as a comprehensive view of the investigated cellular process can be obtained [160,161,162,163,164].

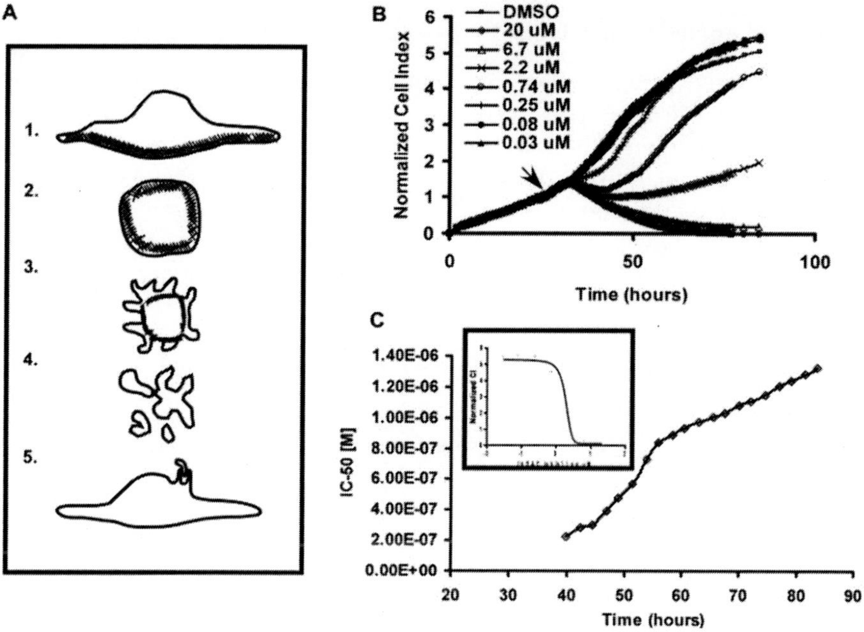

Figure 21. (**a**) Morphological changes associated with apoptosis. (1) shows a cell under normal physiological conditions. The apoptotic response leads to cellrounding (2) followed by membrane blebbing, which occurs due to weak interaction between the membrane and actin cytoskeleton (3), formation of apoptotic bodies (4), and ultimate actin-mediated engulfment of apoptotic bodies by neighboring cells (5); **(b)**Dynamic monitoring of cytotoxicity using the RT-CES system. A549 cells were seeded in microtiter plates containing interdigitated microelectrodes and treated with the HDAC inhibitor Scriptaid at the indicated doses; (**c**) Real-time IC_{50}values for Scriptaid. The inset shows the dose-response to Scriptaid at a single time point. (Reprinted from [165]. © 2008, with permission from John Wiley and Sons).

However, the impedance measurement of a large population of cells attached on one electrode may result in complex and even questionable interpretations, because only an overall level of cellular activities can be detected without differentiating the effect of proliferation, motility, and cell-cell separation. It is difficult to precisely monitor changes in membrane properties of individual cells with the current ECIS systems. The response of a single cell to a given dose of a pharmacological drug or harmful toxin may be more significant than that of multiple cells. In a single-cell sensing system, cells respond to drugs individually, and this effect can be accurately recorded without cell-cell interactions [166,167].

3.5.3. Drug Analysis Using LAPS

By monitoring extracellular parameters after drug treatment, the corresponding drug effect can be directly investigated. Both the microphysiometer and the cell-semiconductor hybrid LAPS device can be used for drug analysis and evaluation, depending on the extracellular parameters to be measured. The microphysiometer has become a utility tool in drug analysis, screening of prospective pharmacological agents, characterizations of dose responses and structure-activity relationships, and investigation of cellular mechanisms.

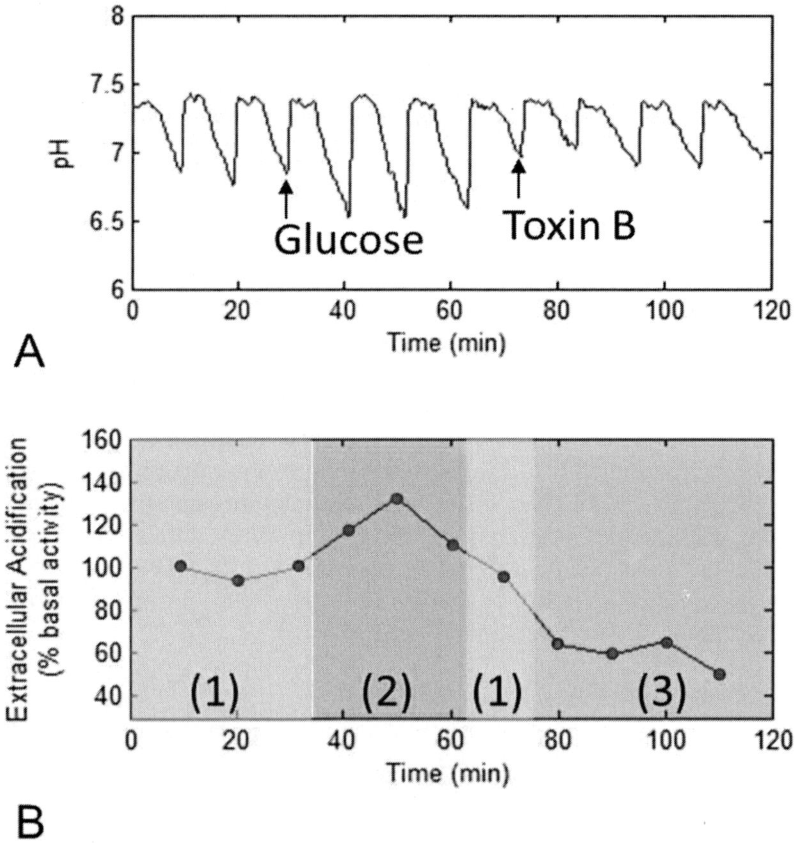

Figure 22. Cellular metabolism detection under drug stimulation. (**a**) Original data shows the change of pH corresponding to the photocurrent; (**b**) The extracellular acidification rate under the blank (1), glucose (2) and clostridium difficile toxin B (3). (Reprinted from [111]. © 2012, with permission from World Scientific Publishing Co.)

In the work of Hu *et al.* [111], the metabolic activities of MCF-7 cells were monitored by the secretion of the cellular acidic metabolites. The extracellular acidification rate was one of the most significant parameters which indicated cellular metabolic rate. Therefore, the extracellular acidification was monitored to determine the performance of the microphysiometer. Firstly, the extracellular acidification rate was recorded in the presence of cell culture medium alone as basal extracellular acidification rate. Subsequently, glucose and clostridium difficile toxin B was injected respectively into the sensor unit, and the change of extracellular acidification was monitored.Figure 22 displays the cell responses in the stage of blank (1), glucose (2), and clostridium difficile toxin B (3). In the absence of drug, the cells secreted the acidic metabolites in the native state. Meanwhile, the glucose induced a sharp increase in the extracellular acidification rate of MCF-7 cells, with a maximum about 138% after 40 min. Then, clostridium difficile toxin B was injected into the culture medium and induced a 40% decrease of the basal extracellular acidification rate after 70 min.

Glucose is the necessary nutrition and supplies the main energy ATP in normal cellular life. With the increase of glucose, intake of glucose was sharply raised by MCF-7 cells, which increased the generation of the protons under aerobic and anaerobic conditions. Therefore, after the cells absorbed more glucose, metabolic activities were enhanced in the form of an increased extracellular acidification rate. Meanwhile, clostridium difficile toxin B is a toxin generated by clostridium difficile. It is usually described as an enterotoxin, but it also has some activities as a cytotoxin [168]. Toxin B is a chromosomally-encoded exotoxin that is produced and secreted by several bacterial organisms. It is often heat-stable, is of low molecular weight and water-soluble. Enterotoxins are generally cytotoxic and kill cells by altering the apical membrane permeability of the epithelial cells. The action of enterotoxins increased chloride ion permeability of the apical membrane of cells. These membrane pores were activated either by the increased cAMP or by increased intracellular calcium ion concentration. Therefore, the toxin B affected the normal metabolic activities of MCF-7 cells and induced a weak extracellular acidification rate.

Multifunction is also essential for basic research as well as for various fields of biomedical applications. The multianalyte function for extracellular ion monitoring, such as Na^+, K^+, Ca^{2+}, may change along with cellular metabolism. In order to analyze simultaneously the relations of the extracellular environmental H^+, Na^+, K^+, Ca^{2+} under the effects of drugs, a novel microphysiometer was developed based on multiparameter-LAPS [29,169].

The LAPS surface was deposited with different sensitive membranes by the silicon microfabrication technique and the PVC membrane technique. The different sensitive membranes (H^+, K^+, Ca^{2+}) are illuminated in parallel with light sources at different frequencies (3 kHz for K^+, 3.5 kHz for Ca^{2+}, 4 kHz for H^+) (Figure 23(a)). The amplitude of each frequency component might be measured online by software FFT analysis (Figure 23(b)). Dilantin, *i.e.*, phenytoin sodium, a sort of anti-epilepsy drug and anti-arrhythmia drug, had significant effects as a tranquilizer, hypnotic and anti-seizure agent. It was proved that dilantin had membrane stabilizing action on neural cells because it reduced the permeability of pericellular membrane ions (Na^+, Ca^{2+}), inhibited Na^+ and Ca^{2+} influx and staved K^+ efflux. Thus, the refractory period were prolonged, pericellular membrane stabilized and excitability decreased (Figure 23(c)).

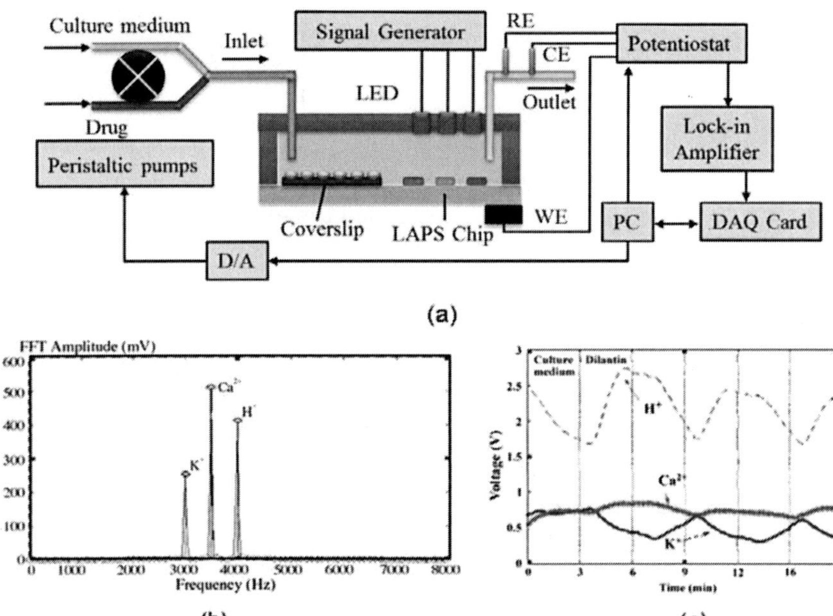

Figure 23. Multianalyte microphysiometer studies based on LAPS. (**a**) The schematic drawing of the multiparameter-LAPS system to different extracellular ions (H^+, K^+, and Ca^{2+}); (**b**) FFT analysis of signals from three sensitive membranes with three light sources; (**c**) Simultaneous analyzing of H^+, K^+, $Ca^{2+}$by multi-LAPS. (Reprinted from [29]. © 2001, with permission from Elsevier).

4. DEVELOPMENT TRENDS OF ELECTROCHEMICAL CELL-BASED BIOSENSORS

4.1. Integrated Sensors for Multi-Parameters Monitoring

Cellular life is characterized by dynamic molecular processes whose highly interconnected regulation is essential for the whole organism. Cells continuously integrate different sources of chemical and physical signals originating from both internal and external environments. The 'output' of this cellular signaling network may be manifested as a decision about growth and mitosis, activation of distinct metabolic pathways, the production and release of proteins or the initiation of programmed cell death. Conventional methods, for instance fluorescence techniques, are most often used to dissect the molecular basis of cell functions, either with fluorescent protein analogs or with synthetic dyes. However, no truly appropriate fluorescence techniques have been developed so far for the analysis of cell electric activity, morphology and metabolic rates. Functional cellular assays are employed to monitor several kinds of signals at the same time. Thus the signal-processing of living cells can be profiled in a more detailed and thorough way.

Brischwein *et al.* [160] first reported on multi-parametric cell based assays with data obtained solely with integrated sensors on silicon chips. Extracellular acidification rates (with ISFETs), cellular oxygen consumption rates (with amperometric electrode structures) and cell morphological alterations (with impedimetric electrode structures, IDES) were monitored simultaneously for up to several days. Ceriotti *et al.* [163] employed a multi-parametric chip-based system to measure cell adhesion, metabolism, and response to metal compounds which was previously classified as cytotoxic in immortalized mouse fibroblasts (BALB/3T3 cell line). In a parallel, online, and in a label-free way, the system measured extracellular acidification rates (with ISFETs), the cellular oxygen consumption (with amperometric electrode structures (Clark-type sensors)), and cell adhesion (with IDESs) (Figure 24(a)). A group from the biosensor national special lab in China has established an automatic multi-parametric platform based on an integrated biochip which measures the cell electric activity (with MEA), the cell morphology (with ECIS) and metabolic rates (with LAPS) (Figure 24(b)). Using the integrated multi-parametric sensors, some cell-based assays are conducted, including cytotoxicity and drug assessment, cell development and stem cell differentiation research.

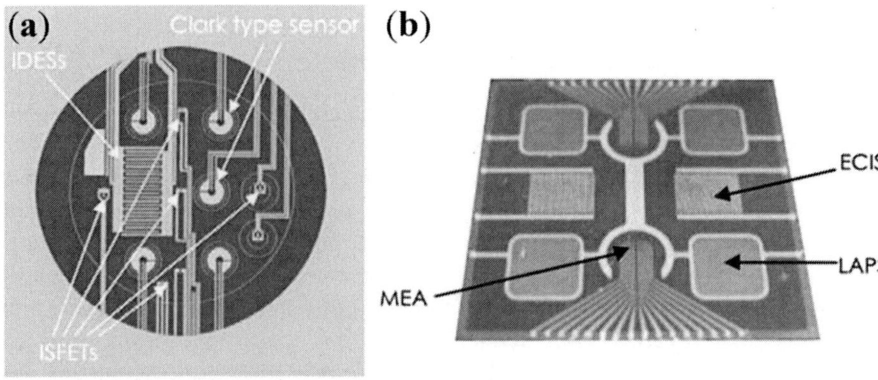

Figure 24. (**a**) A metabolic chip integrated with IDES, ISFET sensors and Clark-type electrodes. (Reprinted from [163]. © 2007, with permission from Elsevier). (**b**) Integrated biochip with MEA, ECIS and LAPS.

From another point of view, extracellular acidification and redox potential are important indicators of cell metabolism activity. As the EIS and EMIS sensor have similar structure and working principle, it makes sense to integrate the two sensors on the same chip. The related work has been carried out by Wang *et al.* [170]. The integrated potentiometric sensor (Figure 25) was constructed by depositing a gold metal layer on a partial surface of silicon dioxide. The sensitivity for pH and redox potential measurement is approximately 41.6 mV/pH and 53.2 mV/log[Fe(III)/Fe(II)], respectively. The latter is in accordance with the work of Adami *et al.* [171]. The integrated sensor was used for nephrotoxicity assay under drug stimulation.

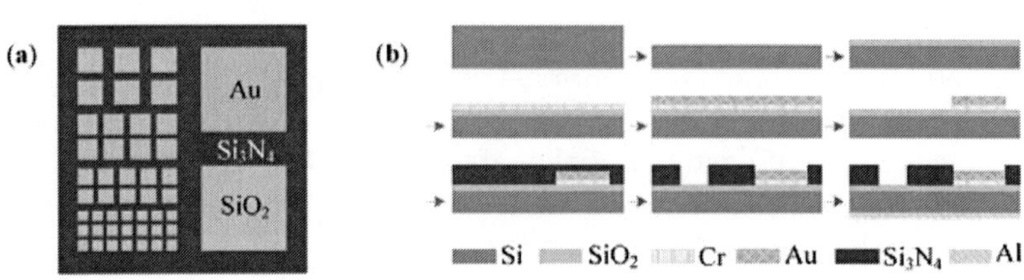

Figure 25. Schematic diagram of an integrated photovoltage-based biosensor used for cellular detection. (**a**) Sensor design; (**b**) Sensor fabrication.

The physiological state shows an acidified extracellular microenvironment and a reduced redox potential. It is revealed from this study that gentamicin inhibits the regular cellular metabolism, and the toxicity enhances with the concentration, but the toxicity is reversible when washed with PBS solution, which is in accordance with reversible or partially reversible kidney failure induced by gentamicin in clinical use, while 5-FU exhibits an anti-metabolism effect both intra and post 5-FU administration. The change of extracellular redox potential is significantly different from that of acidification, which indicates that the drugs are toxic for kidney cells mainly through acidification inhibition but induce no significant change to the concentration of extracellular redox pairs. However, this study was just a preliminary application of integrated photovoltage-based biosensors for cellular toxicity evaluation and a complete dose-response analysis and EC_{50}determination was not presented. Anyhow, the concept of an integrated sensor for acidification and redox potential detection has potential in drug toxicity analysis and can work as a supplementary means for preclinical drug screening.

4.2. Cell-Based Electrochemical Biosensor with Microfluidic Technology

Microfluidic technology is a powerful tool to facilitate the miniaturization and integration of analysis systems and can even address many issues that traditional methods cannot deal with. It has a cell-comparable small size and can precisely confine cells in the specific microchannel. By employing micro pumps and valves, mechanical or chemical stimulus can be introduced to cells and the corresponding cellular level biological responses can be monitored by integrated electrochemical electrodes. Even single cell and small cell group analysis can be achieved. Besides, microfluidics can realize high-throughput parallel biological monitoring and analysis.

The microfluidic chip with specific micro-fabricated electrodes has broad applications in cell morphology research and has many dramatic advantages over traditional assays, such as simple operation, rapid detection, and less invasiveness. Cells are cultured in the microfluidic chip and the culture media solution can be introduced automatically through the microfluidic channel. Using suitable electrochemical detection methods, cell morphology information, such as adhesion, migration, proliferation, and apoptosis, can be monitored in real time. One typical example is the automatic transwell assay, in which two microfluidic chambers separated by a porous membrane are designed to monitor cell migration [172]. As illustrated in Figure 26(a), cells are seeded in the upper chamber through microfluidic channels and an EIS sensor is placed in the lower chamber. Once cells migrate from the upper chamber into the lower one, great impedance changes can be detected. Figure 26(b) shows the impedance changes at different times. In addition, the influence of different ECM components can be also detected by this assay.

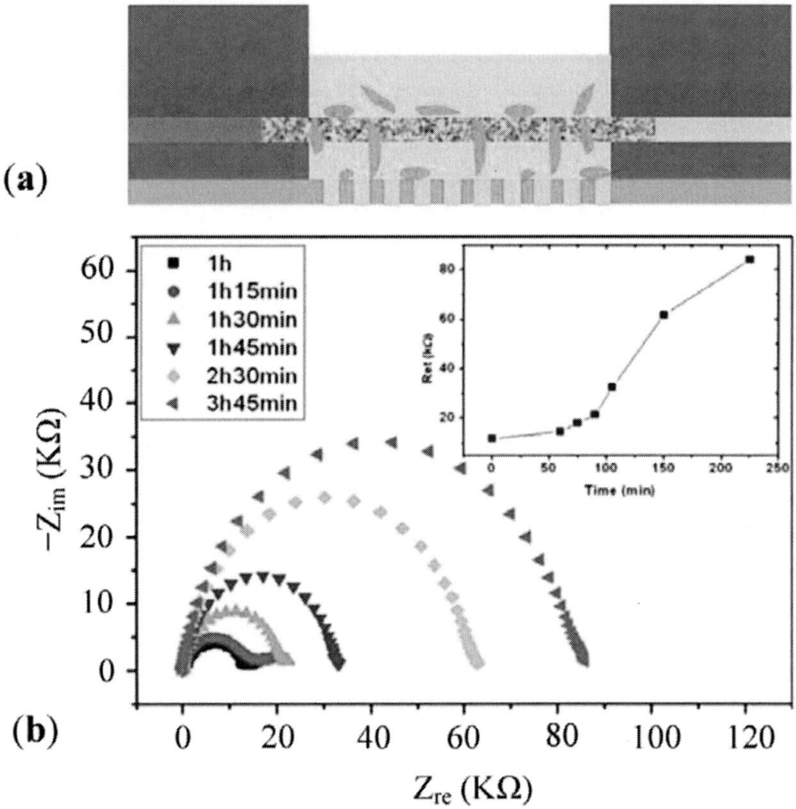

Figure 26. (**a**) Autotranswell cell chip for migration assay; (**b**) Real time monitoring of cell migration. Inset shows the time course of transmigration.

The microfluidic channel also provides great convenience for cellular metabolism analysis, especially in terms of single cells and small cell groups. Cells can be cultured in a tiny space and their physiological and electrical responses will be captured easily. Currently, many researches are covered in analyzing extracellular microenvironment [173] and intracellular components [174], monitoring ligand-induced cell secretion [175], recording the extracellular potential [176] and cell dynamic responses [177,178], *etc.* Besides, the cell patterning technique is an important issue in fundamental cell biology, tissue engineering, and neuron network research. Microfluidic technology plays a great role in this field. The chemical and cellular environment can be controlled precisely through the microchannels modified with diverse chemical molecules [179,180,181].

Assisted with microfluidic technology, cell-based electrochemical biosensors will have broad commercial and medical applications in the future. Low cost, disposable chips can be produced in large amounts and point-of-care diagnosis can be realized in the near future.

4.3. Cell-Based Electrochemical Biosensor with Nanotechnology

In recent years, nanotechnology has been greatly improved and widely applied in the biosensing fields. The most commonly used nanomaterials include nanoparticles, nanotubes, nanowires, and nanobars. Due to attractive electronic, optical, magnetic, thermal and catalytic properties, they have important applications in many interdisciplinary fields.

Nanomaterials play important roles in many aspects of biosensors. Owing to the large surface-volume ratio, nanogold particles are employed to modify the sensitive surface and enlarge the effective working area. Magnetic nanoparticles are usually used to accumulate or separate some important biomolecules conveniently [182]. Since carbon nanotubes (CNTs) have high electrocatalytic effect and fast electro-transfer rate, they are always used in the detection strategy to assist signal detection [183]. As shown in Figure 27(a), an indium tin oxide (ITO) electrode modified with CNTs was used as the working electrode and SU8 was patterned to limit the dimension of the ITO electrode. PC12 cells were cultured in a limited chamber (hundreds of μm^2). Once extracellular stimuli induced the exocytosis of cells, the molecules can be detected by the CNTs-ITO electrode, as illustrated in Figure 27(b). The results demonstrated that the sensitivity of the electrochemical sensor after CNTs surface modification was improved by 2.5–3 orders of magnitude and a typical amperometric response from a single cell was shown in Figure 27(c). The distinct optical performance of nanomaterials is also very attractive in the development of optical biosensors. Many kinds of nanomaterial-based biosensors are investigated to detect DNA, protein, and cellular substance by diverse detection technologies [184].

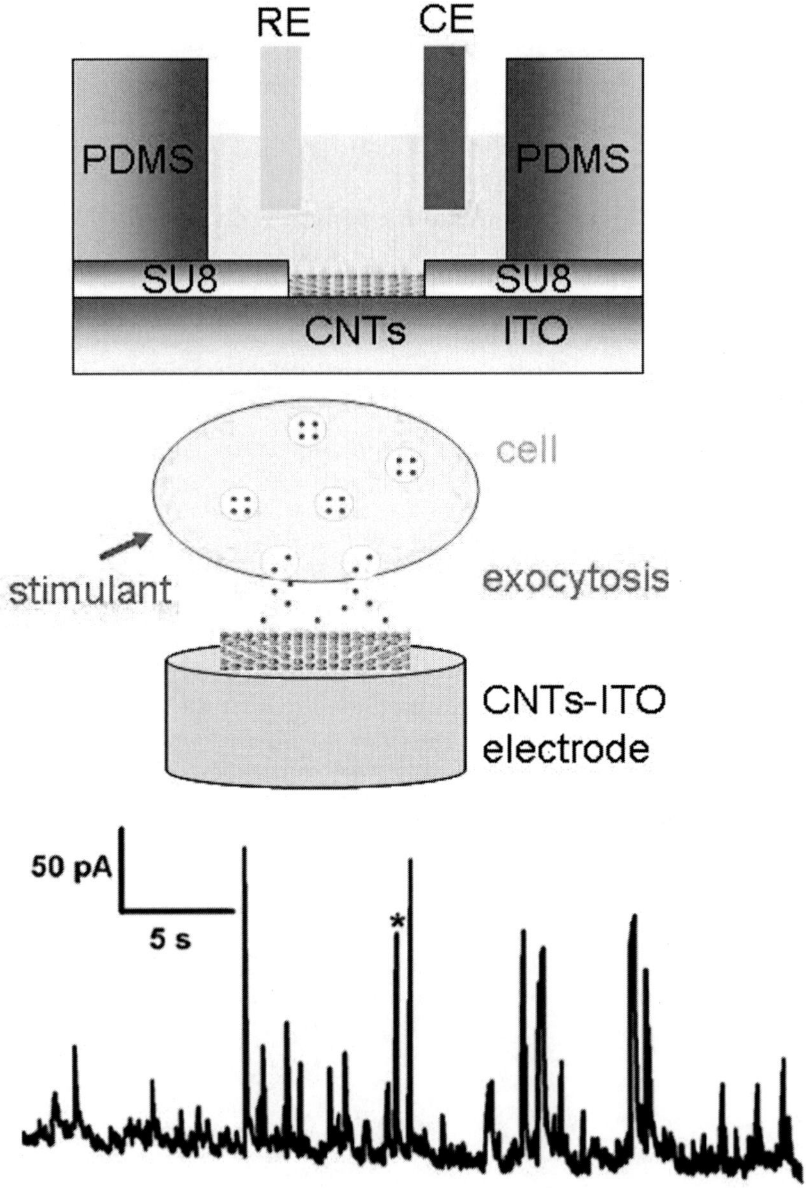

Figure 27. (**a**) Cross-sectional schematic of the electrochemical sensor with CNTs-ITO microelectrode. Ag/AgCl acted as reference electrode (RE) and Pt wire as counter electrode (CE); (**b**) Schematic of the single-cell release monitoring with the CNTs modified ITO electrode; (**c**) The amperometric response to stimulation from a single cell. (Reprinted from [183]. © 2011, with permission from Elsevier).

As an attractive trend, the distinct performance of nanomaterials can contribute much to cellular diagnosis. Nano-dimension electrodes can be used to probe intracellular compounds and monitor cell metabolism and electrophysiological activities. Nanomaterials can help to simplify and miniaturize biosensors for biomolecular trace detection.

5. CONCLUSIONS

This review essentially presents the microfabricated electrochemical cell-based biosensors that are commonly used in living cell analysis. MEA is applicable mostly for the electrophysiological recording of electrically excitable cells, and analysis of signal transduction between different cells. ECIS has become mature in the morphology monitoring of anchorage-dependent cells. LAPS is outstanding in the metabolite detection of the extracellular microenvironment. These microfabricated biosensors have demonstrated good performance in the mechanism study of cellular level, and even in drug analysis and environment monitoring. The biggest challenge in the development of high-precision commercialized cell-based biosensors is the designated growth and strong adhesion between the chip and cells. The introduction of nanotechnology and microfluidic technology may help to improve the performance of biosensor systems mentioned above and accelerate the birth of novel biosensors. It can be expected that the entrapment of cells and construction of single-cell-based biosensors will be facilitated. Another important development is high throughput and multi-parameter detection. Investigations of multi-channel and multi-functional integrated biosensors are gradually carried out, and this will result in a higher demand for supporting measurement systems.

ACKNOWLEDGMENTS

This work was supported by the National Natural Science Foundation of China (Grant No. 60725102, 30970765, 81027003) and the National Basic Research Program of China (Grant No. 2009CB320303).

REFERENCES

1. Bery, M.N.; Grivell, M.B. An electrochemical description of metabolism. In*Bioelectrochemistry of Cells and Tissues*; Walz, D., Berg, H., Milazzo, G., Eds.; Birkhauser Verlag: Basel, Switzerland, 1995; pp. 134–158.

2. Nonner, W.; Eisenberg, B. Electrodiffusion in ionic channels of biological membranes. *J. Mol. Liq.* **2000**, *87*, 149–162.

3. Du, D.; Cai, J.; Ju, H.X.; Yan, F.; Chen, J.; Jiang, X.Q.; Chen, H.Y. Construction of a biomimetic zwitterionic interface for monitoring cell proliferation and apoptosis. *Langmuir***2005**, *21*, 8394–8399.

4. Du, D.; Liu, S.L.; Chen, J.; Ju, H.X.; Lian, H.Z.; Li, J.X. Colloidal gold nanoparticle modified carbon paste interface for studies of tumor cell adhesion and viability. *Biomaterials* **2005**, *26*, 6487–6495.

5. Yan, F.; Chen, J.; Ju, H.X. Immobilization and electrochemical behavior of gold nanoparticles modified leukemia K562 cells and application in drug sensitivity test. *Electrochem. Commun.***2007**, *9*, 293–298.

6. Cheng, W.; Ding, L.; Lei, J.P.; Ding, S.J.; Ju, H.X. Effective cell capture with tetrapeptide-functionalized carbon nanotubes and dual signal amplification for cytosensing and evaluation of cell surface carbohydrate. *Anal. Chem.* **2008**, *80*, 3867–3872.

7. Hao, C.; Ding, L.; Zhang, X.J.; Ju, H.X. Biocompatible conductive architecture of carbon nanofiber-doped chitosan prepared with controllable electrodeposition for cytosensing.*Anal. Chem.* **2007**, *79*, 4442–4447.

8. Hu, H.L.; Jiang, H.; Wang, X.M.; Chen, B.A. Detection and distinguishability of leukemia cancer cells based on Au nanoparticles modified electrodes. *Electrochem. Commun.* **2008**, *10*, 1121–1124.

9. Shen, Q.; You, S.K.; Park, S.G.; Jiang, H.; Guo, D.D.; Chen, B.A.; Wang, X.M. Electrochemical biosensing for cancer cells based on TiO2/CNT nanocomposites modified electrodes.*Electroanalysis* **2008**, *20*, 2526–2530.

10. Jia, X.E.; Tan, L.; Zhou, Y.P.; Jiang, X.F.; Xie, Q.J.; Tang, H.; Yao, S.Z. Magnetic immobilization and electrochemical detection of leukemia K562 cells. *Electrochem. Commun.***2009**, *11*, 141–144.

11. Zhong, X.; Bai, H.J.; Xu, J.J.; Chen, H.Y.; Zhu, Y.H. A reusable interface constructed by 3-aminophenylboronic acid functionalized multiwalled carbon nanotubes for cell capture, release, and cytosensing. *Adv. Funct. Mater.* **2010**, *20*, 992–999.

12. Weng, J.; Zhang, Z.W.; Sun, L.P.; Wang, J.A. High sensitive detection of cancer cell with a folic acid-based boron-doped diamond electrode using an AC impedimetric approach.*Biosens. Bioelectron.* **2011**, *26*, 1847–1852.

13. Zhang, J.J.; Cheng, F.F.; Zheng, T.T.; Zhu, J.J. Design and implementation of electrochemical cytosensor for evaluation of cell surface carbohydrate and glycoprotein. *Anal. Chem.* **2010**,*82*, 3547–3555.

14. Ding, L.; Du, D.; Wu, H.; Ju, H.X. A disposable impedance sensor for electrochemical study and monitoring of adhesion and proliferation of K562 leukaemia cells. *Electrochem. Commun.***2007**, *9*, 953–958.

15. Wang, P.; Liu, Q.J. *Cell-Based Biosensors: Principles and Applications*; Artech House Publishers: Norwood, MA, USA, 2010.

16. Thomas, C.A., Jr.; Springer, P.A.; Loeb, G.E.; Berwald-Netter, Y.; Okun, L.M. A miniature microelectrode array to monitor the bioelectric activity of cultured cells. *Exp. Cell Res.* **1972**,*74*, 61–66.

17. Cheng, Y.; Luo, X.; Tsao, C.Y.; Wu, H.C.; Betz, J.; Payne, G.F.; Bentley, W.E.; Rubloff, G.W. Biocompatible multi-address 3D cell assembly in microfluidic devices using spatially programmable gel formation. *Lab Chip* **2011**, *11*, 2316–2318.

18. Hsiung, L.C.; Yang, C.H.; Chiu, C.L.; Chen, C.L.; Wang, Y.; Lee, H.; Cheng, J.Y.; Ho, M.C.; Wo, A.M. A planar interdigitated ring electrode array via dielectrophoresis for uniform patterning of cells. *Biosens. Bioelectron.* **2008**, *24*, 869–875.

19. Giaever, I.; Keese, C.R. Monitoring fibroblast behavior in tissue-culture with an applied electric-field. *Proc. Natl. Acad. Sci. USA* **1984**, *81*, 3761–3764.

20. Giaever, I.; Keese, C.R. A Morphological biosensor for mammalian-cells. *Nature* **1993**, *366*, 591–592.

21. Ehret, R.; Baumann, W.; Brischwein, M.; Schwinde, A.; Stegbauer, K.; Wolf, B. Monitoring of cellular behaviour by impedance measurements on interdigitated electrode structures.*Biosens. Bioelectron.* **1997**, *12*, 29–41.

22. Wegener, J.; Sieber, M.; Galla, H.J. Impedance analysis of epithelial and endothelial cell monolayers cultured on gold surfaces. *J. Biochem. Biophys. Methods* **1996**, *32*, 151–170.

23. Nguyen, D.D.; Huang, X.Q.; Greve, D.W.; Domach, M.M. Fibroblast growth and H-7 protein kinase inhibitor response monitored in microimpedance sensor arrays. *Biotechnol. Bioeng.***2004**, *87*, 138–144.

24. Bergveld, P. Development of an ion-sensitive solid-state device for neurophysiological measurements. *IEEE Trans. Biomed. Eng.* **1970**, *Bm17*, 70–71.

25. Lin, P.; Yan, F.; Yu, J.; Chan, H.L.W.; Yang, M. The application of organic electrochemical transistors in cell-based biosensors. *Adv. Mater.* **2010**, *22*, 3655–3660.

26. Cheng, Y.; Chen, K.S.; Meyer, N.L.; Yuan, J.; Hirst, L.S.; Chase, P.B.; Xiong, P. Functionalized SnO2 nanobelt field-effect transistor sensors for label-free detection of cardiac troponin.*Biosens. Bioelectron.* **2011**, *26*, 4538–4544.

27. Hafeman, D.G.; Parce, J.W.; Mcconnell, H.M. Light-addressable potentiometric sensor for biochemical systems. *Science* **1988**, *240*, 1182–1185.

28. McConnell, H.M.; Owicki, J.C.; Parce, J.W.; Miller, D.L.; Baxter, G.T.; Wada, H.G.; Pitchford, S. The cytosensor microphysiometer: Biological applications of silicon technology. *Science***1992**, *257*, 1906–1912.

29. Wu, Y.C.; Wang, P.; Ye, X.S.; Zhang, Q.T.; Li, R.; Yan, W.M.; Zheng, X.X. A novel microphysiometer based on MLAPS for drugs screening. *Biosens. Bioelectron.* **2001**, *16*, 277–286.

30. Siu, W.M.; Cobbold, R.S.C. Basic properties of the electrolyte-SiO_2-Si system: Physical and theoretical aspects. *IEEE Trans.Electron Devices* **1979**, *26*, 1805–1815.

31. Bove, M.; Grattarola, M.; Martinoia, S.; Verreschi, G. Interfacing cultured neurons to planar substrate microelectrodes: Characterization of the neuron-to-microelectrode junction.*Bioelectrochem. Bioenerg.* **1995**, *38*, 255–265.

32. Breckenridge, L.; Wilson, R.; Connolly, P.; Curtis, A.; Dow, J.; Blackshaw, S.; Wilkinson, C. Advantages of using microfabricated extracellular electrodes for *in vitro* neuronal recording.*J. Neurosci. Res.* **1995**, *42*, 266–276.

33. Regehr, W.G.; Pine, J.; Cohan, C.S.; Mischke, M.D.; Tank, D.W. Sealing cultured invertebrate neurons to embedded dish electrodes facilitates long-term stimulation and recording. *J. Neurosci. Methods* **1989**, *30*, 91–106.

34. Grattarola, M.; Martinoia, S.; Massobrio, G.; Cambiaso, A.; Rosichini, R.; Tetti, M. Computer simulations of the responses of passive and active integrated microbiosensors to cell activity. *Sens. Actuat. B: Chem.* **1991**, *4*, 261–265.

35. Heuschkel, M.O.; Fejtl, M.; Raggenbass, M.; Bertrand, D.; Renaud, P. A three-dimensional multi-electrode array for multi-site stimulation and recording in acute brain slices. *J. Neurosci. Methods* **2002**, *114*, 135–148.

36. Koo, K.; Chung, H.; Yu, Y.; Seo, J.; Park, J. Fabrication of pyramid shaped three-dimensional 8 × 8 electrodes for artificial retina. *Sens. Actuat. A: Phys.* **2006**, *130*, 609–615.

37. Gross, G.W.; Wen, W.Y.; Lin, J.W. Transparent indium-tin oxide electrode patterns for extracellular, multisite recording in neuronal cultures. *J. Neurosci. Methods* **1985**, *15*, 243–252.

38. Pine, J. Recording action potentials from cultured neurons with extracellular microcircuit electrodes. *J. Neurosci. Methods* **1980**, *2*, 19–31.

39. Wegener, J.; Keese, C.R.; Giaever, I. Electric cell-substrate impedance sensing (ECIS) as a noninvasive means to monitor the kinetics of cell spreading to artificial surfaces. *Exp. Cell Res.* **2000**, *259*, 158–166.

40. Giaever, I.; Keese, C.R. Use of electric fields to monitor the dynamical aspect of cell behavior in tissue culture. *IEEE Trans. Biomed. Eng.* **1986**, *33*, 242–247.

41. Xiao, C.; Lachance, B.; Sunahara, G.; Luong, J.H.T. An in-depth analysis of electric cell-substrate impedance sensing to study the attachment and spreading of mammalian cells.*Anal. Chem.* **2002**, *74*, 1333–1339.

42. Wang, L.; Zhu, J.; Deng, C.; Xing, W.; Cheng, J. An automatic and quantitative on-chip cell migration assay using self-assembled monolayers combined with real-time cellular impedance sensing. *Lab Chip* **2008**, *8*, 872–878.

43. Chang, B.W.; Chen, C.H.; Ding, S.J.; Chen, D.C.H.; Chang, H.C. Impedimetric monitoring of cell attachment on interdigitated microelectrodes. *Sens. Actuat. B: Chem.* **2005**, *105*, 159–163.

44. Parce, J.W.; Owicki, J.C.; Kercso, K.M.; Sigal, G.B.; Wada, H.; Muir, V.C.; Bousse, L.J.; Ross, K.L.; Sikic, B.I.; McConnell, H.M. Detection of cell-affecting agents with a silicon biosensor.*Science* **1989**, *246*, 243–247.

45. Piras, L.; Adami, M.; Fenu, S.; Dovis, M.; Nicolini, C. Immunoenzymatic application of a redox potential biosensor. *Anal. Chim. Acta* **1996**, *335*, 127–135.

46. Bousse, L. The chemical Sensitivity of Electrolyte/Interface/Silicon Structures. Ph.D. Dissertation, Twente University of Technology, Enschede, The Netherlands. 1982.

47. Stern, E.; Klemic, J.F.; Routenberg, D.A.; Wyrembak, P.N.; Turner-Evans, D.B.; Hamilton, A.D.; LaVan, D.A.; Fahmy, T.M.; Reed, M.A.

Label-free immunodetection with CMOS-compatible semiconducting nanowires. *Nature* **2007**, *445*, 519–522.

48. Cheng, Y.; Xiong, P.; Yun, C.S.; Strouse, G.; Zheng, J.; Yang, R.; Wang, Z. Mechanism and optimization of pH sensing using SnO2 nanobelt field effect transistors. *Nano Lett.* **2008**, *8*, 4179–4184.

49. Hafner, F. Cytosensor® Microphysiometer: Technology and recent applications. *Biosens. Bioelectron.* **2000**, *15*, 149–158.

50. Bard, A.J.; Faulkner, L.R. *Electrochemical Methods: Fundamentals and Applications*; John Wiley & Sons: New York, NY, USA, 1980.

51. Rabinowitz, J.D.; Vacchino, J.F.; Beeson, C.; McConnell, H.M. Potentiometric measurement of intracellular redox activity. *J. Am. Chem. Soc.* **1998**, *120*, 2464–2473.

52. Ward, K.; Hawkridge, F.; Brahim, S.; Lewis, K.L.; Rhoten, M.C. Non-biofouling, universal redox electrode and measurement system. U.S. Patent Application 20090134043, May 2009.

53. Natarajan, A.; Stancescu, M.; Dhir, V.; Armstrong, C.; Sommerhage, F.; Hickman, J.J.; Molnar, P. Patterned cardiomyocytes on microelectrode arrays as a functional, high information content drug screening platform. *Biomaterials* **2011**, *32*, 4267–4274.

54. Chen, M.Q.; Whittington, R.H.; Day, P.W.; Kobilka, B.K.; Giovangrandi, L.; Kovacs, G.T.A. A device for separated and reversible co-culture of cardiomyocytes. *Biotechnol. Prog.* **2010**, *26*, 1164–1171.

55. Gholmieh, G.; Soussou, W.; Han, M.; Ahuja, A.; Hsiao, M.C.; Song, D.; Tanguay, A.R., Jr.; Berger, T.W. Custom-designed high-density conformal planar multielectrode arrays for brain slice electrophysiology. *J. Neurosci. Methods* **2006**, *152*, 116–129.

56. Hofmann, F.; Bading, H. Long term recordings with microelectrode arrays: Studies of transcription-dependent neuronal plasticity and axonal regeneration. *J. Physiol.-Paris* **2006**,*99*, 125–132.

57. Kiryushko, D.; Kofoed, T.; Skladchikova, G.; Holm, A.; Berezin, V.; Bock, E. A synthetic peptide ligand of neural cell adhesion molecule (NCAM), C3d, promotes neuritogenesis and synaptogenesis and modulates presynaptic function in primary cultures of rat hippocampal neurons. *J. Biol. Chem.* **2003**, *278*, 12325–12334.

58. David-Pur, M.; Shein, M.; Hanein, Y.; Balasubramanian, K. Carbon nanotube-based neurochips. *Methods Mol. Biol.* **2010**, *625*, 171–177.

59. Charvet, G.; Rousseau, L.; Billoint, O.; Gharbi, S.; Rostaing, J.P.; Joucla, S.; Trevisiol, M.; Bourgerette, A.; Chauvet, P.; Moulin, C. BioMEA^{TM}: A versatile high-density 3D microelectrode array system using integrated electronics. *Biosens. Bioelectron.* **2010**, *25*, 1889–1896.

60. Rajaraman, S.; Choi, S.O.; Shafer, R.H.; Ross, J.D.; Vukasinovic, J.; Choi, Y.; DeWeerth, S.P.; Glezer, A.; Allen, M.G. Microfabrication technologies for a coupled three-dimensional microelectrode, microfluidic array. *J. Micromech. Microeng.* **2007**, *17*, 163–171.

61. Chu, H.Y.; Kuo, T.Y.; Chang, B.; Lu, S.W.; Chiao, C.C.; Fang, W. Design and fabrication of novel three-dimensional multi-electrode array using SOI wafer. *Sens. Actuat. A: Phys.* **2006**,*130*, 254–261.

62. Morin, F.O.; Takamura, Y.; Tamiya, E. Investigating neuronal activity with planar microelectrode arrays: Achievements and new perspectives. *J. Biosci. Bioeng.* **2005**, *100*, 131–143.

63. Streit, J.; Tscherter, A.; Heuschkel, M.O.; Renaud, P. The generation of rhythmic activity in dissociated cultures of rat spinal cord. *Eur. J. Neurosci.* **2001**, *14*, 191–202.

64. Robinson, H.; Kawahara, M.; Jimbo, Y.; Torimitsu, K.; Kuroda, Y.; Kawana, A. Periodic synchronized bursting and intracellular calcium transients elicited by low magnesium in cultured cortical neurons. *J. Neurophysiol.* **1993**, *70*, 1606–1616.

65. Pan, L.; Xiang, G.; Huang, L.; Yu, Z.; Cheng, J.; Xing, W.; Zhou, Y. Automatic positioning and sensing microelectrode array (APSMEA) for multi-site electrophysiological recordings.*J. Neurosci. Methods* **2008**, *170*, 123–129.

66. Jing, G.; Yao, Y.; Gnerlich, M.; Perry, S.; Tatic-Lucic, S. Towards a multi-electrode array (MEA) system for patterned neural networks. *Procedia Chem.* **2009**, *1*, 329–332.

67. Berdondini, L.; Chiappalone, M.; Van Der Wal, P.; Imfeld, K.; De Rooij, N.F.; Koudelka-Hep, M.; Tedesco, M.; Martinoia, S.; Van Pelt, J.; Le Masson, G. A microelectrode array (MEA) integrated with clustering structures for investigating *in vitro* neurodynamics in confined interconnected sub-populations of neurons. *Sens. Actuat. B: Chem.* **2006**, *114*, 530–541.

68. Pearce, T.M.; Wilson, J.A.; Oakes, S.G.; Chiu, S.Y.; Williams, J.C. Integrated microelectrode array and microfluidics for temperature clamp of sensory neurons in culture. *Lab Chip* **2004**,*5*, 97–101.

69. Dworak, B.J.; Wheeler, B.C. Novel MEA platform with PDMS microtunnels enables the detection of action potential propagation from isolated axons in culture. *Lab Chip* **2008**, *9*, 404–410.

70. Parviz, M.; Gross, G.W. Quantification of zinc toxicity using neuronal networks on microelectrode arrays. *NeuroToxicology* **2007**, *28*, 520–531.

71. Kamioka, H.; Maeda, E.; Jimbo, Y.; Robinson, H.P.C.; Kawana, A. Spontaneous periodic synchronized bursting during formation of mature patterns of connections in cortical cultures. *Neurosci. Lett.* **1996**, *206*, 109–112.

72. Hescheler, J.; Halbach, M.; Egert, U.; Lu, Z.J.; Bohlen, H.; Fleischmann, B.K.; Reppel, M. Determination of electrical properties of ES cell-derived cardiomyocytes using MEAs. *J. Electrocardiol.* **2004**, *37*, 110–116.

73. Illes, S.; Fleischer, W.; Siebler, M.; Hartung, H.P.; Dihné, M. Development and pharmacological modulation of embryonic stem cell-derived neuronal network activity. *Exp. Neurol.* **2007**, *207*, 171–176.

74. Taketani, M.; Baudry, M. *Advances in Network Electrophysiology: Using Multi-Electrode Arrays*; Springer: Berlin/Heidelberg, Germany, 2006; pp. 321–331.

75. Liu, Q.; Ye, W.; Xiao, L.; Du, L.; Hu, N.; Wang, P. Extracellular potentials recording in intact olfactory epithelium by microelectrode array for a bioelectronic nose. *Biosens. Bioelectron.***2010**, *25*, 2212–2217.

76. LeBaron, R.G.; Athanasiou, K.A. Extracellular matrix cell adhesion peptides: Functional applications in orthopedic materials. *Tissue Eng.* **2000**, *6*, 85–103.

77. Cuvelier, D.; Théry, M.; Chu, Y.S.; Dufour, S.; Thiéry, J.P.; Bornens, M.; Nassoy, P.; Mahadevan, L. The universal dynamics of cell spreading. *Curr. Biol.* **2007**, *17*, 694–699.

78. Luong, J.H.T.; Xiao, C.; Lachance, B.; Leabu, S.M.; Li, X.; Uniyal, S.; Chan, B. Extended applications of electric cell-substrate impedance sensing for assessment of the structure–function of α2β1 integrin. *Anal. Chim. Acta* **2004**, *501*, 61–69.

79. Bouafsoun, A.; Helali, S.; Othmane, A.; Kerkeni, A.; Prigent, A.F.; Jaffrézic-Renault, N.; Bessueille, F.; Léonard, D.; Ponsonnet, L. Evaluation of endothelial cell adhesion onto different protein/gold electrodes by EIS. *Macromol. Biosci.* **2007**, *7*, 599–610.

80. Kataoka, N.; Iwaki, K.; Hashimoto, K.; Mochizuki, S.; Ogasawara, Y.; Sato, M.; Tsujioka, K.; Kajiya, F. Measurements of endothelial cell-to-cell and cell-to-substrate gaps and micromechanical properties of endothelial cells during monocyte adhesion. *Proc. Natil. Acad. Sci. USA* **2002**, *99*, 15638–15643.

81. Atienza, J.M.; Yu, N.; Wang, X.; Xu, X.; Abassi, Y. Label-free and real-time cell-based kinase assay for screening selective and potent receptor tyrosine kinase inhibitors using microelectronic sensor array. *J. Biomol. Screen.* **2006**, *11*, 634–643.

82. Yu, N.; Atienza, J.M.; Bernard, J.; Blanc, S.; Zhu, J.; Wang, X.; Xu, X.; Abassi, Y.A. Real-time monitoring of morphological changes in living cells by electronic cell sensor arrays: An approach to study G protein-coupled receptors. *Anal. Chem.* **2006**, *78*, 35–43.

83. Cho, R.J.; Huang, M.; Campbell, M.J.; Dong, H.; Steinmetz, L.; Sapinoso, L.; Hampton, G.; Elledge, S.J.; Davis, R.W.; Lockhart, D.J. Transcriptional regulation and function during the human cell cycle. *Nat. Genet.* **2001**, *27*, 48–54.

84. Tapon, N.; Moberg, K.H.; Hariharan, I.K. The coupling of cell growth to the cell cycle. *Curr. Opin. Cell Biol.* **2001**, *13*, 731–737.

85. Wang, L.; Yin, H.; Xing, W.; Yu, Z.; Guo, M.; Cheng, J. Real-time, label-free monitoring of the cell cycle with a cellular impedance sensing chip. *Biosens. Bioelectron.* **2010**, *25*, 990–995.

86. Bieberich, E.; Guiseppi-Elie, A. Neuronal differentiation and synapse formation of PC12 and embryonic stem cells on interdigitated microelectrode arrays: Contact structures for neuron-to-electrode signal transmission (NEST). *Biosens. Bioelectron.* **2004**, *19*, 923–931.

87. Maercker, C.; Rogge, T.; Mathis, H.; Ridinger, H.; Bieback, K. Development of live cell chips to monitor cell differentiation processes. *Eng. Life Sci.* **2008**, *8*, 33–39.

88. Diederichs, S.; Riechers, D.; Sempf, F.; Kall, S.; Kasper, C.; Griensven, M.; Scheper, T. Investigation of the effect of mechanical strain on the osteogenic differentiation of mesenchymal stem cells. *Cells Cult.* **2010**, 579–589.

89. Hediger, S.; Fontannaz, J.; Sayah, A.; Hunziker, W.; Gijs, M. Biosystem for the culture and characterisation of epithelial cell tissues. *Sens. Actuat. B: Chem.* **2000**, *63*, 63–73.

90. Asami, K. Dielectric spectra of biological cells and tissues simulated by three-dimensional finite difference method. *IFMBE Proc.* **2007**, *17*, 98–101.

91. Frampton, J.; Hynd, M.; Williams, J.; Shuler, M.; Shain, W. Three-dimensional hydrogel cultures for modeling changes in tissue impedance around microfabricated neural probes. *J. Neural Eng.* **2007**, *4*, 399–409.

92. Han, A.; Yang, L.; Frazier, A.B. Quantification of the heterogeneity in breast cancer cell lines using whole-cell impedance spectroscopy. *Clin. Cancer Res.* **2007**, *13*, 139–143.

93. Linderholm, P.; Vannod, J.; Barrandon, Y.; Renaud, P. Bipolar resistivity profiling of 3D tissue culture. *Biosens. Bioelectron.* **2007**, *22*, 789–796.

94. Williams, J.C.; Hippensteel, J.A.; Dilgen, J.; Shain, W.; Kipke, D.R. Complex impedance spectroscopy for monitoring tissue responses to inserted neural implants. *J. Neural Eng.***2007**, *4*, 410–423.

95. Atienza, J.M.; Zhu, J.; Wang, X.; Xu, X.; Abassi, Y. Dynamic monitoring of cell adhesion and spreading on microelectronic sensor arrays. *J. Biomol. Screen.* **2005**, *10*, 795–805.

96. Webb, D.J.; Zhang, H.Y.; Horwitz, A.F. Cell migration. In *Cell Migration: Developmental Methods and Protocols*; Guan, J.-L., Ed.; Humana Press Inc.: Totowa, NJ, USA, 2005; Volume 294, pp. 3–11.

97. Noiri, E.; Hu, Y.; Bahou, W.F.; Keese, C.R.; Giaever, I.; Goligorsky, M.S. Permissive role of nitric oxide in endothelin-induced migration of endothelial cells. *J. Biol. Chem.* **1997**, *272*, 1747–1752.

98. Keese, C.R.; Wegener, J.; Walker, S.R.; Giaever, I. Electrical wound-healing assay for cells in vitro. *Proc. Natl. Acad. Sci. USA* **2004**, *101*, 1554–1559.

99. Shaw, L.M. Tumor cell invasion assays. *Methods Mol. Biol.* **2005**, *294*, 97–105.

100. Keese, C.R.; Bhawe, K.; Wegener, J.; Giaever, I. Real-time impedance assay to follow the invasive activities of metastatic cells in culture. *Biotechniques* **2002**, *33*, 842–851.

101. Wang, G.; Zhang, W. The signaling network of tumor invasion. *Histol. Histopathol.* **2005**, *20*, 593–602.

102. Ren, J.; Xiao, Y.; Singh, L.S.; Zhao, X.; Zhao, Z.; Feng, L.; Rose, T.M.; Prestwich, G.D.; Xu, Y. Lysophosphatidic acid is constitutively produced by human peritoneal mesothelial cells and enhances adhesion, migration, and invasion of ovarian cancer cell. *Cancer Res.* **2006**, *66*, 3006–3014.

103. Saxena, N.K.; Sharma, D.; Ding, X.; Lin, S.; Marra, F.; Merlin, D.; Anania, F.A. Concomitant activation of the JAK/STAT, PI3K/AKT, and

ERK signaling is involved in leptin-mediated promotion of invasion and migration of hepatocellular carcinoma cells. *Cancer Res.* **2007**, *67*, 2497–2507.

104. Chen, J.; Ye, L.; Zhang, L.; Jiang, W.G. Placenta growth factor, PLGF, influences the motility of lung cancer cells, the role of Rho associated kinase, Rock1. *J. Cell. Biochem.* **2008**, *105*, 313–320.

105. Earley, S.; Plopper, G.E. Phosphorylation of focal adhesion kinase promotes extravasation of breast cancer cells. *Biochem. Biophys. Res. Commun.* **2008**, *366*, 476–482.

106. Saxena, N.K.; Taliaferro-Smith, L.T.; Knight, B.B.; Merlin, D.; Anania, F.A.; O'Regan, R.M.; Sharma, D. Bidirectional crosstalk between leptin and insulin-like growth factor-I signaling promotes invasion and migration of breast cancer cells via transactivation of epidermal growth factor receptor. *Cancer Res.* **2008**, *68*, 9712–9722.

107. Sgambato, A.; De Paola, B.; Migaldi, M.; Di Salvatore, M.; Rettino, A.; Rossi, G.; Faraglia, B.; Boninsegna, A.; Maiorana, A.; Cittadini, A. Dystroglycan expression is reduced during prostate tumorigenesis and is regulated by androgens in prostate cancer cells. *J. Cell. Physiol.***2007**, *213*, 528–539.

108. Taliaferro-Smith, L.; Nagalingam, A.; Zhong, D.; Zhou, W.; Saxena, N.; Sharma, D. LKB1 is required for adiponectin-mediated modulation of AMPK–S6K axis and inhibition of migration and invasion of breast cancer cells. *Oncogene* **2009**, *28*, 2621–2633.

109. Patani, N.; Douglas-Jones, A.; Mansel, R.; Jiang, W.; Mokbel, K. Tumour suppressor function of MDA-7/IL-24 in human breast cancer. *Cancer Cell Int.* **2010**, *10*, 29.

110. Mandel, L.J. Energy metabolism of cellular activation, growth, and transformation. *Curr. Top. Membr. Transp.* **1986**, *27*, 261–291.

111. Hu, N.; Ha, D.; Wu, C.X.; Cheng, G.; Yu, H.; Wang, T.X.; Wu, J.Y.; Cai, H.; Liu, Q.J.; Wang, P. Design of microphysiometer based on multiparameter cell-based biosensors for quick drug analysis. *J. Innov. Opt. Health Sci.* **2012**, *5*, 1150005–1.

112. Deuticke, B. Monocarboxylate transport in red blood cells: Kinetics and chemical modification. *Methods Enzymol.* **1989**, *173*, 300–329.

113. Wada, H.; Owicki, J.; Bruner, L.; Miller, K.; Raley-Susman, K.; Panfili, P.; Humphries, G.; Parce, J. Measurement of cellular responses to toxic agents using a silicon microphysiometer. *Altern. Anim. Test. Exp. (AATEX)* **1992**, *1*, 154–164.

114. Fujii, R.; Hosoya, M.; Fukusumi, S.; Kawamata, Y.; Habata, Y.; Hinuma, S.; Onda, H.; Nishimura, O.; Fujino, M. Identification of neuromedin U as the cognate ligand of the orphan G protein-coupled receptor FM-3*. *J. Biol. Chem.* **2000**, *275*, 21068–21074.

115. Wille, K.; Paige, L.A.; Higgins, A.J. Application of the Cytosensor microphysiometer to drug discovery. *Recept. Channels* **2003**, *9*, 125–131.

116. Kramarenko, I.I.; Bunni, M.A.; Morinelli, T.A.; Raymond, J.R.; Garnovskaya, M.N. Identification of functional bradykinin B_2 receptors endogenously expressed in HEK293 cells. *Biochem. Pharmacol.* **2009**, *77*, 269–276.

117. Needham, J.; Needham, D.M. The oxidation-reduction potential of protoplasm: A review.*Protoplasma* **1926**, *1*, 255–294.

118. Allyn, W.P.; Baldwin, I. Oxidation-reduction potentials in relation to the growth of an aerobic form of bacteria. *J. Bacteriol.* **1932**, *23*, 369–398.

119. Ward, W.E. The apparent oxidation-reduction potentials of bright platinum electrodes in synthetic media cultures of bacteria. *J. Bacteriol.* **1938**, *36*, 337–355.

120. Kwong, S.C.; Randers, L.; Rao, G. On-line assessment of metabolic activities based on culture redox potential and dissolved oxygen profiles during aerobic fermentation. *Biotechnol. Prog.***1992**, *8*, 576–579.

121. Cater, D.; Phillips, A.; Silver, I. The measurement of oxidation-reduction potentials, pH, and oxygen tension in tumours. *Proc. R. Soc. Lond. Ser. B Biol. Sci.* **1957**, *146*, 382–399.

122. Eyer, K.; Heinzle, E. On-line estimation of viable cells in a hybridoma culture at various DO levels using ATP balancing and redox potential measurement. *Biotechnol. Bioeng.* **1996**, *49*, 277–283.

123. Higareda, A.E.; Possani, L.D.; Ramirez, O.T. The use of culture redox potential and oxygen uptake rate for assessing glucose and glutamine depletion in hybridoma cultures. *Biotechnol. Bioeng.* **1997**, *56*, 555–563.

124. Relchart, O.; Szakmar, K.; Jozwiak, A.; Felfoldi, J.; Baranyai, L. Redox potential measurement as a rapid method for microbiological testing and its validation for coliform determination.*Int. J. Food Microbiol.* **2007**, *114*, 143–148.

125. Zheng, J.D.; Xu, G.; Chu, J.; Wang, Y.; Zhuang, Y.; Zhang, S. Using oxidation-reduction potential to optimize lactic acid production by Lactobacillus paracasei. *Chin. J. Bioprocess Eng.* **2008**, *6*, 73–77.

126. Koo, B.S.; Gong, Y.J.; Kim, S.Y.; Kim, C.W.; Lee, H.C. Improvement of coenzyme Q(10) production by increasing the NADH/NAD($^+$) ratio in Agrobacterium tumefaciens. *Biosci. Biotechnol. Biochem.* **2010**, *74*, 895–898.

127. Hwang, C.; Sinskey, A. The role of oxidation-reduction potential in monitoring growth of cultured mammalian cells. In *Production of Biologicals from Animal Cells in Culture*; Spier Raymond G.J.B. Meignier, B., Ed.; Butterworth-Heinemann: Oxford, UK, 1991; pp. 548–569.

128. Carelli, S.; Ceriotti, A.; Cabibbo, A.; Fassina, G.; Ruvo, M.; Sitia, R. Cysteine and glutathione secretion in response to protein disulfide bond formation in the ER. *Science* **1997**, *277*, 1681–1684.

129. Hwang, C.; Sinskey, A.J.; Lodish, H.F. Oxidized Redox State of Glutathione in the Endoplasmic-Reticulum. *Science* **1992**, *257*, 1496–1502.]

130. Bellomo, G.; Palladini, G.; Vairetti, M. Intranuclear distribution, function and fate of glutathione and glutathione-S-conjugate in living rat hepatocytes studied by fluorescence microscopy. *Microsc. Res. Tech.* **1997**, *36*, 243–252.

131. Natarajan, A.; Molnar, P.; Sieverdes, K.; Jamshidi, A.; Hickman, J. Microelectrode array recordings of cardiac action potentials as a high throughput method to evaluate pesticide toxicity. *Toxicol. in vitro* **2006**, *20*, 375–381.

132. Borkholder, D.A. Cell Based Biosensors Using Microelectrodes. Ph.D. Thesis, Stanford University, Stanford, CA, USA. 1998.

133. Halbach, M.D.; Egert, U.; Hescheler, J.; Banach, K. Estimation of action potential changes from field potential recordings in multicellular mouse cardiac myocyte cultures. *Cell. Physiol. Biochem.* **2003**, *13*, 271–284.

134. Reppel, M.; Igelmund, P.; Egert, U.; Juchelka, F.; Hescheler, J.; Drobinskaya, I. Effect of cardioactive drugs on action potential generation and propagation in embryonic stem cell-derived cardiomyocytes. *Cell. Physiol. Biochem.* **2007**, *19*, 213–224.

135. Gilchrist, K.H. Characterization and validation of cell-based biosensors. Ph.D. Thesis, Stanford University, Stanford, CA, USA. 2003.

136. Yeung, C.; Sommerhage, F.; Wrobel, G.; Offenhäusser, A.; Chan, M.; Ingebrandt, S. Drug profiling using planar microelectrode arrays. *Anal. Bioanal. Chem.* **2007**, *387*, 2673–2680.

137. Tsai, C.T.; Tseng, C.D.; Hwang, J.J.; Wu, C.K.; Yu, C.C.; Wang, Y.C.; Chen, W.P.; Lai, L.P.; Chiang, F.T.; Lin, J.L. Tachycardia of atrial myocytes induces collagen expression in atrial fibroblasts through transforming growth factor β1. *Cardiovasc. Res.* **2011**, *89*, 805–815.

138. Stett, A.; Egert, U.; Guenther, E.; Hofmann, F.; Meyer, T.; Nisch, W.; Haemmerle, H. Biological application of microelectrode arrays in drug discovery and basic research. *Anal. Bioanal. Chem.* **2003**, *377*, 486–495.

139. Meyer, T.; Leisgen, C.; Gonser, B.; Günther, E. QT-screen: High-throughput cardiac safety pharmacology by extracellular electrophysiology on primary cardiac myocytes. *Assay Drug Dev. Technol.* **2004**, *2*, 507–514.

140. Gross, G.W.; Harsch, A.; Rhoades, B.K.; Göpel, W. Odor, drug and toxin analysis with neuronal networks *in vitro*: Extracellular array recording of network responses. *Biosens. Bioelectron.* **1997**, *12*, 373–393.

141. Xiang, G.; Pan, L.; Huang, L.; Yu, Z.; Song, X.; Cheng, J.; Xing, W. Microelectrode array-based system for neuropharmacological applications with cortical neurons cultured *in vitro.Biosens. Bioelectron.* **2007**, *22*, 2478–2484.

142. Gross, G.W.; Rhoades, B.; Jordan, R. Neuronal networks for biochemical sensing. *Sens. Actuat. B: Chem.* **1992**, *6*, 1–8.

143. Eytan, D.; Minerbi, A.; Ziv, N.; Marom, S. Dopamine-induced dispersion of correlations between action potentials in networks of cortical neurons. *J. Neurophysiol.* **2004**, *92*, 1817–1824.

144. Keefer, E.W.; Gramowski, A.; Gross, G.W. NMDA receptor-dependent periodic oscillations in cultured spinal cord networks. *J. Neurophysiol.* **2001**, *86*, 3030–3042.

145. Harsch, A.; Ziegler, C.; Gopel, W. Strychnine analysis with neuronal networks in vitro: Extracellular array recording of network responses. *Biosens. Bioelectron.* **1997**, *12*, 827–835.

146. Gross, G.W.; Rhoades, B.K.; Azzazy, H.M.E.; Wu, M.C. The use of neuronal networks on multielectrode arrays as biosensors. *Biosens. Bioelectron.* **1995**, *10*, 553–567.

147. Keefer, E.W.; Norton, S.J.; Boyle, N.A.J.; Talesa, V.; Gross, G.W. Acute toxicity screening of novel AChE inhibitors using neuronal networks on microelectrode arrays. *NeuroToxicology* **2001**, *22*, 3–12.

148. Xia, Y.; Gross, G.W. Histiotypic electrophysiological responses of cultured neuronal networks to ethanol. *Alcohol* **2003**, *30*, 167–174.

149. Scarlatos, A.; Cadotte, A.; DeMarse, T.; Welt, B. Cortical networks grown on microelectrode arrays as a biosensor for botulinum toxin. *J. Food Sci.* **2008**, *73*, E129–E136.

150. Guenther, E.; Herrmann, T.; Stett, A. The retinasensor: An in vitro tool to study drug effects on retinal signaling. In *Advances in Network Electrophysiology: Using Multi-Electrode Arrays*; Taketani, M., Baudry, M., Eds.; Springer Science+Business Media Inc.: New York, NY, USA, 2006; pp. 321–331.

151. Suyama, K.; Daikoku, S.; Funabashi, T.; Kimura, F. Effects of GABA and bicuculline on the electrical activity of rat olfactory placode neurons derived at E13. 5 and cultured for 1 week on multi-electrode dishes. *Endocr. J.* **2004**, *51*, 171–176.

152. Omori, T.; Saijo, K.; Kato, M.; Hayashi, M.; Itagaki, H.; Miyazaki, S.; Ohno, T.; Sugawara, H.; Teramoto, N.; Tanaka, N. Validation study on five cytotoxicity assays by JSAAE II. Statistical analysis. *Altern. Anim. Test. Exp.* **1998**, *5*, 39–58.

153. Clemedson, C.; Barile, F.; Chesne, C.; Cottin, M.; Curren, R.; EKWALL, B.; Ferro, M.; Gomez-Lechon, M.; Imai, K.; Janus, J. MEIC evaluation of acute systemic toxicity. Part VII. Prediction of human toxicity by results from testing of the first 30 reference chemicals with 27 further *in vitro* assays. *ATLA. Altern. Lab. Anim.* **2000**, *28*, 161–200.

154. Botham, P.A. Acute systemic toxicity. *ILAR J.* **2002**, *43*, 27–30.

155. Botham, P. Acute systemic toxicity—Prospects for tiered testing strategies. *Toxicol. in vitro* **2004**, *18*, 227–230.

156. Xing, J.Z.; Zhu, L.; Jackson, J.A.; Gabos, S.; Sun, X.J.; Wang, X.; Xu, X. Dynamic monitoring of cytotoxicity on microelectronic sensors. *Chem. Res. Toxicol.* **2005**, *18*, 154–161.

157. Keese, C.; Karra, N.; Dillon, B.; Goldberg, A.; Giaever, I. Cell-substratum interactions as a predictor of cytotoxicity. *In Vitro Mol. Toxicol.* **1998**, *11*, 183–192.

158. Xiao, C.; Luong, J.H.T. On-line monitoring of cell growth and cytotoxicity using electric cell-substrate impedance sensing (ECIS). *Biotechnol. Prog.* **2003**, *19*, 1000–1005.

159. Xing, J.Z.; Zhu, L.; Gabos, S.; Xie, L. Microelectronic cell sensor assay for detection of cytotoxicity and prediction of acute toxicity. *Toxicol. in vitro* **2006**, *20*, 995–1004.

160. Brischwein, M.; Motrescu, E.; Cabala, E.; Otto, A.; Grothe, H.; Wolf, B. Functional cellular assays with multiparametric silicon sensor chips. *Lab Chip* **2003**, *3*, 234–240.

161. Otto, A.M.; Brischwein, M.; Niendorf, A.; Henning, T.; Motrescu, E.; Wolf, B. Microphysiological testing for chemosensitivity of living tumor cells with multiparametric microsensor chips. *Cancer Detect. Prev.* **2003**, *27*, 291–296.

162. Geisler, T.; Ressler, J.; Harz, H.; Wolf, B.; Uhl, R. Automated multiparametric platform for high-content and high-throughput analytical screening on living cells. *IEEE Trans. Autom. Sci. Eng.* **2006**, *3*, 169–176.

163. Ceriotti, L.; Kob, A.; Drechsler, S.; Ponti, J.; Thedinga, E.; Colpo, P.; Ehret, R.; Rossi, F. Online monitoring of BALB/3T3 metabolism and adhesion with multiparametric chip-based system. *Anal. Biochem.* **2007**, *371*, 92–104.

164. Becker, B.; Lob, V.; Janzen, N.; Grundl, D.; Ilchmann, F.; Wolf, B. Automated multi-parametric label free 24 channel real-time screening system. *IFMBE Proc.* **2008**, *20*, 186–189.

165. Xi, B.; Yu, N.; Wang, X.; Xu, X.; Abassi, Y. The application of cell-based label-free technology in drug discovery. *Biotechnol. J.* **2008**, *3*, 484–495.

166. Asphahani, F.; Wang, K.; Thein, M.; Veiseh, O.; Yung, S.; Xu, J.; Zhang, M. Single-cell bioelectrical impedance platform for monitoring cellular response to drug treatment. *Phys. Biol.* **2011**, *8*, 015006.

167. Wang, W.; Foley, K.; Shan, X.; Wang, S.; Eaton, S.; Nagaraj, V.J.; Wiktor, P.; Patel, U.; Tao, N. Single cells and intracellular processes studied by a plasmonic-based electrochemical impedance microscopy. *Nat. Chem.* **2011**, *3*, 249–255.

168. Poulain, B.; De Paiva, A.; Deloye, F.; Doussau, F.; Tauc, L.; Weller, U.; Dolly, J. Differences in the multiple step process of inhibition of neurotransmitter release induced by tetanus toxin and botulinum neurotoxins type A and B atAplysia synapses. *Neuroscience* **1996**, *70*, 567–576.

169. Wu, Y.C.; Wang, P.; Ye, X.S.; Zhang, G.Y.; He, H.Q.; Yan, W.M.; Zheng, X.X.; Han, J.H.; Cui, D.F. Drug evaluations using a novel

microphysiometer based on cell-based biosensors. *Sens. Actuat. B: Chem.* **2001**, *80*, 215–221.

170. Wang, J.; Yu, H.; Cai, H.; Du, L.; Liu, Q.; Wang, P. A photovoltage-based integrated sensor for nephrotoxicity evaluation under drug stimulation. In Proceeding of the 16th International Solid-State SensorsActuators and Microsystems Conference (TRANSDUCERS), Beijing, China, 5–9 June 2011; pp. 2122–2125.

171. Adami, M.; Martini, M.; Piras, L. Characterization and enzymatic application of a redox potential biosensor based on a silicon transducer. *Biosens. Bioelectron.* **1995**, *10*, 633–638.

172. Primiceri, E.; Chiriacò, M.S.; Dioguardi, F.; Monteduro, A.G.; D'Amone, E.; Rinaldi, R.; Giannelli, G.; Maruccio, G. Automatic transwell assay by an EIS cell chip to monitor cell migration. *Lab Chip* **2011**, *11*, 4081–4086.

173. Wu, C.C.; Lin, W.C.; Fu, S.Y. The open container-used microfluidic chip using IrOx ultramicroelectrodes for the *in situ* measurement of extracellular acidification. *Biosens. Bioelectron.* **2011**, *26*, 4191–4197.

174. Chao, T.C.; Ros, A. Microfluidic single-cell analysis of intracellular compounds. *J. R. Soc. Interface* **2008**, *5*, S139–S150.

175. Gao, Y.; Bhattacharya, S.; Chen, X.; Barizuddin, S.; Gangopadhyay, S.; Gillis, K.D. A microfluidic cell trap device for automated measurement of quantal catecholamine release from cells. *Lab Chip* **2009**, *9*, 3442–3446.

176. Werdich, A.A.; Lima, E.A.; Ivanov, B.; Ges, I.; Anderson, M.E.; Wikswo, J.P.; Baudenbacher, F.J. A microfluidic device to confine a single cardiac myocyte in a sub-nanoliter volume on planar microelectrodes for extracellular potential recordings. *Lab Chip* **2004**, *4*, 357–362.

177. Gao, C.; Rilo, H.; Myneni, P.; Ahn, C. A new on-chip insulin biosensor for monitoring dynamic reponse of human islet cells. *Spec. Publ.-R. Soc. Chem.* **2004**, *297*, 581–583.

178. Pereira Rodrigues, N.; Sakai, Y.; Fujii, T. Cell-based microfluidic biochip for the electrochemical real-time monitoring of glucose and oxygen. *Sens. Actuat. B: Chem.* **2008**, *132*, 608–613.

179. Fan, C.Y.; Tung, Y.C.; Takayama, S.; Meyhöfer, E.; Kurabayashi, K. Electrically programmable surfaces for configurable patterning of cells. *Adv. Mater.* **2008**, *20*, 1418–1423.

180. Kaji, H.; Hashimoto, M.; Sekine, S.; Kawashima, T.; Nishizawa, M. Patterning adherent cells within microchannels by combination of electrochemical biolithography technique and repulsive dielectrophoretic force. *Electrochemistry* **2008**, *76*, 555–558.

181. Kaji, H.; Camci-Unal, G.; Langer, R.; Khademhosseini, A. Engineering systems for the generation of patterned co-cultures for controlling cell-cell interactions. *Engineering systems for the generation of patterned co-cultures for controlling cell-cell interactions* **2011**, *1810*, 239–250.

182. Li, Q.; Tang, D.; Tang, J.; Su, B.; Chen, G.; Wei, M. Magneto-controlled electrochemical immunosensor for direct detection of squamous cell carcinoma antigen by using serum as supporting electrolyte. *Biosens. Bioelectron.* **2011**, *27*, 153–159.

183. Shi, B.X.; Wang, Y.; Zhang, K.; Lam, T.L.; Chan, H.L.W. Monitoring of dopamine release in single cell using ultrasensitive ITO microsensors modified with carbon nanotubes. *Biosens. Bioelectron.* **2011**, *26*, 2917–2921.

184. Merkoçi, A. Nanoparticles-based strategies for DNA, protein and cell sensors. *Biosens. Bioelectron.* **2010**, *26*, 1164–1177.

Index